The Rational Universe Evolving for Humans

The Rational Universe Evolving for Humans

Exploring the Rational Design of the Universe

DING, ZHAO

Foreword by Kari Carlisle

RESOURCE *Publications* • Eugene, Oregon

THE RATIONAL UNIVERSE EVOLVING FOR HUMANS
Exploring the Rational Design of the Universe

Copyright © 2025 Ding, Zhao. All rights reserved. Except for brief quotations in critical publications or reviews, no part of this book may be reproduced in any manner without prior written permission from the publisher. Write: Permissions, Wipf and Stock Publishers, 199 W. 8th Ave., Suite 3, Eugene, OR 97401.

Resource Publications
An Imprint of Wipf and Stock Publishers
199 W. 8th Ave., Suite 3
Eugene, OR 97401

www.wipfandstock.com

PAPERBACK ISBN: 978-1-7252-9284-0
HARDCOVER ISBN: 978-1-7252-9285-7
EBOOK ISBN: 978-1-7252-9286-4
VERSION NUMBER 05/05/25

Sincerely, this book is dedicated to the memory
of the great scientist and thinker,
Albert Einstein!

"Quantum mechanics is certainly imposing. But an inner voice tells me that it is not yet the real thing. The theory says a lot, but does not really bring us any closer to the secret of the 'Old One.' I, at any rate, am convinced that He is not playing at dice."
—Albert Einstein

"The religion of the future will be a cosmic religion. It should transcend personal God and avoid dogma and theology. Covering both the natural and the spiritual, it should be based on a religious sense arising from the experience of all things natural and spiritual as a meaningful unity."
—Albert Einstein

"Science without religion is lame, religion without science is blind."
—Albert Einstein

"My religiosity consists in a humble admiration of the infinitely superior spirit that reveals itself in the little that we, with our weak and transitory understanding, can comprehend of reality."
—Albert Einstein

"Human beings, vegetables, or cosmic dust, we all dance to a mysterious tune intoned in the distance by an invisible player."
—Albert Einstein

"I am satisfied with the mystery of the eternity of life and with the awareness and a glimpse of the marvelous structure of the existing world, together with the devoted striving to comprehend a portion, be it ever so tiny, of the Reason that manifests itself in nature."
Albert Einstein

"To know that what is impenetrable for us really exist and manifest itself as the highest wisdom and the most beauty..."
—Albert Einstein

Contents

Foreword by Kari Carlisle | xi

Preface | xiii

Acknowledgement | xv

Prefatory Note | xvii

Introduction | xxiii
        The Concept of Natural Philosophy | xxiii
        The Natural Philosophy of Albert Einstein | xxiv
        The Two Principles of Natural Philosophy | xxvii
        The Relationship between the Two Principles and the Reason in the Development of Nature | xxviii

I.1   The Rationality of the Earth's Development in Time: The Time Effect of Natural Development on Earth | 1
        The Mysterious Time Effect | 2

I.2   The Rationality of the Earth's Development in Time: The Natural Development Chain | 31
        The Origin of the Planetary Wind System | 32
        Another Viewpoint: The Natural Contributing Factors | 45
        Ancient Seawater | 51
        The Illustration of Natural Development Chains | 59

II.1   The Rationality of the Earth System: The Composition of the Earth System | 64
        Astronomical Factors | 65

CONTENTS

        Geological Factors | 78
        Factors Related to Earth's Surface | 79
        Biological Factors | 93
        Factors Related to Civilization | 96
        The Definition of the Earth System | 104

II.2 The Rationality of the Earth System:
      The Structure of the Earth System | 107

        Earth's Core | 108
        The Mantle and the Crust | 109
        The Plate and its Materials | 109
        How to Process the Earth's Surface | 110
        The Formation of Plates and the Movement of Plates | 114

II.3 The Rationality of the Earth System:
      Who Predetermined Our Human World? | 117

        Introduction to Ancient Civilizations | 117
        A Very Logical Distribution of Land and Sea | 130
        The Mysterious Distribution of Islands in the World | 150
        The Mysterious Plate Movement | 159
        The Ancient Civilization Sites on the Collision Boundary | 175
        The Uniqueness of the Islands of Eastern Asia | 182
        North and South America | 183
        Overview | 185
        Nature Set the Way for the Development
          of Human Civilization | 187
        Nature, the Chief Director of the Human World | 201
        Why the Natural World Supports Humanity | 203

III The Mysterious Periodic Table | 209

        The Magical Order of Elements in the Periodic Table | 209
        Understanding of the Periodic Table from
          Natural Philosophy | 211
        The Mysterious Abundance of Elements on the Table | 218

CONTENTS

IV   What Exactly Is the Earth? | 252
        Earth's Uniqueness in the Universe | 252
        What Exactly is the Earth? | 264

V   The Remote Future of Humanity | 290
        Humankind—the Greatest Secret of the Universe | 290
        The Remote Future of Humanity | 299

VI   The Universe Evolved for Humans | 314
        Nature's Most Fundamental Creations | 315
        Setting the Stage for Reverse Reasoning | 320
        The Reverse Reasoning for the Evolution of the Universe | 330
        Back to Einstein and Hawking | 344
        A Simple Summary of Cosmic Evolution | 351

Epilogue (On the human brain) | 361
Comprehension Exercises | 364
Reflection Questions | 370
Bibliography | 373

Foreword

It is an honor to have been selected as the editor of *The Rational Universe Evolving for Humans*. This project has been both a challenging and rewarding journey, one that has deepened my love and understanding of science in ways I never anticipated. When I first approached Ding's manuscript, I knew this would be a unique opportunity—not just to refine a work on the natural world, but to help shape a narrative that bridges science, philosophy, and our broader human experience.

My lifelong interest in science allowed me to understand much of the material in Ding's writing, and I have done my best to bring his ideas to life for an English-speaking audience while preserving his original intent. Collaborating with Ding has been one of the most enriching aspects of this project. We've shared knowledge not only about the content of the book but about our different experiences and perspectives, which has added layers to the final product.

In addition to my scientific interests, my background in theology, anthropology, and history allowed me to draw connections between the physical universe and deeper philosophical and religious questions. *The Rational Universe Evolving for Humans* doesn't just explore the origins of the cosmos—it invites readers to consider how scientific insights might resonate with and enhance their own worldviews. Whether one approaches this book from a purely scientific lens or through the framework of personal beliefs, there is something to gain.

For me, this work has been transformative. It has not only deepened my scientific knowledge but also broadened my perspective on the world around us—how science, current events, geography, and international relations are all interwoven into the same fabric of understanding. This book has influenced the way I think about humanity's place in nature and how we relate to one another across time and space.

I believe readers, no matter their worldview, will find in these pages a profound exploration of how the universe's origin and development connect to larger questions of existence. I invite you to join Ding on this journey of discovery and reflection. It's an experience that may change the way you see the world, as it has for me.

Kari Carlisle, MA

Preface

Einstein's "The Reason in Nature" is highly abstract and very hard to grasp in its essence, but it is the only bridge for humanity to have an insight into the nature of Nature. The material of this bridge is the deep understanding of Nature both in time and in space; the exploration of nature is an essential means.

Once all the chemical elements were born in the universe, the process of creating civilization was logically defined by the principles of natural philosophy, and there was no alternative. The periodic table is "the constitution" of the universe, but it is created by "the highest wisdom" guided by "the Reason in nature."

This is an exciting story of natural development from the Big Bang to the emergence of humanity and modern civilization. Based on time limitations and principles of natural philosophy, natural development is bound to be an extremely complex nonlinear system in time and space. It can only be regulated by "the Reason in nature" that manifests itself as "the highest wisdom" in the cosmic cycles.

Since nature is of the Reason and rational for humanity, then logically, the universe must have evolved for humans, and that is just the logic of natural development. Humans, with their creative brain as the product of Nature, must follow the logic to release their wisdom fully. Therefore, the extensive demonstration of the rationality of nature is a profound subject for human cognition of the nature of Nature, but the formation of the rationality is another subject, which is much more complex and profound than the former.

The main points of this book are to demonstrate in detail that the real existence of Einstein's "Reason in nature" with "the highest wisdom" and that the progressive trend of human society brought about by the Renaissance is in line with the logic of natural development (cosmic evolution) and that human beings will move towards a brilliant future and complete the sacred natural mission given to them by the awesome universe. These three points are closely linked by cosmic logic.

Understanding the profound complexity of nature and its formation is vital to grasping its inherent rationality. As children of the universe, only through a deep understanding of this remarkable attribute can humans navigate the universe's logic and embark on a promising future, fulfilling our sacred natural mission!

Important note: This book uses a significant amount of data, primarily sourced from popular science websites on the Chinese Internet and several Chinese-language textbooks, rather than from monographs. While the figures presented may vary slightly across different sources, the focus of this book as a work of natural philosophy is not on the exact accuracy of these figures but on understanding the relative relationships between numerical values. Since the sources are primarily in Chinese, a direct translation or citation in English is not feasible. For figures that are not critical to the argument, no footnotes are provided.

After reading this book, you may understand why the logic of natural development is always consistent. From the atom to the human brain to intelligence in the form of energy, it's all about keeping intelligence throughout the cosmic cycles. A universe without humans would be meaningless. Humanity, the cosmic pride, you must respect your sacred natural mission.

DING, Zhao

Acknowledgement

This research book is based on popular science knowledge made available by the Chinese Internet and the Encyclopedia of China. I would like to thank the authors of all kinds of knowledge on the Chinese Internet (generally referred to as Internet encyclopedias) and the Encyclopedia of China. Many thanks to Ms. Kari Carlisle for her excellent editing. I am not a native English speaker and cannot write fluent and beautiful English, for which she has made great efforts to both retain my ideas and improve my level of written expression. Most of the images quoted in the book are from the Internet, and I would like to thank their authors. Finally, I would like to thank my wife, Ms. He, Yimei (何义梅 1937–2023), who had created a quiet environment for me to learn to think and write for more than 20 years, and provided me with some important references.

Prefatory Note

Is there a God? I don't know. Albert Einstein, however, believed that God does not "play at dice"[1] with the universe and that nature is "rational,"[2] nature is of "the Reason"[3] (rationality), "the profoundest reason and the most radiant beauty,"[4] "the highest wisdom and the most radiant beauty."[5] The basis of his views was his familiarity with physics. These ideas are very abstract, and people cannot understand them in a concrete way. Exploring the "rationality" and "the highest wisdom" hidden in various natural phenomena is the only way to understand his view concretely. Therefore, the author devoted a book to a comprehensive exploration of the "rationality" and "the highest wisdom" and found that the common feature of these rationalities is that they are all necessary for the eventual emergence of humanity and civilization in the universe. In other words, *this book demonstrates the true existence of the Reason in nature*. It must be emphasized here that Einstein's natural rationality is expressed from the standpoint of human beings. Since nature is rational in time and space for humans, it is entirely logical that humans would eventually emerge in the universe through natural development, that is, the universe evolved for humans.

The study of the rationality of a natural factor in the Earth System, such as the atmosphere or the tilt of the Earth's axis, and how that rationality came into being are two different things. This book deals only

1. Einstein, *Letter to Max Born*
2. Einstein, *Ideas and Opinions*
3. Einstein, *Letters to Solovine*, 102
4. Einstein, *The World*
5. Einstein, *The World*

with the rationality of natural factors. We have neither the ability nor the need to study the formation process of a rational natural factor like the atmosphere, which certainly involves many natural laws and scientific principles, including mathematics, physics, chemistry, etc., known and unknown. Therefore, behind the rationality of every natural factor, there is a complicated story, and this is where the mystery of nature lies!

I was born and raised in January 1935 in Qingdao, Shandong Province, China. Qingdao is a beautiful modern coastal city initially developed and built by the Germans. My ancestors (our family name is Ding 丁 in Chinese) are from Taoluo Town, Rizhao City, Shandong Province. My great-grandfather, Ding Shou-cun (丁守存 1812–1883), was a famous ordnance manufacturer in China. He excelled in rocket manufacturing, and the rocket he developed ranged from 1968 to 2624 feet (600 to 800 meters). He was also interested in Nature and wrote his book, *Researching for the Origins of Nature*. Sadly, it has been lost for years. My grandfather, Ding Lin-nian (丁麟年 1870–930), was the curator of the Library of Shandong Province, and my father, Ding Ru-yi (丁汝彝 1889–1944), majored in modern radio at a university founded by Germany in Qingdao. He was intensely interested in radio technology and modern popular science. My mother, Yang Zhong-yi (杨仲怡 1906–1988), raised me and my brothers with great difficulty.

I graduated from Tsinghua University in Beijing in 1958 with a five-year educational system. Tsinghua University and Peking University are the top Chinese universities, and both are in Beijing. My preschool name was Ding Wie-zhao. My major was river engineering, the department most closely related to nature. However, due to the lack of availability then, I had no opportunity to continue my studies for a doctorate after graduation.

Thinking about nature is my lifelong passion, an interest in my soul. My research owes much to my own experience. I grew up near the sea, and after graduating from university, I went to work in the mountains with rivers and different kinds of weathered rocks for twenty-three years. I love to watch what commonly occurs in the natural world and like to give some thought to them. I have always believed that it is necessary to understand nature with one's own heart to feel the relationship between various natural phenomena. This has been a long-term process of accumulation. Without such a long time, nature cannot be understood with your mind.

PREFATORY NOTE

After I retired in 1995, I devoted myself to the subject of "understanding nature" and gradually transformed into a scholar of natural philosophy to exclusively study Einstein's "the Reason in nature." In this sense, even if I got a doctorate in 1958, it would have nothing to do with the study of Einstein's natural philosophy that I am engaged in today, let alone the fact that there is no course on his natural philosophy in universities. On the contrary, my rich field experience and personal feelings on the various changes in nature have been the best basis for my study of the natural philosophy of Einstein.

On the other hand, my understanding of nature is based on popular science. I use this mature knowledge to understand nature, which indicates that Einstein's natural philosophy exists extensively in nature rather than just in relativity. For example, you enjoy seeing the sea, but you may not know the functions of the sea and the size and volume of the sea required for those functions, and you may not know why the salinity of the seawater is necessary.

My understanding of nature began with the origin of the planetary wind system, a subject that has never been studied. My interest in this subject was sparked by the huge difference between trees hundreds of millions of years ago and trees today in environmental conditions of survival and their height. I examined these conditions from a soil-mechanical point of view, thus, the mystery of the origin of planetary wind system was gradually revealed. In this study, mechanical knowledge and field experience play a great role.

The planetary wind system is of great significance to the Earth's environment and the development of human beings. Still, its origin (formation) is based on some natural conditions, and nature seems to have taken billions of years to form these natural conditions on the Earth. The problem closely related to the planetary wind system is the process of seawater becoming salty. They are two fundamental factors affecting the Earth's environment. Then, I developed a deep understanding of the rationality (necessity) of every element of the Earth System (examples include Pangaea, polyploidy, and clay minerals). In particular, given the great significance of the Mediterranean Sea as well as the Great Rift Valley in East Africa to human beings, I spent a lot of space logically arguing their formation process, a process that convincingly demonstrates the existence of the Reason (rationality) in nature, that is, natural development for the eventual emergence of humans and civilization. Finally, I argue the relationship between natural development and the development of human society. As a child of nature, the trend of human social development since the Renaissance conforms to the logic of natural development, which is an irresistible giant trend in the world.

Furthermore, I argue that the Earth is a perfect combination of numerous factors (elements) that are precise, subtle in time, space, and quantity. The formation of each of these factors has a complex process. But just how many factors are involved, modern science has not been able to determine. The greatest function of this combination of our planet is the evolution from chemical elements into life, humans, and modern civilization. The universe has taken billions of years to form this extraordinary planet composed of numerous factors, which is an extremely complex process in time and space. A completely random process will never produce any meaningful results or produce meaningful results needing endless time, but natural development has a strict time limitation; therefore, this extremely complex process with a time limitation can only have a requirement for timing, and thus only be regulated by the Reason advanced by Einstein. A universe without intelligence is meaningless.

Many eternal questions about the mysterious relationship between the universe, the Earth, humans, and civilization cannot be answered by modern science, and we may never be able to answer them. Einstein's natural philosophy is the only way to understand these weighty problems. In other words, Einstein's natural philosophy is embodied through the exploration of nature, and that's what this book is all about.

In today's Internet age, through extensive searching, I believe this is the first book in the world to argue overall the real existence of the Reason in nature.

Over the past nearly 30 years, as my understanding of nature has deepened, I have written four books, three in Chinese and one in English. They are all available in the Library of Congress and some famous university libraries in the United States and the United Kingdom. These four books represent the gradual deepening of the author's understanding of nature. This current one, the fifth, represents the author's comprehensive understanding of nature, which, of course, contains a lot of new content and a lot of content from my previous books, embodying the development process of my understanding nature.

In fact, as early as 2001, I put my ideas into a booklet in English, which is a very simple prototype of my understanding of nature; the title is *In the Beginning, Understanding Nature*. My good Australian friend,

PREFATORY NOTE

Mr. Keith Montague (passed away), edited my manuscript and submitted it to Athena Press. Soon after, the publisher accepted it and made a Readers Report by Mark Sykes, Chief Editor, on Jan. 3, 2002, of which the first sentence is: *"This is certainly a remarkable document."* However, many new arguments had been formed in my mind, and Tsinghua University Press agreed to publish it, so I naturally gave up the publication of the English booklet, and my first book in Chinese was published in 2004. Mr. Ni, Weidou (倪维斗), a prominent professor at Tsinghua University, first supported the publication with his preface for the book. After the publication of my first book, a prominent American professor, Donald E. Worster, read the condensed version in English and replied to me with the following comments in 2005: *"This is a grand project, highly ambitious, and it is always good to see someone taking on the big question that continue to be important. I see something of the Humboldt in you as well—a big scale explorer!"* In 2010, an American geologist Eric R. Force read my manuscript and made the following comment: *I am struck by your view of the earth's processes having evolved gradually AND UNIDIRECTIONALLY.* From July 2017 to December 2018, the American Philosophical Society conducted a detailed review of my manuscript, finding it to be "very thorough."

In addition, the course I taught at Tsinghua University (2007–2016) was an elective course with the topic of understanding nature, which was welcomed by students from all departments in the school. They listened to this course with curiosity and interest. Hundreds of people attended each class and filled the classroom.

Nature hides endless mysteries. Humans study physics, mathematics, and chemistry and acquire very profound scientific knowledge. But we can say that this knowledge (laws of nature) existed in nature through time before humans discovered them and that they form the basis of natural existence. Natural philosophy does not study specific laws of nature; it is a sublimation of natural science from the macro understanding of the profound relationship between humans and nature. Einstein saw "the marvelous structure of the existing world"[6] from the relationship between time, space, and matter revealed by his theory of relativity. Then he proposed the idea of "the Reason in nature," or that he was aware of the existence of "the Reason in nature."

---

6. Einstein, *Letters to Solovine*, 102

PREFATORY NOTE

Stephan Hawking once expressed a meaningful passage, which now ends this preface:

> Up to now, most scientists have been too occupied with the development of new theories that describe *what* the universe is to ask the question *why*. On the other hand, the people whose business it is to ask *why*, the philosophers, have not been able to keep up with the advance of scientific theories . . .[7]

7. Hawking, *A Brief History*

Introduction

THE CONCEPT OF NATURAL PHILOSOPHY

God does not "play at dice" with the universe, then logically, *what will God do with the universe?*

Modern science cannot answer this weighty question and may never be able to answer it. It is essentially within the realm of natural philosophy.

Simply put, the natural development from the Big Bang to the emergence of humanity and modern civilization is undoubtedly a random process, whereas completely random processes do not produce meaningful results. In other words, it takes endless time to produce meaningful results.

However, there are three key points that seem to control the random process of natural development.

1. *Time limitation.* Natural development is subject to time limitation, resulting in the requirement for timing. For example, the evolution of life on Earth could only occur during the optimal period of solar radiation. If there were no time requirements for various natural processes, natural development would be disorderly, and humans would not be able to emerge. This is easy to understand.

2. *Two common sense principles.* The consistency of structure and function and the fact that everything happens conditionally are two common sense principles of natural philosophy.

3. *"The Reason in nature."* Deep comprehension of Einstein's "the Reason in nature" is essential concerning the fate of humanity. Einstein

believed that nature has "the highest wisdom" and "the Reason." However, these ideas at that time mainly referred to physics and life. Today, "the Reason" and "the highest wisdom" through modern popular science can extend to the whole of nature in time and space. Now we can ask nature why, billions of years ago, you "knew" that your choices about the Earth's mass, size, and the distance between the Earth and the Sun were appropriate and necessary for the formation of a civilized planet. Does it embody "the Reason" (rationality) in nature? This book will answer this and similar questions.

Now, we will explain points two and three further.

THE NATURAL PHILOSOPHY OF ALBERT EINSTEIN

Generally speaking, there are two options for natural development. One is that it is entirely random, and the emergence of humans is an accident in the vast universe. The second is that some mysterious force created human beings. This book will argue the third option, that humans came into being based on the Reason revealed by Einstein, which exists in natural random processes.

The term "natural philosophy" was first used in the title of Newton's work, *The Mathematical Principles of Natural Philosophy (Philosophiæ Naturalis Principia Mathematica)*. At that time, natural philosophy referred to the natural sciences. Is there a philosophy for nature today? Yes, this is the philosophy that Einstein advocated; we might call it Einstein's natural philosophy. Is there a course on his natural philosophy in universities? Perhaps not. Why?

To answer this question, we must understand the characteristics of Einstein's natural philosophy.

Einstein, like Leibniz, Hegel, Spinoza, and Kant, did a lot of philosophical thinking about nature, but strictly speaking, he was very different from the others. Einstein was not a scientist who liked to use mathematical reasoning to prove his own concepts only.

Einstein's logical thinking was even more abstract than mathematics, with this high level of abstract thinking dominating his mind. Due to his extraordinary depth of thought, "the marvelous structure of the existing world" revealed by his theory of relativity naturally led to his unique philosophy of nature (with "marvelous" being a keyword here). Einstein sought to understand the rationality of natural structures through a philosophical lens during his arduous exploration of nature, aiming to comprehend the essence of nature itself. This was Einstein's philosophical approach to understanding nature. Broadly speaking, this is an endless process. The more we explore nature, the more rationality we discover, and the deeper our understanding of the nature of nature becomes. As Einstein himself stated:

> But whoever has undergone the intense experience of successful advances made in this domain is moved by profound reverence for the rationality made itself in existence.[8]

Note that the rationality that Einstein mentioned here is, of course, from the human point of view, and this rationality can also be understood as what he referred to as the Reason in nature.

Therefore, Einstein's philosophy has a dual meaning: the scientific side and the philosophical side. In other words, Einstein's philosophy of nature is based on his exploration of nature. To understand the truth of Einstein's natural philosophy, you must personally explore the rationality of natural phenomena (natural processes). This is the distinguishing characteristic of Einstein's philosophy of nature. Please keep this point in mind. It seems that Einstein's natural philosophy does not belong to pure philosophy, and thus does not form an independent philosophical system in higher education.

Einstein's core philosophy of nature revolves around his belief that "God does not play dice with the universe" and his quest to understand "the Reason that manifests itself in nature." In his view, the development of nature is not blind but rational. Einstein explicitly questioned the entire materiality of the world. In addition to matter, "the Reason in nature" exists, and the physical world seems to operate under its rule. He expressed this idea many times. For example,

> A knowledge of the existence of something we cannot penetrate, our perceptions of the profoundest reason and the most radiant beauty, which only in their most primitive forms are accessible to our minds--it is this knowledge and this emotion that constitute true religiosity; in this sense, and in this alone, I am a deeply religious man.[9]

---

8. Einstein, *Ideas and Opinions*
9. Einstein, *The World*

INTRODUCTION

Human beings, vegetables, or cosmic dust, we all dance to a mysterious tune intoned in the distance by an invisible player.[10]

So, if you comprehend nature deeply, *the profoundest reason and the most radiant beauty* can be *accessible to your mind*. Therefore, it is not only a philosophical question, but a religious one. Einstein never acknowledged the existence of a God in the religious sense:

> I cannot conceive of a God who rewards and punishes his creatures, or has a will of the kind that we experience in ourselves.[11]

> I cannot prove to you that there is no personal God, but if I were to speak of him I would be a liar.[12]

As we build on the above, it becomes clear that Einstein's natural philosophy is firmly rooted in physics. It is inconceivable that the philosophy of a scientist with such a profound understanding of the structure of nature could be merely a figment of his imagination. This would be illogical. However, we must recognize that comprehending Einstein's natural philosophy is neither easy nor straightforward. The truth of his ideas—"God does not play dice with the universe" and "the Reason that manifests itself in nature"—can only be grasped by thoroughly studying the rationality of the existing world and the development of nature. This endeavor is a comprehensive, multidisciplinary process.

Without such an in-depth study, proving the truth of his philosophical thoughts is impossible. Understanding the Reason (rationality) in nature is far more complex than simply observing a natural phenomenon.

Essentially, Einstein's natural philosophy is a profound legacy to the world. But suppose no one in the world extensively explores the actual existence of the Reason (rationality) to demonstrate its truth. In that case, this precious legacy can only be empty and meaningless. Therefore, it will be difficult for humanity to understand their natural mission and the meaning of their existence in the wondrous universe.

The problem is people tend to think that the development of nature is entirely random. The question is, if nature is not entirely random, who is manipulating nature? However, such a question reduces extremely complicated natural development into a simple binary choice: random or rational. The answer to such a profound question can only be Einstein's

10. Einstein, *The World*
11. Einstein, *Der Einstein-Gutkind Brief*
12. Hermanns, *Einstein*, 132

natural philosophy combined with the exploration of nature, or the sublimation of scientific thought. Is there any other way except for combining philosophy with natural exploration to comprehend the nature of nature? No!

The mysterious relationship between humanity and nature has always been an eternal question in the human mind. Some people understand it from a religious perspective; others understand it from a secular perspective. According to Einstein, the development of nature is undoubtedly a stochastic process, but "the Reason that manifests itself in nature" seems to control the stochastic process.

It should be noted that Einstein was not satisfied with the term "religion" because it is often misleading. Still, he could not find another widely agreed-upon word to express his devotion to the rationality of nature.

Einstein was the only great natural explorer who integrated science and philosophy. His philosophy grew out of his scientific achievements. His philosophical ideas, emerging from the same brain and mind, deserve as much attention as his scientific achievements. However, in the past 100 years, how many books have exclusively demonstrated the truth of his words? In today's Internet age, it seems there are none. Notably, the truth of relativity can be proved by instrumental observation, but the truth of natural philosophy cannot.

This book presents a wealth of scientific information and demonstrates numerous examples of natural rationality (the Reason), each subject worthy of further investigation. If the multitude of rationalities that have manifested in nature since the Big Bang were studied in depth, it could provide a solid foundation for humanity to establish Einstein's cosmic religion, unifying humanity into a cohesive force.

THE TWO PRINCIPLES OF NATURAL PHILOSOPHY

All processes in nature are subject to two principles of natural philosophy, which, however, is common sense, and no one denies:

1. Consistency of structure and function. Any change in a structure will affect its function no matter how slight the difference.
2. Everything happens (forms) with conditions, or everything happens (forms) conditionally.

According to these principles, the conditions for forming a structure can be one or many, some visible and some invisible. The higher the function, the more complex and refined the structure required, and the more conditions are necessary to form that structure. This applies to both natural structures, like cells, and artificial structures, like clocks. The complexity mentioned here does not imply disorder; rather, it is logical and ordered, though there are differences between various parts of the structure. Fineness refers to the subtlety and sensitivity of organizational structure. Life evolves with changing conditions, albeit very slowly. However, the conditions necessary for the origin of life remain an ongoing subject of study.

Note that although the Big Bang theory and Darwin's theory are widely accepted, they're just theories. Today, humans still do not know what happened in nature precisely over the past billions of years. Humans can only obtain some relevant information from astronomical observations and archaeological research to verify the truth of the theories. However, if you read this book carefully, you will understand that nature, particularly the Earth System, is a highly complex and precise assembly of countless factors. The two principles of natural philosophy govern each of these factors (for example, the atmosphere). Through these two principles, we can further infer that the process from the Big Bang to the emergence of modern civilization is bound to be extremely complex. This is a significant point for understanding natural development.

THE RELATIONSHIP BETWEEN THE TWO PRINCIPLES AND THE REASON IN THE DEVELOPMENT OF NATURE

Nature is undoubtedly full of randomness, but natural structures and development are rational and orderly. According to Einstein's natural philosophy, this reflects natural rationality, or the work of the Reason. We can say, therefore, that the entire process of natural development operates under the control of Reason (rationality). Nature successively creates new conditions for the formation of increasingly complex material structures with higher functions, culminating in the emergence of humans and modern civilization. Can you conceive of a more thoughtful way for evolution to proceed from chemical elements to human beings? On Earth, structure determines function, forming diverse environments over time—an essential precondition for the evolutionary process.

This entire progression must occur during the optimal period of solar radiation.

Understanding the natural conditions necessary to form each natural structure is a complex subject of research. For example, what conditions (causes) led to the Earth's current axial tilt? And the precise composition of the atmosphere? And mass extinctions in the evolutionary process of life? Modern science struggles to answer these questions comprehensively. However, it is undeniable that everything always happens under specific conditions in both time and space.

Randomness cannot produce meaningful results; for example, a monkey randomly tapping a keyboard will never form a poem. In other words, the time it takes for the monkey to create a poem by randomly tapping a keyboard will be infinitely long.

In natural development, the time limitation and the requirement for timing dictate that progress must be rational and logical. The principles of natural philosophy shape the course of natural development, while the Reason appears to guide natural selection. These two principles of natural philosophy act like guardrails, directing the path of this highly complex process. The direction of this path is natural rationality, or the Reason, which has consistently guided natural development toward the eventual emergence of humanity and modern civilization.

There may be little doubt about the two principles of natural philosophy, and the Reason is the hypothesis this book devotes a great deal of space to arguing. You will see the reality of the Reason and how it controls natural development, the process of evolution from chemical elements into human beings. The Reason is an attribute of nature.

Chapter I.1

The Rationality of the Earth's Development in Time

The Time Effect of Natural Development on Earth

"The Reason in nature" or rationality of nature is a core idea of Einstein's natural philosophy. In highly complex natural development, humans could not have emerged without rationality.

In this chapter, we will discover the rationality of time in the Earth's development, which is a primary characteristic of natural development, and we call it the time effect. This rationality also shows the logic of natural development. We cannot imagine natural development without any timing requirement.

The development of the Earth System occurs during the optimal period of solar radiation, which is the prerequisite for the development of the Earth System. Briefly, the time effect manifests itself in close coordination between the inanimate and animate worlds. Every stage, transition, or factor involved shows a strict time requirement, neither too early nor too late but on time, thus leading to the evolution of life until the formation of a civilization planet.

Note that in nature, only life processes, from the fertilized egg to the end of life, have timeliness.

The appearance of civilization on a planet is far from easy!

Every physical factor involved in the development process of the Earth System requires timing, which we may call "timing factors." When should a physical factor such as the polar ice sheets, for example, begin to appear? When should a factor-forming process conclude? And what is the relationship between various factors when it comes to timing? The final formation of every factor involved in the Earth System requires its appropriate condition and timing. If the geographical conditions necessary for the emergence of humans were not created in time, humans would not appear on this planet because everything happens conditionally.

The need for timing is easily understood. What is difficult is for us to imagine the appearance of factors along the time axis being in disorder with no need for precise timing. Without timing, all the factors involved in the development of the Earth System could not possibly line up.

Did the development of the Earth System truly occur in an orderly fashion over the past billions of years? Let's see...

THE MYSTERIOUS TIME EFFECT

There are typically three questions involved in a factor that makes up a part of the Earth System:

1. What is its role to play in the Earth System?
2. What is the factor-forming process?
3. What is the requirement for its timing?

The first question is not difficult to answer. The second one is difficult to answer because factor-forming processes occurred in the past millions of years or deep in the Earth. The answer to the third question reveals the timing needed for the factor to play its role in the Earth's development.

Now, let's turn to the second and third questions. From the perspective of the eventual appearance of humans and civilization on our planet, the "time effect" means nature always did what it should in due time, neither earlier nor later, just on time. This is because natural development is time-limited. For example, it must occur in the expanding phase of the universe, not indefinitely. Therefore, natural development has a requirement for timing, and time effect is only a means of meeting this time requirement in natural development.

The following illustrates how nature performed a series of programs, one by one, in its due time on the Earth to develop the Earth System toward the eventual appearance of humans and civilization.

The Timely Appearance of the Earth

The intensity of solar radiation received on the Earth's surface is critical to life's existence. The Sun's size and lifespan, the Earth's size, and the distance between the Sun and the Earth all determine the intensity of solar radiation received on the Earth (see Chapter VI, Is the Solar System Designed for Earth).

The Sun's radiation has been growing stronger for 5 billion years and will eventually weaken in the distant future. Scientists think the Sun's lifespan is about 10 billion years, so we can logically imagine our planet was born and developed during a particular time when the intensity of the Sun's radiation was conducive to evolutionary processes on the Earth.

Besides the Sun's radiation, many natural factors could cause a significant change in the global environment, impacting the very existence of life on the Earth, such as an asteroid collision, ultra-strong solar storms, supernova outbursts, chromospheric eruptions, etc. All these factors would have had the potential for a massive transformation of the Earth's development.

In a sense, our planet's global environment results from a combination of many factors. Fortunately, over time, Earth formed at the appropriate distance from the Sun during a suitable period. With the necessary solar radiation, these combinations allowed the evolutionary process to unfold over 3.5 to 3.8 billion years successfully and provided civilization with sufficient time to begin and develop into the distant future.

But why did the formation of the Sun and the Earth correspond so well in size, distance, and time? Why were the natural factors mentioned above in harmony on the Earth over the past 4.6 billion years? We may consider that this synchronization results from the time effect manifesting itself in the Earth's processes.

The Timely Birth of the Moon

The Moon is an ideal and necessary companion for the Earth. It significantly impacts the Earth's environment (see Chapter II.1, Our Mysterious Moon), but what was the best time for its appearance?

A few hypotheses have been suggested about the origin of the Moon. Relative to the Earth, the Moon is considerably larger than the satellites of other planets in the solar system, so we may logically surmise that the way the Moon formed might differ from other moons in the solar system. Since we landed on the Moon and brought back some geologic samples, scientists seem to be increasingly more interested in the hypothesis of a giant impact on the Earth about 4.6 billion years ago. If true, that collision would be tremendous: a celestial body of the right size and quality, with the right velocity, in the right direction, and at the right angle, ran into the Earth. Whether this hypothesis is true or not, the Moon appeared at the best time. If the collision had occurred at another time, all life would have been destroyed.

From the viewpoint of probability, the Moon's size, quality, and distance (orbit) meeting the Earth's requirements is practically impossible. If the Moon's density were roughly the same as the Earth's (according to the hypothesis above, the Moon would have come from the Earth), its mass would break the dynamic balance with the Earth. In fact, its density is only 0.121 lb/in^3 (3.35 g/cm^3), which makes its larger size possible. This density is much less than that of Earth (0.199/5.51), Mercury (0.197/5.45), Venus (0.190/5.26), and Mars (0.143/3.96). Although speculative theories have suggested that the Moon might be a relatively "hollow" celestial body, scientific evidence does not support these theories, and the hypothesis above remains strong.

No matter how the Moon was born, why was it formed just at the beginning of the Earth rather than another time? The timely appearance of the Moon, a perfect satellite both in size and in mass for the Earth, is indeed inconceivable! We may consider that the Moon's appearance at the right time is again the time effect manifesting itself in the Earth's processes.

The Timely Differentiation of the Earth and the Timely Appearance of the Oceans

When the planetesimal developed into the primitive Earth, what naturally followed was the gravitational differentiation inside the Earth (the

THE RATIONALITY OF THE EARTH'S DEVELOPMENT IN TIME

hypothesis of gravitational differentiation). This was caused by the tremendous amount of heat energy inside the planet, accompanied by massive eruptions of volcanoes containing plenty of steam as the primary source of seawater. Thus, the following seems to be a reasonable process:

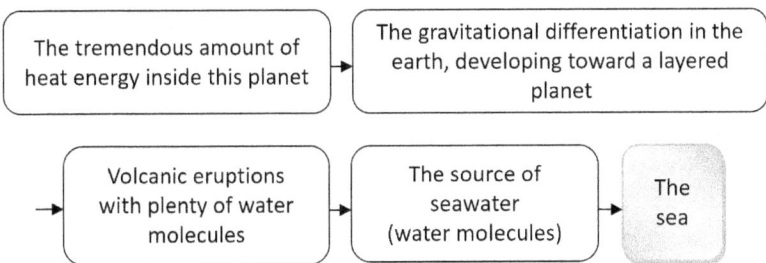

Correspondingly, the crust-forming process is as follows:

Stages	Time
Astronomical stage	4.6 to 3.85 billion years ago
The continental nucleus forming process	3.85 to 2.5 billion years ago
The plate forming process	2.5 billion to 800 million years ago
The Pangea forming process	800 to 208 million years ago
Continental drifting (massive orogeny)	208 million years ago to the present

An essential natural chain holds the world together and without any link in this chain, the world as we know it would not exist. Correspondingly, the geomagnetic field naturally formed after the Earth's layers were established.

Where did the tremendous amount of heat energy and the immense volume of water molecules inside the Earth come from? They are key to forming the layered planet and the sea. Unfortunately, scientists have learned only a little so far. Without them, the development of the Earth could not possibly have transpired. The Moon, gravitational differentiation (and the geomagnetic field), and the primitive sea have formed the basis of the Earth's environment.

But why were these natural events arranged so well in time? We may consider that these events are the time effect manifesting itself in the Earth's processes.

The Timely Birth of Photosynthesis on Earth and the Timely Cambrian Explosion

When all was ready, what was needed next for the creation of life? The establishment of biological energy!

Nature uses only water molecules, sunlight, and carbon dioxide—substances found almost everywhere—to initially create biological energy in the form of glucose through photosynthesis.

Photosynthesis is a complex chemical reaction that occurs in the cells of leaves. The key to photosynthesis lies in chloroplasts (5–7μm or 196.850–275.591 microinches in size) (Figure 1-1), where water molecules are split into hydrogen and oxygen. The oxygen produced is then utilized by living organisms, supporting the evolutionary process from the earliest days up to the present. Another crucial part of photosynthesis is the dark reaction, or CO_2 assimilation. Both reactions occur within the cells, and without the cell's structural support, the complete photosynthesis process (from photons to usable biological energy) would not be possible. The glucose produced by photosynthesis is stored within the cell, making cells essential for converting biological energy into a form that sustains life. Thus, photosynthesis inherently defines the cell as an ideal form for the existence of life, an important point to note.

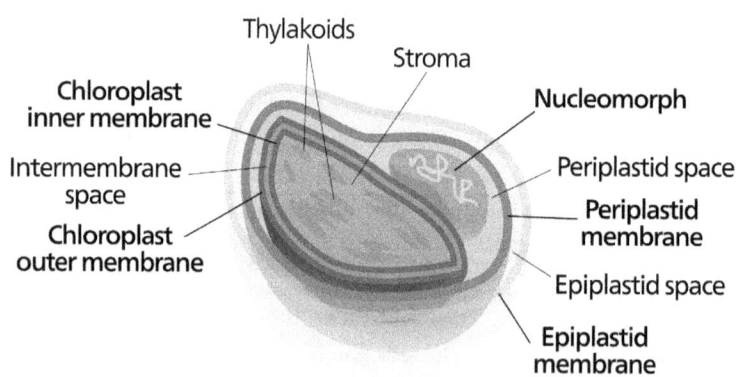

Figure 1-1. Plant cell chloroplast (Source: Kelvinsong, CC BY SA 4.0).

Remember, the disintegration of pure water molecules is difficult. Only 1–2 percent of water molecules can be split at temperatures exceeding 3632°F (2000°C) with very high pressure, indicating that a significant amount of energy is required for this breakdown process. However, chloroplasts—fine and complex in structure—can complete this process relatively easily at normal temperatures. Due to their immense significance in science and the economy, many scientists have studied the remarkable function of chloroplasts for years. We must continuously learn from nature.

The first life forms emerged in an environment with little to no oxygen, that is, anaerobic conditions. Over a long period, the first blue-green algae, or prokaryocytes, with primitive photosynthetic abilities appeared. Their presence led to a gradual increase in atmospheric oxygen content. Subsequently, eukaryotic cells emerged, and life began to gradually adapt to using oxygen for respiration.

But when did these natural events first appear? Scientists continue to research this question, but they believe the occurrence of all these natural events in the initial stages of evolution ranged from 1.6 billion to 3.8 billion years ago, and the first blue-green algae appeared 3.5 billion years ago.

Photosynthesis is the primary natural process that provides biological energy, which seems to be the only solution in the periodic table for life to use solar energy (see Chapter III, Understanding the element carbon).

Amazingly, nature created the small, intricate, and complex structure of chloroplasts and the process of photosynthesis. However, scientific inquiry has revealed little about how nature accomplished this. Chloroplasts are primarily composed of proteins, which hints at the origins of life itself.

The evolutionary process of cells in the ocean was relatively slow during the first 3 billion years. If the evolution of life had proceeded at this same rate, living organisms could not possibly have evolved into what they are today. But an extraordinary natural event, the Cambrian Explosion, took place about 520 million years ago, lasting about 2 million years. A wonderful natural arrangement! During this unusual period, diverse primitive species of animals "suddenly" appeared, composing a "blueprint" that led the evolutionary process to the eventual appearance of humans on Earth. Further research is needed on the cause of the Cambrian Explosion (Figure 1–2).

Figure 1–2: An imagination of Ediacaran Sea life around 600 million years ago (Source: Ryan Somma, CC BY-SA 2.0).

We do believe that without the Cambrian Explosion occurring in its due time, without the formation of photosynthesis in the initial stage of the evolutionary process, the natural world could not have evolved into what we have today.

But why did photosynthesis and the Cambrian Explosion take place at that time? We may consider that their occurrence is the time effect manifesting itself in the Earth's processes.

The Timely Appearance of Life on Land

After the Cambrian Explosion, what comes next for life's continued evolution? Logically, life should transition from water to land, expanding its space for further evolution. If life remained confined to the ocean forever, it would lose its significance on the Earth and even in the universe.

Over the past millions of years, there have been two different kinds of natural development processes on the Earth. One is the evolution of life in the sea, and the other is crustal movement. Since the appearance of photosynthesis, life gradually became varied. Then, eukaryotic cells began to develop in two different ways, autotrophy and heterotrophy, evolving toward the appearance of primitive plants and animals. On the other hand, plate movement (beginning in the Caledonian orogeny in the Early Paleozoic) and sedimentation caused the first regression on a large scale about 400 million years ago. But most landmasses then were scattered in the middle and low latitudes and the equator (Figure 1-3). Both development processes are independent of each other.

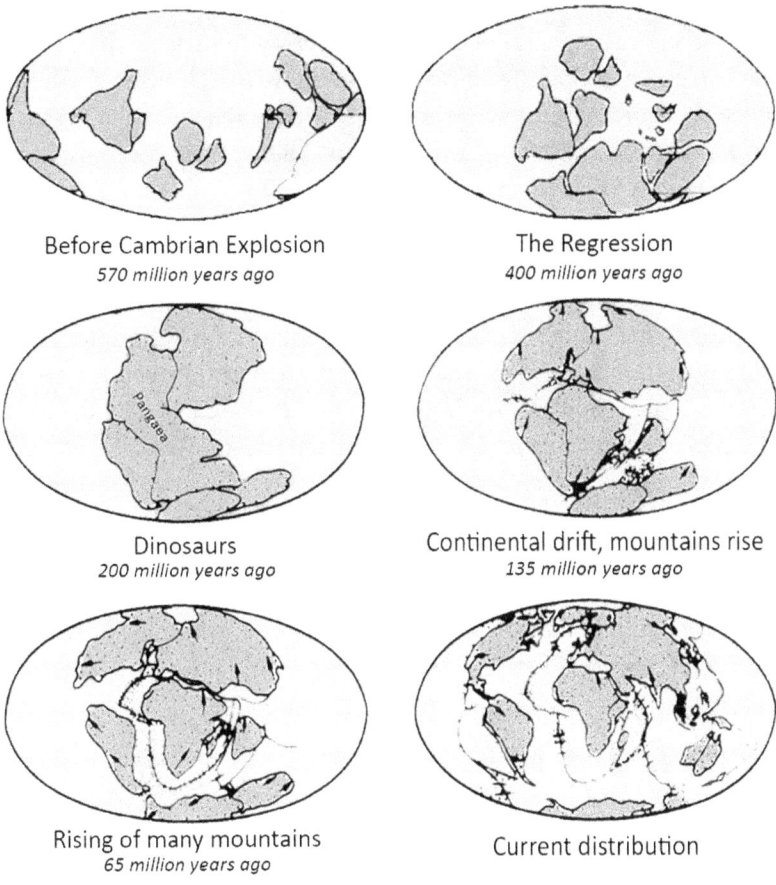

Figure 1-3: The development of ancient continents (Source: Benpei Liu, et al. *Geohistory*. 3rd ed. Beijing: Geology Press, 1996.).

Note that this set of pictures is for reference only. In fact, similar geohistorical maps are found in different documents, with some differences. Plate tectonics (seafloor spreading and continental drift) is one of the great discoveries of modern science. The concept of Pangaea has been widely accepted. Here, we focus on the concept of continental change rather than its exact location.

We do not know why these processes take so long to occur. On the one hand, life in the ocean (through a long period of evolving from prokaryotic cells to primitive plants and animals) could adapt to the land environment about 400 million years ago. On the other hand, over a long geological process, the first regression appeared about the same time frame (from the late part of the Silurian Period to the early part of the Devonian Period), causing life to "have to" move onto land. So, these two processes coincide with each other. Subsequently, life thrived in the shallows and marshland left by the regression, making them the transitional environments for life to extend to the land.

If there had been no regression about 400 million years ago, life landing would have waited for the second regression, which occurred in the late Permian Period 250 million years ago! If life landing had happened later, could humans have been born at the right time? None other than the first regression allowed for the opportunity in due time for primitive plants and animals in the ocean to evolve further on the land. The degree of accuracy of the time point 400 million years ago requires further study.

So, the landing of life from the ocean and the first regression taking place in due time is indeed a surprising natural arrangement in the development of our planet. But why did these two processes coincide? We may consider that their timing is the time effect manifesting itself in the Earth's processes.

The Timely Change in Ancient Land

What came next after life transitioned from the sea to land? The changing environmental conditions on land! According to the theory of evolution, this is easily understood.

The most significant factor in the development of the Earth is plate movement, the primary cause that spurred the other processes, affecting the results of evolutionary processes on the Earth.

Fortunately, the timely change in ancient land is so favorable that it made the evolutionary process possible, from the earliest life in the ocean to the landing of life to the wide variety of species to the emergence of humans and civilization.

THE RATIONALITY OF THE EARTH'S DEVELOPMENT IN TIME

The process is as follows: the complete formation of plates took place 800 million years ago, the scattered ancient landmasses were increasing in area (the first regression corresponding to life landing, 400 million years ago), and they moved toward gathering into one landmass, becoming Pangea, about 208 million years ago, which, however, was "soon" disintegrated, breaking into separate drifting continents (Figure 1–3) and continuing up to the present. All these significant natural events were caused by plate movement.

In the development of the Earth, Pangea seems to be an essential stage or a geographically natural arrangement that resulted in two crucial sub-processes:

First, Pangea made the global exchange of species possible. For example, dinosaurs once existed throughout Pangea, and their remains have been found in every continent but Antarctica (see Chapter II.3).

Second, according to the science of plate tectonics, the disintegration of Pangea through continental drift inevitably brought vast diversity both in topography (many high mountain ranges arose, that is, orogeny) and in the distribution of continents, promoting the variety of species in evolutionary processes (from Old Alpedic in the Late Mesozoic to Young Alpedic, the current Cenozoic). What continental drifting brought are not only high mountains but also Greenland and Antarctica, as well as many by-products, such as rivers, alluvial plains, etc. (see Chapter II.3).

The change in ancient land indicated above seems to be an inevitable and logical process; with the formation of plates, the landmasses formed from continental nuclei individually like islands and then came together to form Pangea. Subsequently, the seafloor spreading with continental drift formed the current global distribution of land and sea and various geographical environments. Life evolved into land-based life throughout the process as the land area grew from a continental nucleus into landmasses. A wonderful natural arrangement (Figure 1–3 and table above)! Can there be another way? Probably not.

Therefore, timely change in ancient land has contributed to the formation of the global climatic pattern and the direction of evolutionary processes.

But why didn't Greenland drift to the North Polar Region as the Antarctic continent did in the southern polar region? We do not know. If it had, logically, the development of the global climatic pattern would certainly be radically different (at the very least, temperatures at the

North Pole would be much lower than they are today), deviating from the environment humans have experienced for millions of years.

If the plates had been stationary without any movement for billions of years, if there had been no Pangea in the Earth's development, if the distribution of continents had occurred in another way (say, if most landmasses were situated in other latitudes rather than the temperate zone), could humanity and civilization have emerged (refer to the theory of Wilson cycle in another source)?

Please note a particular condition is a primary requirement for a specific occurrence. Without the relevant conditions, the wide variety of organisms could not possibly have emerged in the evolutionary process, so the following process seems necessary:

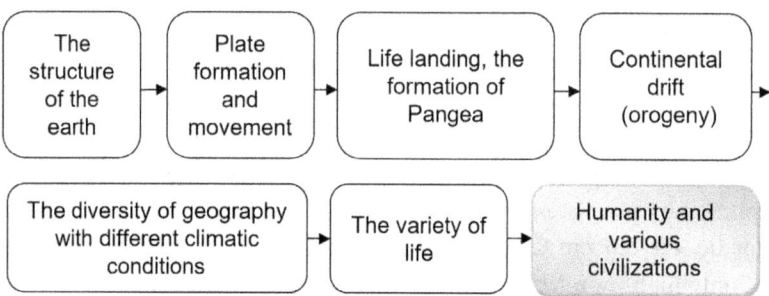

Obviously, plate movement is the primary factor in the development of the Earth, which, as it were, successfully directed the evolution of life. This tells us again that the development of civilization on a planet is far from simple!

But why did the timing of the change in ancient land occur the way it did rather than another way? Why did the logical process occur in due time? We may consider that this is the time effect manifesting itself in the Earth's processes. (see Chapter II.3).

The Timely Appearance of the Atmosphere

After exploring the events on the Earth's surface, let's shift our focus to the story unfolding in the sky.

Our atmosphere possesses significant functions and is a necessary component of the Earth System. But how is the atmosphere formed? Scientists continue to be puzzled.

THE RATIONALITY OF THE EARTH'S DEVELOPMENT IN TIME

The Earth's atmosphere is a mixture of gases formed by various natural factors, including volcanic eruptions, the gas exchange of plants and animals, the oceans absorbing carbon dioxide, and combustion consuming oxygen and producing carbon dioxide. Each factor acts as an independent variable. The air is primarily composed of oxygen and nitrogen, with trace amounts of other gases like carbon dioxide, methane, and inert gases. These components are crucial for maintaining the global average temperature, supporting all forms of life, and enabling combustion, which is essential for the development of civilization.

The atmosphere's composition is believed to have gradually changed over billions of years. This gradual change is called atmosphere development. In the initial stages of our planet's development, the composition of the primitive atmosphere was significantly different from today, almost entirely made up of carbon dioxide, methane, steam, etc., and no oxygen. How did the primitive atmosphere gradually transform into the existing atmosphere over billions of years? We know only a little. Understanding this process is a big problem. Today, global warming has resulted from a slight increase in carbon dioxide in the air. This fact shows how delicate the atmosphere is!

The primitive atmosphere turning into the present layered atmosphere must have been a complex and subtle process, lasting billions of years, and this reveals a time effect during the Earth's development (see Figure 2–12 in Chapter II.1). The ozone sphere is believed to have formed earlier than 400 million years ago, providing a layer of protection for the first life on the land about 400 million years ago (time effect!).

Today, the layered atmosphere creates a desirable environment for all life to exist and for us to develop civilization on our planet. Do you realize how ideal our atmosphere is in terms of temperature, humidity, precipitation, combustion, filtration of ultraviolet rays, photosynthesis, the colorful world, etc.? Without all the upper layers, the troposphere could not have formed. Each upper layer has a unique role in creating the troposphere, realizing all the functions mentioned above. The properties of oxygen and nitrogen define the formation of the layered atmosphere. In a long, complicated natural process, a tremendous amount of oxygen and nitrogen accumulated above the Earth's surface, forming layers under the power of solar radiation. The resulting layered atmosphere includes the ideal proportion of gases.

The stochastic natural process, from the properties of the gases to the accumulation of the gases to the formation of the layered atmosphere, is an incredible natural miracle!

The following diagram shows us how solar radiation through the atmosphere provides a colorful environment to support our lives.

Remember that today's desirable, layered atmosphere was not formed until the long, changing natural process reached this stage. The time effect manifests itself in this natural process.

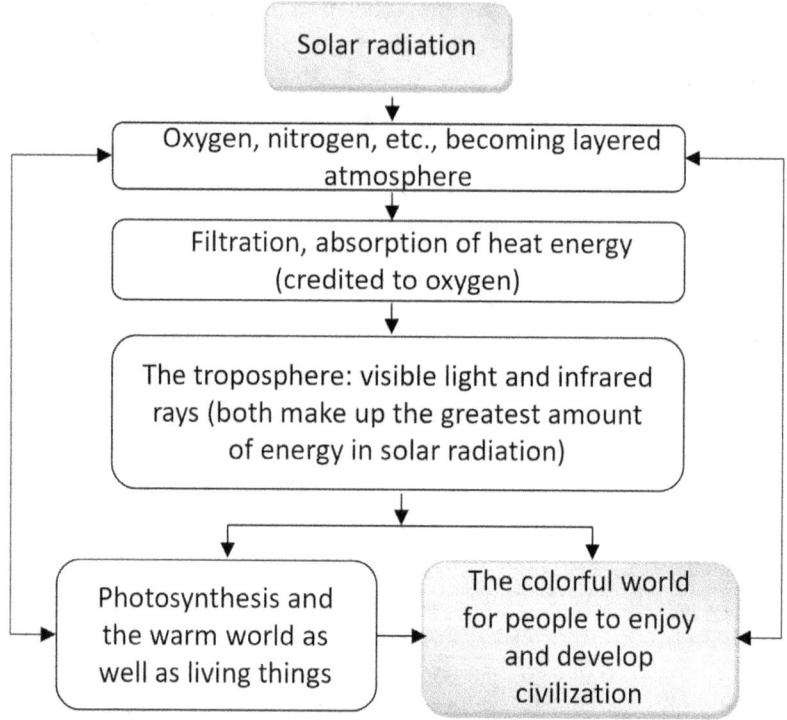

Solar radiation is "filtered" by oxygen. Meanwhile, UV rays harmful to life are absorbed (including the ozone sphere), allowing only visible and infrared light to pass through the troposphere. The former brings light; the latter brings heat. A wonderful natural arrangement!

The Timely Change in Temperature and the Timely Appearance of the Planetary Wind System

Global atmospheric temperature is a crucial factor that influences the evolution of life. The temperature is constantly changing with a stochastic process, subject to many factors such as astronomical, geological, geographical, content of gases in the air, marine temperatures, seawater salinity, the extent of plant cover, etc. Surprisingly, the changing temperature over billions of years has been favorable rather than adverse to the evolutionary process.

During the late Paleozoic and Mesozoic Eras, the global temperature was in a higher range (except for the two great ice ages; according to modern Earth history, there was no ice sheet in the North Polar Region, including Greenland, until the Pleistocene), while organisms with a weak vitality began to live on land and dinosaurs and gymnosperms prevailed. But the global temperature dropped sustainably in the Cenozoic Era (Figure 1-4). Suppose the temperature drop had occurred in the Mesozoic Era. In that case, the evolutionary process could not have happened as it did because a higher temperature is an essential condition for dinosaurs to exist.

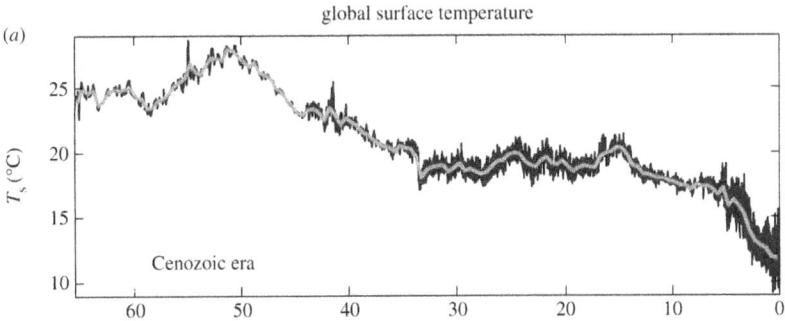

Figure 1-4: The process of the temperature dropping in the Cenozoic Era (Source: James Hansen, Makiko Sato, Gary Russell, and Pushker Kharecha, CC BY 3.0).

The causes of the temperature drop remain a complex problem that needs further research.

Scientists widely accept that a range of high mountains, from the Alps to the Himalayas, began to ascend (see Chapter II.3, The Mysterious Plate Movement) and resulted in a series of effects, such as the increase of rainfall, stronger weathering, etc., bringing the decrease of carbon dioxide in the air and the temperature drop accordingly. If this theory holds, it implies that rainfall before that time would have been much less than afterward. Regardless, the temperature drop took place.

The global temperature drop in the Cenozoic Era eventually resulted in the expansion of a thick ice sheet in the polar region (Greenland and the Antarctic continent are the base of the ice sheet). This ice sheet was responsible for the formation of the substantial difference in temperature between the equator and the polar region, a primary condition of the establishment of the planetary wind system (the other condition being the Earth's rotation and tilt axis) and the global climate zones (see Figures 2–4 and 2–5 in Chapter II.1).

The expansive polar ice sheet could not have formed without the sustainable temperature drop in the Cenozoic Era. As the difference in temperature between the equator and the two Polar Regions gradually increased, the planetary wind system progressively formed. The timely formation of the planetary wind system accelerated various environmental conditions, promoting the evolution of life.

In the plant kingdom, angiosperms started to dominate. Specifically, herbs began to thrive from the mid-Tertiary Period, and grasses, including monocotyledons, began to flourish in the late Tertiary Period. Correspondingly, herbivores appeared during this era. Subsequently, grasses evolved into wild wheat and other edible plants, and herbivores evolved into various animals, such as primitive horses.

A planetary wind system brings wide precipitation, the best way to form water circulation worldwide. Without regular precipitation, the world would become a desert. Therefore, the dropping of global temperatures in the Cenozoic Era can never be undervalued (see Chapter II.1, Water, Precipitation, and Rivers).

But why were global temperatures higher during the late Paleozoic and Mesozoic Eras and began dropping in the Cenozoic Era (no matter the causes), gradually forming the planetary wind system? We can consider that the changes are the time effect manifesting in the Earth's processes.

Given the significance of the planetary wind system to humanity, we will examine in detail the time of its forming in the next chapter.

THE RATIONALITY OF THE EARTH'S DEVELOPMENT IN TIME

The Timely Appearance of the Great Ice Age during the Quaternary Period

The Great Ice Age of the Quaternary Period is the most significant fruit of the Earth System's development, which began about 2.5 million years ago and remains ongoing. The global temperature drop, continental drift, and the formation of the polar ice sheet (continental drift led to the appearance of the Antarctic continent and Greenland, a significant condition) all occurred in the late Cenozoic Era. It is a wonderful natural arrangement; in the absence of any of them, today's world would not be possible. However, scientists have not found a generally accepted theory to explain the occurrence of the Great Ice Age of the Quaternary Period.

The Great Ice Age is characterized by a climatic alternation of ice ages (glacial epoch) and warm ages (interglacial epoch), each lasting about a few ten thousand years (refer to the Milankovitch Theory below). This alternation gradually created various environmental conditions, including glaciers, rivers, lakes, the expanse of plains with thick layers of soil, etc., which are favorable to the wide presence of life and evolutionary processes. The most significant fruits brought by the Great Ice Age are as follows:

Although paleo-polyploidy appeared millions of years ago, the massive appearance of angiosperm was in the Quaternary Period, which may have been induced by climatic alternation (see below). Polyploid plants have strong vitality and have made green cover almost everywhere, including the high mountains, high latitudes, and even some areas averse to plant growth. Grasses are essential polyploid plants, including many useful species critical for the subsistence of early humans, such as wild wheat, rice, and cotton, as well as various crops such as vegetables and fruits.

The expansion of grasslands significantly accelerated the evolution of herbivores. As a result, many animals, such as wild horses, oxen, donkeys, camels, pigs, sheep, silkworms, dogs, cats, chickens, and ducks, began to thrive. Thus, we can assert that all species essential to humans are directly or indirectly connected to polyploidy. Polyploidy is crucial for human survival and development! The initial development of civilization would have been impossible without grasses and herbivores, which provided beneficial resources for food, power, clothing, and transportation in early human history. We might consider these as gifts from nature, or natural arrangements. How thoughtful nature is!

All these resources can be credited to the massive appearance of polyploid plants. However, evolutionary processes could not have fully matured without the Great Ice Age. The process seems reasonable:

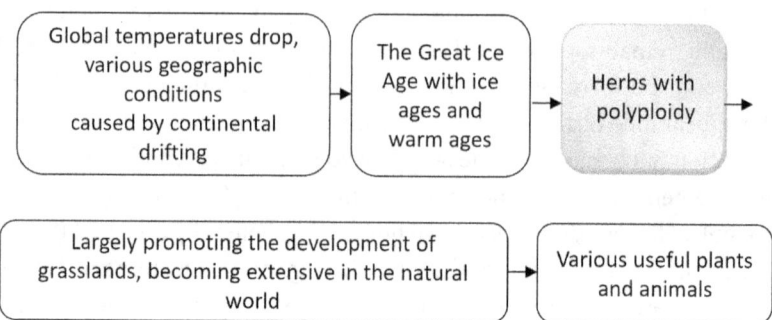

Perhaps the greatest miracle of all time occurred during the Quaternary Period. The convergence of the planetary wind system, soil, rain, wild wheat, horses, oxen, a suitable range of atmospheric temperatures, ideal oxygen content in the air, the discovery of shallow copper and tin ore deposits, and various plants through numerous evolutionary and geological processes ensured human development. This allowed people to gather and store food, use fuel, and employ animals to assist with labor. With the necessary materials and implements, humans had the perfect environment to live and develop early civilization.

So, the great natural meeting between the animate and the inanimate was just right for early humans to exploit, and human intelligence was developed during this time (see Chapter II.1, Four Seasons in a Year, 24 Hours in a Day, and Utilization of the Two Primary Substance Chains).

Grasses are small but great! A wonderful natural arrangement:

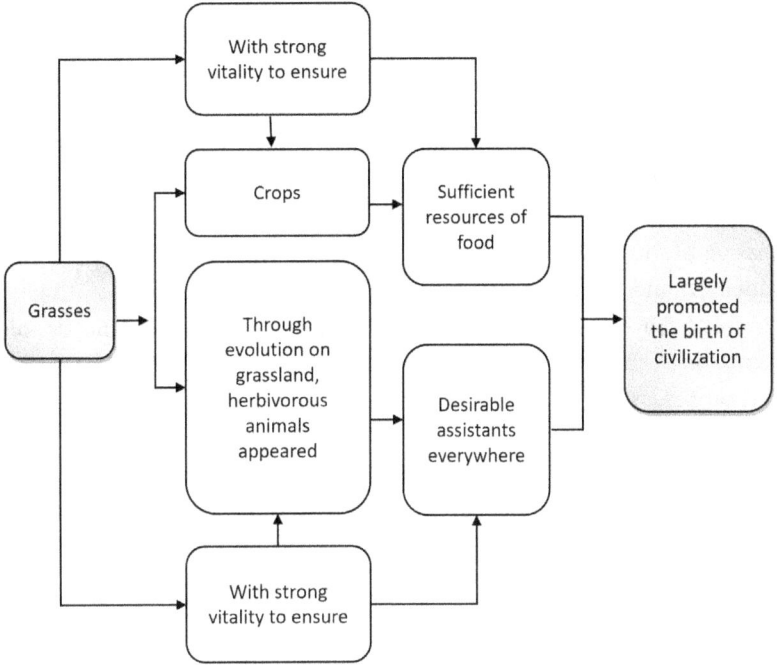

Though the origin of polyploid plant life remains a scientific mystery, it's important to note the significant emergence of polyploid plants during the Quaternary Period. The ice age's cold weather caused many plants to wither away, but some survived due to the transition from diploidy to polyploidy. This adaptation likely resulted from a long process of natural selection. Polyploid plants withstood the harsh, cold climate and became more resilient than their diploid counterparts.

In the natural world, polyploidy is widespread in the plant kingdom, especially among herbs and grasses. Without polyploid plants, our green world would lack its current vibrance, the evolution of many useful species (both animals and plants) would have faced significant obstacles, and the emergence of early civilization would have been challenging and uncertain. Polyploidy acts as a power booster in the evolutionary process of plants, essential for the richness and diversity of the natural world.

The evolutionary process of horses suggests the existence of expansive grasslands. It wasn't until the Pleistocene that *Equus Linnaeus*, grass-eating equine inhabitants of semi-arid regions worldwide, first appeared. They could run at great speeds for long distances, although the earliest primitive horse originated about 50 million years ago. Thus, we may conclude there would be no modern horses without extensive grasslands in the late Tertiary and the Quaternary. Nature created many varied species long before the Quaternary, but not until the appearance of the Quaternary Ice Age did they reach maturity, including human beings.

Here, the most critical factor is the climatic alternation of ice ages and warm ages caused by slight changes in the three orbital elements of the Earth, namely the obliquity of the ecliptic, eccentricity, and the precession of the equinoxes. The thermal transformation between the Sun's radiation and the Earth's surface, the Milankovitch Theory, is a highly complex process (see Figure 2–2 in Chapter II.1). Whether climatic alternation occurred throughout the Great Ice Age is a question that needs further research. Regardless, climatic alternation is significant and cannot be overrated.

Why was the Quaternary Ice Age with the climatic alternation formed in due time? Why did the great natural meeting occur just in time for early humans to exploit and enjoy? We may consider that this is the time effect manifesting itself in the Earth's processes.

The Timely Process of Seawater Becoming Salty

We cannot neglect any natural factor involved in the Earth System. The salinity of seawater is a natural factor that seems minor but can impact global climate, precipitation, temperatures, and humidity. This section, however, focuses on the timing of the saltiness of the sea, the process we will explore in detail in the next chapter.

Were the oceans always salty? If not, when did they start to become salty?

Nature faces two inevitable missions during its development on our planet: the transition of life from the sea to land and the salinization of seawater. These two processes are closely interconnected. Modern science explains that salts, such as sodium chloride and magnesium chloride, are carried into the ocean by rivers, requiring several essential conditions, including sufficient precipitation. The sequence of this process is as follows (see Chapter II.1, Water, Precipitation, and Rivers):

THE RATIONALITY OF THE EARTH'S DEVELOPMENT IN TIME

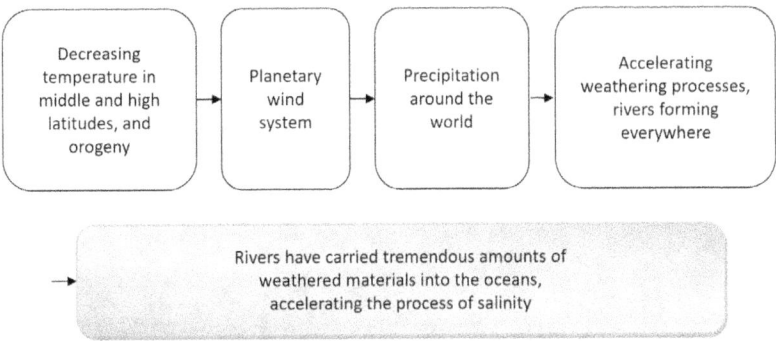

Few rivers may have existed before the Cenozoic Era, and they may have originated due to special topographical conditions rather than heavy rainfall. However, the massive appearance of rivers undoubtedly began with plenty of precipitation caused by the planetary wind system and the orogeny caused by continental drift (see Chapter II.1, Mountains). So, the best hypothesis regarding the two natural missions of life on land and seawater salinity would be that life landed first, and much later, rivers massively appeared and accelerated the process of seawater becoming salty.

If seawater had become salty before the emergence of life onto land, say, in the Paleozoic Era, all the shallows and marshland left by the regression (see Chapter I.1, The Timely Appearance of Life on Land) would have been salty water, and all life would have adapted to saltwater. As it was, there are no signs to suggest plants and animals on land were saltwater life in the Paleozoic Era.

On our planet, saltwater life, with few exceptions, cannot survive in a freshwater environment, so the sequence of these two missions could not have been reversed. For example, mangroves (angiosperms) have developed complex physiological traits to adapt to the high salinity of seawater (Figure 1–5). In contrast, plants from the Paleozoic Era (such as pteridophytes) did not exhibit such complex adaptations.

Figure 1-5: Mangroves with many brace roots (prop roots) exist in the ooze of tidal areas. The physiological functions of the leaves are much different from those of land plants (Source: Hans NYC, CC BY 2.0).

But why were the two natural missions arranged so well in due time? We may consider that the order is the time effect manifesting itself in the Earth's processes. Given the significance of seawater to humanity, the timing of seawater becoming salty will be examined in detail in the next chapter.

The Timely Mass Extinction of Dinosaurs

The extinction of dinosaurs has been a big puzzle in science. The current hypothesis for the mass extinction of dinosaurs is that a collision on the Earth by a small celestial body occurred about 65 million years ago. If the hypothesis is true, the collision would have been an extraordinary event of precise magnitude and timing.

From a scientific perspective, dinosaurs are a crucial link in the evolutionary process. They may have served as transitional animals from oviparity (amniotic eggs of dinosaurs) to viviparity, with some species evolving into primitive birds and other animals. The mass extinction of dinosaurs also paved the way for the evolution of mammals. Regardless of the factors that ultimately led to the permanent extinction of dinosaurs, they were removed from the evolutionary process after completing their natural missions.

But if the extinction of dinosaurs had occurred much later, or if dinosaurs could have survived through the present day, could humans and civilization have emerged? What the modern world would be like if these incredible animals were still present everywhere today is difficult to imagine.

Dinosaur extinction is indeed a miracle in the evolutionary process! We may consider that the dinosaur extinction event is the time effect manifesting itself in the Earth's processes. Nature has always done what it should in due time, even when it is seemingly an accident.

The Timely Mass Extinctions

The evolution of life on our planet would seem to be a magical process that was not merely random but reasonable. The timely mass extinctions appear to support this statement.

The first stage of the evolutionary process, which took about 3 billion years, occurred in the ocean before the Cambrian Explosion. Had the process continued at this gradual pace, life as we know it might not have emerged before the Sun's radiation began to weaken (see Chapter I.1, The Timely Appearance of the Earth). How did nature address this challenge? The answer may lie in mass extinctions, which could promote natural selection or elimination. Throughout the long evolutionary process, there were five significant mass extinction events: the Ordovician-Silurian (440 million years ago), the Late Devonian (365 million years ago), the Permian-Triassic (230 million years ago), the Triassic-Jurassic (195 million years ago), and the Cretaceous-Paleogene (65 million years ago). Remarkably, certain species consistently survived these mass extinctions and continued to evolve and thrive.

According to the evolutionary record, these survivors are invertebrates in the ocean → fish → amphibians → reptiles → mammalians and birds. Besides those mass extinctions, relatively normal extinctions occurred in the evolutionary process, also furthering natural selection.

Extinctions are like steps, promoting evolutionary processes to reach more complex levels. As for the causes of extinction, there are many, such as an impact by an asteroid, a supernova outburst near the solar system, massive volcanic eruptions, etc., but they all are no more than hypotheses.

Why were living organisms not all destroyed during a mass extinction? Why did some "seeds" always survive each extinction and become new species prevailing on the Earth? Why did mass extinctions occur in due time, respectively? We may consider that is the time effect manifesting itself in the Earth's processes.

The Timely Appearance of the Great Rift Valley in the East Africa Plateau

The origin of humans, indeed a complex and even subtle natural process, is still a puzzle in science. But there is no doubt the key to this natural process is the geographic state of a "womb" that could be pregnant with the "fetus" of human beings.

The Great Rift Valley (GRV) in East Africa is just such a womb, geologically a graben, an extensive area formed by strata subsidence between two faults (Figures 1–6 and 1–7 below, and Figure 2–28 in Chapter II.3). The geological process forming that area resulted from the crust cracking caused by rising hot magma, accompanied by massive volcanic eruptions. Scientists believe the process began in the Oligocene era (30 million years ago) and then, through a lengthy weathering process and other natural factors, gradually formed special environmental conditions with a fertile soil layer, expanses of forests, and plenty of water resources, favorable to the evolutionary development of primates.

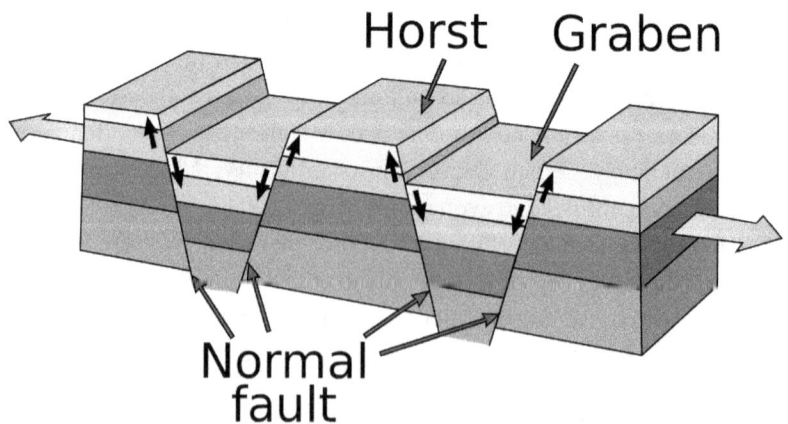

Figure 1–6: A graben (Source: Public domain).

THE RATIONALITY OF THE EARTH'S DEVELOPMENT IN TIME

Figure 1-7: The Great Rift Valley, a natural graben (Source: Roma Neus, CC by 3.0).

Generally speaking, the primate appeared in the world as early as in the Eocene (56 million years ago), but only the timely appearance of the GRV led to the evolution of bipedalism about 3–4 million years ago. The well-known skeletal fossils of Lucy, belonging to *Hominidae*, once lived 3.2 million years ago in that area. The GRV is the only place in the world to have completed this unique evolutionary process, possibly because of a transition from forests into the expanse of savanna (a prairie islanded with woods) in that area. What was once the expanse of forests became the expanse of savanna (see Chapter II.1, Forests and Grasslands).

As stated above, massive volcanic eruptions produced a lot of volcanic ash, making the soil layer more fertile than other savannas. Such fertile soils resulted in the plants thriving with plenty of fruits, contributing to the subsistence of primates, an advantage over other areas at the time. In this process, a fascinating fact is that the strange behavior of dung beetles, which existed widely in that area, contributed to the soil's fertility.

The geographic conditions in the GRV were distinct from those in other parts of the world. Was it necessary for the GRV formation process to take around 30 million years? Yes, it was essential, as the process involved multiple contributing factors, including geological activities, weathering, water resource formation, forest development, savanna expansion, and primate evolution. The timing of the GRV formation seems to have been perfect. The completion of the GRV formation, including the savanna's expansion, coincided with Australopithecus's appearance in this "womb" about four to one million years ago.

Research on the origin of humans is complex. Various hypotheses remain unproven, but the role of the GRV in this subject is generally accepted. Many findings, including the fossils of footprints and bones, support the GRV as the birthplace of humanity. All evidence indicates primates existed in this area at different times, representing a progressive evolutionary process. Although few deny the importance of the GRV in the origin of *Hominidae*, the views on the emergence of modern humans are contradictory. Most but not all researchers believe modern humans also originated from that area. Without that "womb," the birth of humanity on our planet would have been less likely because there seems to be no other place where scientists have found so many traces or signs of humans' origin. Consequently, humanity could not have appeared without the unique role of the GRV.

The series of significant changes in geographical conditions occurring in due time and in the same location within a certain area across the equator and not far from the ocean would seem to be purposely arranged by nature to construct a "womb" for becoming pregnant with the "fetus" of human beings. Don't you think so? Although we do not know the exact background of the natural processes mentioned above, we can still imagine that if the development of the GRV had not been in that location or not in due time, what would have happened?

Isn't the process in time and space amazing? Why was the GRV formed just in due time? Why does the GRV only occur near the equator in East Africa and not anywhere else? Why were the transitional environmental conditions from forests to grasslands realized just in time to promote the completion of the evolutionary process from early to bipedal primates? We may consider this the time effect manifesting itself in the Earth's processes.

Given the importance of GRV, its formation will be further examined below from different perspectives (see Chapter II.3, The Mysterious Formation of the Great Rift Valley and the Red Sea).

The Timely Formation of the Mediterranean

The formation of the Mediterranean region is a fascinating natural process that unfolded over space and time. It is well-known that modern science originated in this area, particularly in the large peninsulas of Greece and Italy, where the Renaissance took place from the fourteenth to the seventeenth century. What is surprising is the geologically complex process behind the Mediterranean's formation.

Briefly, while the African and Indian plates each drifted respectively toward the north, their drift eventually resulted in the collision between them and the Europe-Asia plate, consisting of two geological processes, plate collision and fold (Figure 1–8). The force of the collision was so tremendous that it caused the rise of a range of high mountains from the Alps to the Himalayas (both are fold mountains) and closed the ancient Tethys Sea (Figure 1–3: "65 ma ago," "the existing distribution"). But the ancient Mediterranean surprisingly remained intact after the collision during the Young Alpedic in the Cenozoic Era. Later, the Mediterranean filled with seawater from the Atlantic Ocean through the Strait of Gibraltar about 5 million years ago. Before that, the Mediterranean was a desert expanse.

Figure 1–8: Folded gneiss (Source: The American Museum Journal, c.1900–18, public domain).

Another surprise is that the two peninsulas, Greece and Italy, also remained intact after the collision. From the viewpoint of probability, the formation of these two peninsulas was not simple, although we do not know the exact geological processes that formed them.

A notable fact is that the Mediterranean lies on the seismic belt today, near the tectonic plate boundaries (see Figure 2-19 in Chapter II.2 and Figure 2-38 in Chapter II.3). If the ancient Mediterranean was closed, we might naturally take it for granted. Furthermore, the geological process that formed the Mediterranean occurred on the three-dimensional, spherical surface of the globe rather than a flat plane, which could bring more complexities. So, don't you think the existence of the Mediterranean and the two peninsulas is a natural miracle?

Why did the ancient Mediterranean region remain for roughly millions of years despite the collision? And why did seawater pour into the Mediterranean through the open Strait of Gibraltar in due time? Don't you think the flow with sufficient width from the Strait of Gibraltar to the Mediterranean is a clever natural arrangement?

We may consider that the Mediterranean's development is the time effect manifesting itself in the Earth's processes. Given the importance of the Mediterranean, its formation will be explored again in detail in other chapters from different perspectives (see Chapter II.3, The Mysterious Plate Movement).

The Timely Formation of Two Primary Substance Chains

The formation of mineral deposits on our planet in space and time is exciting and thought-provoking (see Chapter II.1). Logically, there are two primary substance chains in the natural world, a natural chain in energy and a natural chain in metal, which led to the development of civilization from ancient times to the present.

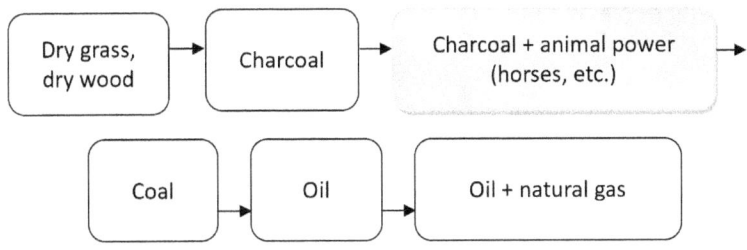

The Natural Chain in Energy

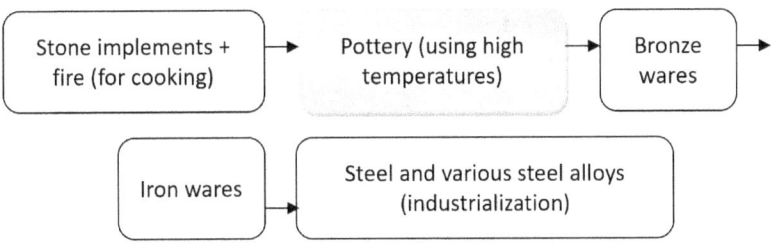

The Natural Chain in Metal

But how did nature make those two chains available? Nature began with crustal movement to make mineral deposits of those two chains long before the appearance of humans on Earth. For example, the natural process of forming coal deposits can be traced to the Carboniferous Period, about 300 million years ago.

In that geologic period, Pteridophyta forests prevailed. Nature used wood as raw materials through complex geological processes with low probability, forming the coal deposits we use today. Surprisingly, with the limited space of sedimentary layers, a limited amount of coal, oil, and natural gas should be stored in the ground. Still, there is enough for the development of civilization from the early days to now and into the future. The large quantity of natural fossil energy trapped in our planet is an incredible miracle! We must be aware of the great significance of this resource and not take it for granted. Likewise, metal deposit-forming processes began with the birth of the Earth. A series of geological processes accompany their formation, such as crustal movement, eruptions, metamorphism, weathering processes, and sedimentation.

The "seeds" of mineral deposits, including nonmetallic raw materials, were "sown" before the Cambrian Period. However, through various complicated geological processes of mineralization until the Quaternary Period, the "fruit" of all kinds of mineral deposits matured, becoming available for humans to mine productively.

Today, when you use natural gas or oil or coal for cooking, heating, lighting, etc., do you realize they are a generous gift from nature? Scientists think nature will provide sufficient natural energy for another few hundred years. How thoughtful nature is! Pay close attention here. The probability of forming fossil energy is extremely low. Isn't this a thought-provoking fact?

But why did nature deposit minerals as early as billions of years ago, available in sufficient quantities for our needs? We may consider those deposits the time effect manifesting itself in the Earth's processes.

Above, we have examined the Earth's development from a time perspective. Step by step, follow the timeline from the planet's birth to the Quaternary Period; every time section of this process is necessary and timely. Therefore, the Earth's development is orderly and a gradual, logical process. Without the required timing, any essential factor would be unable to fulfill its natural mission.

As science advances, scientists will find more and more natural events that occurred in due time in the Earth's development. For example, scientists reportedly discovered that bryophytes appeared in massive numbers on land about 470 million years ago and produced a great deal of oxygen in the air, reaching today's oxygen content about 400 million years ago.[1] Without the prevailing of bryophytes, humanity could not have appeared!

These two primary substance chains will be examined in detail in the next section. After reading this chapter, can you understand the presence of the Reason in nature?

---

1. Lenton, *Earliest Land Plants*

Chapter I.2

The Rationality of the Earth's Development in Time
The Natural Development Chain

Above, we string together several related natural events in chronological order to show the timeliness of the Earth's development (the requirement for timing), forming a complete natural development chain. Among them, the emergence of the planetary wind system is a turning point in the formation of the existing global environment. With the emergence of the planetary wind system on a large scale, the problem of the seawater salinity process will inevitably follow. These are two significant factors affecting the global environment. To show the natural development chain convincingly, the timeliness of these two processes (planetary wind system, seawater salinity) is argued as follows. Natural development on Earth is likely in chains and not some other way.

Wind is a natural phenomenon that could have occurred randomly in any geological era. However, the atmospheric circulation we experience today, known as the planetary wind system, is distinct. It happens regularly on a global scale and is governed by strict conditions: the Earth's rotation with the obliquity of the ecliptic and the presence of ice sheets in the Polar Regions.

The polar ice sheet is an essential condition that forms the climatic zones, seasonality, planetary wind system, and regular global precipitation. But when was the polar ice sheet fully formed? Generally, there are three recognized periods of significant glaciation, known as the Ice Ages, during the Earth's development. The first is the Sinian Ice Age (about 600 million years ago, when all life existed in the sea). The second is the Carboniferous-Permian Ice Age (about 290 million years ago, occurring in some parts of the Southern Hemisphere only), and the third is the ongoing Quaternary Period Ice Age. Before the late Cenozoic Era, the North Polar Region would have had little to no ice sheet. However, the specific causes behind the formation of these Ice Ages have yet to be fully understood.

The process of the polar ice sheet's expansion is a lengthy and intricate geological event that likely began around 14 million years ago, with the formation of the south polar ice sheet predating that of the north. However, the precise timeline remains uncertain. This process appears to be closely linked to continental drift, global cooling, and specific astronomical factors, such as the Earth's tilted axis. However, the exact interplay between these factors is not yet fully understood.

It is believed that the development of the Antarctic continent and Greenland, resulting from continental drift, played a crucial role in forming extensive polar ice sheets there (see Chapter II.3, The defining of the Antarctic and South American Continents). Consequently, understanding the formation of the polar ice sheets is integral to addressing the two related subjects below.

THE ORIGIN OF THE PLANETARY WIND SYSTEM

As a common natural phenomenon, the planetary wind system (wind) has a vital role in the evolutionary process of life and the development of human civilization. We may say that without such a natural phenomenon as wind, there would not be civilization as we know it.

We must go back a long way to understand what causes wind and when this phenomenon began. We should look briefly at the planet's creation and then seek some understanding of the nature of wind itself. There is no firm proof of how Earth was formed, but for the sake of our discussion, it is sufficient to accept that Earth was different on the surface then from what we can see today in satellite images.

In the earliest times, as cooling and condensation took place, a crust was formed, the extent of which is unknown. Later sub-surface

THE RATIONALITY OF THE EARTH'S DEVELOPMENT IN TIME

movements likely caused mountains to form. The continental landmasses were not as we now know them, nor were the oceans. Collisions between tectonic plates have caused surface movement throughout the Earth's development.

With this simplistic understanding of the Earth's creation, we may move on now to establish some knowledge of winds and how they are produced today. As the Earth rotates on its inclined axis, the Sun unequally heats the Earth's atmosphere at different altitudes and latitudes, resulting in air movement. Wind distribution is also related to atmospheric pressure, so both differing atmospheric pressures and unequal heating of the atmosphere must be considered.

The topography of the landmass also alters wind patterns. Near the equator lies the equatorial calm belt, often called the doldrums. Prevailing winds in the northern sub-tropical belt are northeast trade winds, while in the southern sub-tropical belt, they are southeast. Westerlies predominate between latitudes 30–60° north and south of the equator, while easterlies dominate in the north and south Polar Regions (see Figure 2–6 in Chapter II.1).

There are also seasonal variations of wind because of the changing position of the Sun relative to the Earth, and similar variations are caused by physical differences between landmass and water. Further, we have yet to consider local atmospheric movement caused by offshore and inshore temperature differences and consequential airflow.

This brief look into the causes and effects of wind differences gives a small insight into the complexity of the formation of wind and weather prediction. It is an abstract question even to speculate whether winds were present in the earliest times, as little is known about the formation of the Earth's atmosphere and crust. We will, however, expand our logic for the development of early winds by first examining the recognized eras in time.

The geologic scale of time since the Earth's creation is defined in four main eras (not including the present time to about 11,000 years ago):

Cenozoic Era	11 thousand to 65 million years ago
Mesozoic Era	65 to 250 million years ago
Paleozoic Era	250 to 570 million years ago
Pre-Cambrian Era	570 million to 4.6 billion years ago

For this exercise, we will exclude the Pre-Cambrian era due to the scarcity of fossils. Life during that period lacked shells or bones as indicators of its existence, despite the widespread discovery of rocks from this ancient era. So, we will begin from the middle Paleozoic Era, in which geologic period life began to exist on the land.

Paleozoic Era (250 to 570 million years ago)

The rocks of the Paleozoic Era contain the first abundant fossil records. This era is listed below from youngest to most ancient, in some references extending from 200 to 500 million years ago, and is further subdivided into six periods:

- Permian
- Carboniferous (pteridophyte abundant)
- Devonian
- Silurian (earliest land plants appeared)
- Ordovician
- Cambrian (large fauna of marine invertebrates)

First, let us examine how a tree can stand firmly in the wind while it continues to grow.

From a mechanical perspective, a tree's stability relies on its root system, which intertwines with and adheres to the soil. The extensive root network and the soil it bonds with counteract the force of wind on the tree's canopy, preventing it from tipping over at the base (Figure 1–9). Thus, the cohesion between soil and root fibers and the weight of the soil above the root system are crucial for the tree's stability. The tree species and the soil's moisture level surrounding its roots determine the depth and breadth of the root system, making certain trees more resilient to wind. In areas with shallow groundwater, trees tend to have shallow root systems, along with large leaves and canopies that offer poor stability against wind. Conversely, trees in desert environments develop deep root systems, accompanied by small leaves and canopies.

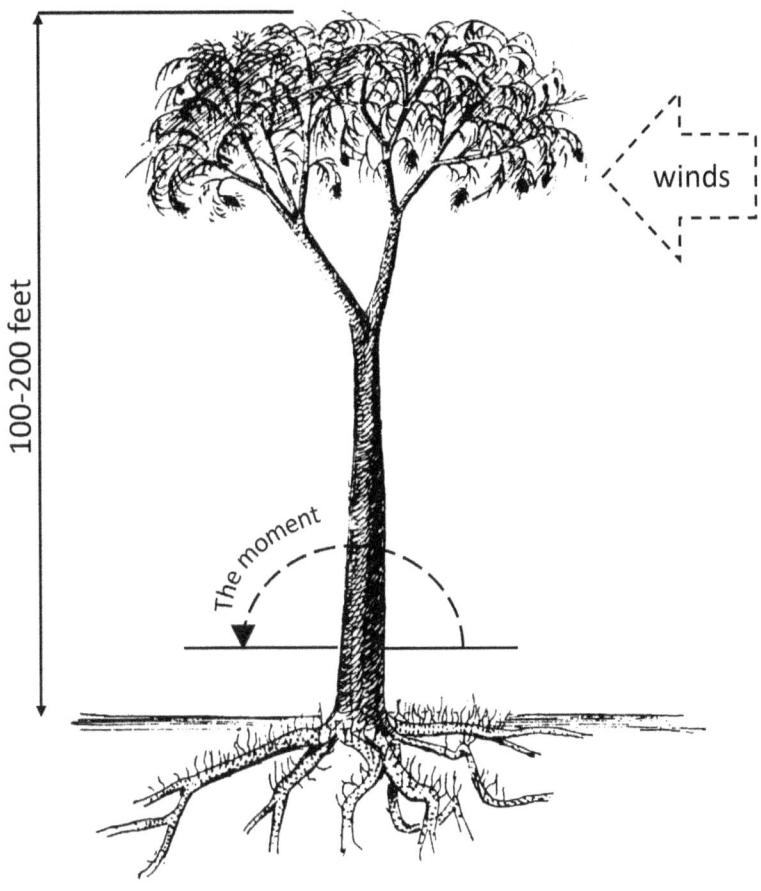

Figure 1-9: A tall tree (Lepidodendron) in strong wind (not to scale). Mechanically, the moment is defined as the product of the force exerted by the wind and the distance from the point of application of the force to the point of rotation, typically the base of the tree. The moment influences how the tree bends or potentially uproots under wind stress. (Source: Eli Heimans, 1911, public domain).

The earliest plant life discovered in fossils dates to 350–400 million years ago, during the late Silurian and Devonian periods of the Paleozoic Era. These pteridophytes had to emerge from shallow waters or marshlands due to their reliance on water for fertilization (hydrophily). The earliest pteridophytes, known as psilophytes (Figure 1-10), were small, slender plants, about 4 in (10 cm) tall, with stems less than 0.08 in (2 mm) in diameter, devoid of leaves or roots. Found during the Silurian and Devonian periods, these plants had their lower stems submerged underwater, with only hair-thin rhizoids at the stem base for material assimilation and attachment (forming a rhizome on the ground). They mark a crucial evolutionary stage, signifying the transition of life from sea to land. However, due to their small size and slim structure, these plants had poor stability, akin to a child learning to stand. Thus, their survival required conditions devoid of strong winds and frost.

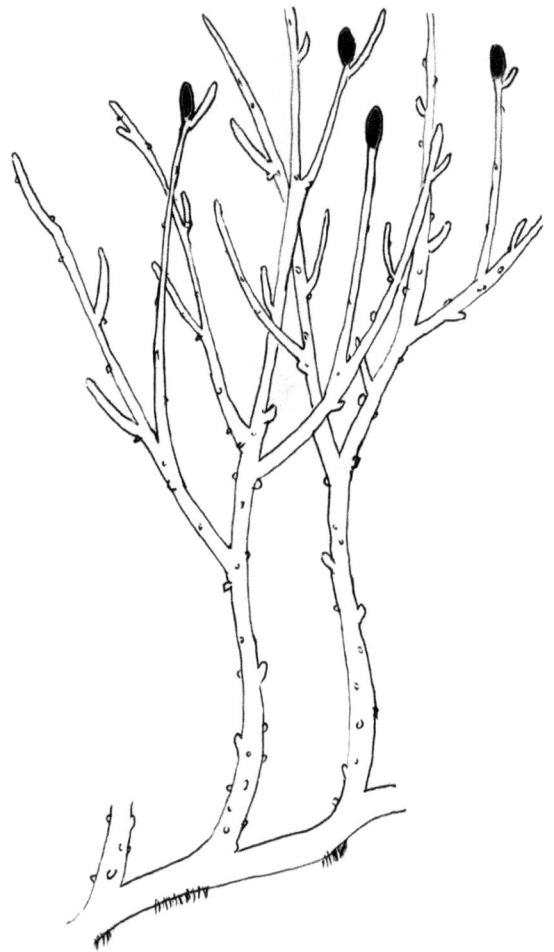

Figure 1-10: The earliest plant on land existed about 400 million years ago. The *Rhynia gwynne-vaughanii* was 7–20 in (18–50 cm) high (not to scale) (Source: Griensteidl, CC BY-SA 3.0).

Psilophytes existed for about 50 million years. Many fossils of various psilophytes are found in Europe, North America, and China. Most landmasses then were scattered in the middle and low latitudes and the equator (Figure 1–3).

In the evolutionary course, what followed psilophytes were Lycopsida, Sphenosida, and Filicopsida, all found in the Carboniferous and Permian Periods (about 300 million years ago). *Lepidodendron* (Figure 1–9), *Sigillaria* (Lycopsida), *Calamites* (Sphenopsids), and *Psaronius* (Filicopsida) were high trees with nearly horizontal root systems in the soil layer underwater that prevailed in these two periods (Carboniferous and Permian). The *Lepidodendron* were 100–165 ft (30–50 m) high, 3–6.5 ft (1–2 m) in diameter, the *Sigillaria* 100 ft (30 m) high, 6.5–10 ft (2–3 m) in diameter while *Calamites* were also 100 ft (30 m) high. *Psaronius* were about 32 ft (10 m) high but only 4 in (10 cm) in diameter. All these high trees formed vast stretches of pteridophyte forests at different times and in various parts of the world, propagating from generation to generation and lasting about 100 million years.

During this era, the soil condition beneath the shallows or marshland, which sustained the forests, is believed to have been relatively loose with poor cohesion. However, deeper soils may have been more compacted (even today, most soil layers at the bottom of rivers, lakes, and seas remain relatively loose). Researchers generally perceive that the stability of these towering trees, reaching heights of 100–165 ft (30–50 m) with diameters of about 3–10 ft (1–3 m), and shallow root systems submerged underwater, was not as secure as that of tall trees today. Consequently, a prerequisite for their survival would have been the absence of strong winds; otherwise, the force of strong winds on their large crowns would likely have caused them to topple (Figure 1–9). We can imagine a tree higher than a twenty-story building with shallow roots growing in the water. Can it be stable?

Let's look now at the appearance of mangrove trees growing from silt in the shallow sea today; each has many prop roots supporting the trunk to effectively resist attacks from strong winds and waves acting on them (Figure 1–5). However, the pteridophyte in the Paleozoic Era was comparatively simple in its structure, without any prop root to support its trunk. The striking contrast between these trees indicates the significant difference in the environment of the two eras. The Paleozoic Era environment was likely calmer to safely support those high trees growing in the shallows and marshland.

Mesozoic Era (65 to 250 million years ago)

Within the Mesozoic Era, three recognized periods separate and define the appearance of various animal and plant life.

Cretaceous	extinction of dinosaurs
Jurassic	dinosaurs' zenith
Triassic	appearance of dinosaurs

During this geologic era, much of the shallows and marshland gradually disappeared due to the movement in the Earth's crust, which removed the conditions for supporting pteridophyte forests, leading to their disappearance in the late Paleozoic Era. Gymnospermae followed them, adapting to the changing environment and once prevailing over the whole Mesozoic Era, lasting about 150 million years. Gymnospermae will be examined in more detail later. Turning to the animal world, we find many reptiles began to appear, of which the dinosaur was the most remarkable. Some species of dinosaurs weighed up to an estimated 70 to 100 tons.

A particular feature of dinosaurs is their large feet. We can imagine that the pressure of their footing on the ground could reach as high as more than 10 tons, even 20 tons under each foot. Considering their momentum, the action on the ground would be much greater than that of the static pressure alone. The question arises as to what kind of ground could bear such heavy animals and allow them to move freely on the land and the shallows, including the shoal where plant life abounded for their survival.

Although dinosaur fossils have been discovered on all continents except Antarctica, scientists conclude that they lived mainly in the lowlands with some shallow waters where ground surfaces were relatively flat and food was abundant.

If the soil layers at the time were entirely composed of silt and clay, the soil under the shallows and in the proximity of water cover would become mud due to the high moisture content, adversely affecting the mobility of these heavy reptilian creatures. From a mechanical viewpoint, the soil on the land would probably be composed of sand and clay in a denser state that would provide a higher load-bearing capacity, whereas, under the water, a comparatively looser soil state would prevail.

Dinosaur fossils are generally found in sandstone and siltstone layers, which may support the above observation.

Another fact worth noticing is that no bushes or herbs existed in that geologic period. These began appearing in the Cenozoic Era. The root system of bushes and herbs can significantly and effectively strengthen the soil layer. From a mechanical viewpoint, we can imagine that the soil layer in the Mesozoic Era with high moisture would be composed of relatively coarse grains lacking cohesion. Under this condition, the soil layer could have a higher load-bearing capacity to support a dinosaur's weight (Figure 1-11).

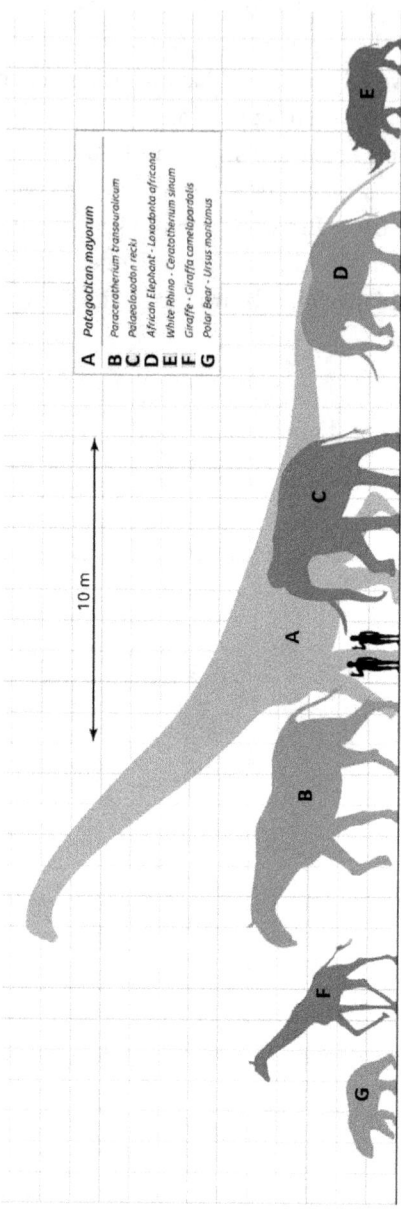

Figure 1–11: A comparison of the vegetarian dinosaurs with other animals. The soil layer must have enough load-bearing capacity to support dinosaur weight, whether they moved about on land, in swamps, or in shallow waters (Source: Steveoc 86, CC BY-SA 4.0).

In China's Sichuan Province, fossils of *Mamenchisaurus hochuanensis* were discovered in alternating layers between siltstone and thick sandstone layers. These dinosaurs were about 70 ft (22 m) long, 11 ft (3.5 m) high, and weighed an estimated 50 tons, with four feet over a foot or two long, which lived about 130 million years ago. We might infer that the soil on the land at that time must have been composed of dense sand and silt with some clay that supported the free movement of the dinosaur and plant life. Another example is from footprint fossils about 5 ft (1.5 m) long, 4 ft (1.2 m) wide each, on a thick sandstone layer, discovered in Dampier Peninsula, Western Australia (Figure 1–12). The estimated weight of the dinosaur is about 50 tons. The footprints' formation suggests the underground water level reached the ground surface.

Figure 1–12: Footprint of a sauropod in Dampier Peninsula, Western Australia. Dinosaur footprint fossils (trace fossils) also appear in sandstone stratums worldwide (Source: Salisbury, Steven W., Anthony Romilio, Matthew C. Herne, Ryan T. Tucker, and Jay P. Nair. "The Dinosaurian Ichnofauna of the Lower Cretaceous (Valanginian–Barremian) Broome Sandstone of the Walmadany Area (James Price Point), Dampier Peninsula, Western Australia." *Journal of Vertebrate Paleontology* 36, no. S1 (2016): 1–152. https://doi.org/10.1080/02724634.2016.1269539.).

A massive concentration of dinosaur footprints was found at an excavation site in Shanghang County, southeast China's Fujian Province, in 2021. The stratum of the footprints is a large area of siltstone. Dinosaurs lived in the Late Cretaceous.

Let's return to examining plant life and Gymnospermae. Cycadopsida, Ginkgopsida, and Coniferopsida were tall trees once prevailing in the Mesozoic Era. The root systems of these trees, growing in such a soil layer with shallow groundwater levels as indicated above, would not be as deep as today's trees. With the necessity for anemophily (wind pollination) for these trees, winds were undoubtedly blowing over the land in the Mesozoic Era. However, as these tall trees had shallow root systems and large crowns, strong winds were not likely present, with the sandy soil layer having little cohesion to support the trees.

Anemophily for pollination requires merely a breeze to carry the pollen. In this era, on the one hand, dinosaurs prevailed; on the other hand, the vitality of Gymnospermae, which was the food resource for the enormous animals, was relatively weak and slow in its growth compared with Angiospermae today. The existence of dinosaurs would have been an adverse factor to the plant's existence. So, to maintain the weak ecological balance between these two sides, neither strong winds nor low temperatures would likely exist in this era. Low temperatures would lead to deciduous leaves of some kinds of plants, such as Ginkgopsida, cutting off the food sources for dinosaurs. The diagram below shows the weak ecological balance between these two sides:

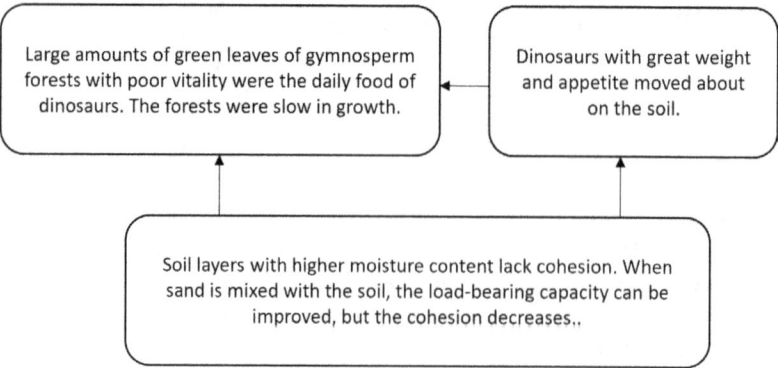

During the Earth's development, the various natural factors for forming the soil to support plant life in the Paleozoic and Mesozoic Eras included the presence of wind. From an evolutionary perspective, this stage was an early developing one, still far from maturity with the relevant soil-forming factors. Organisms such as plant life, micro-organisms, and animals were much fewer on Earth in varieties of species and in their abundance if compared with the more recent.

Dinosaur footprints worldwide are found in sandstone or siltstone, indicating that the soil at the time lacked soil cohesion. As soil quality and cohesion improved over the eras, the stability of the trees also improved. This slow accumulation process includes the accumulation of clay minerals in the soil (see Chapter II.1, The Soil).

Furthermore, the vitality of plants in the Mesozoic and Paleozoic Eras was relatively poor. It is logical to assume that if thunderstorms, driven by strong atmospheric circulation, had frequently occurred during that period, the evolution of pteridophyte and gymnosperm forests would have faced significant obstacles.

Cenozoic Era (up to 65 million years ago)

The Cenozoic Era commenced about 65 million years ago, remains ongoing, and is represented by two main periods, namely the Quaternary period (2.5 million years ago to the present) and the Tertiary period (65 million to 2.5 million years ago).

The plant kingdom has two main groups: lower plant species and higher plant species. Among the higher plants, Spermatophyta is the most prevalent. The appearance of seeds in plants resulted from their evolution on land and adaptation to changing environments. Spermatophyta includes the sub-phyla Gymnospermae and Angiospermae. Gymnospermae were dominant during the Mesozoic Era, but most have become extinct and later turned into coal fossils. Only about 700 species, including Cycadopsida, Ginkgopsida, Coniferopsida, and Yewopsia, have survived to the present day. These plants are sources of timber and raw materials for producing fiber, resin, tannin, and other products.

In comparison, Angiospermae are more advanced in the evolutionary process, characterized by the presence of flowers and ovaries, setting them apart from other extinct plant types. Today, approximately 250,000 species of Angiospermae are found almost everywhere worldwide. Among them, herbs generally exhibit great vitality and are considered more evolutionarily advanced than woody plants.

The first appearance of primitive Angiospermae in the form of woody plants can date back to the early part of the Cretaceous, but herbs did not massively appear on the land until the mid-Cenozoic Era. And some species of herbs, through further evolution, became monocotyledoneae because of the changing environment, including grasses with shallow, fibrous root systems. In the Quaternary Period, among the monocotyledoneae, primitive wheat, rice, etc., began appearing, which is of extreme importance for the development of human civilization.

Let's return now to the subject of wind. Compared with woody plants, there are two principal natural conditions under which herbs can exist and propagate. The water necessary for herbs comes entirely from topsoil water, originating from precipitation, particularly rainfall stored at the top of the soil layer, a limited depth from the ground surface ≤20 in (50 cm) deep. In contrast, water for woody plants comes from deeper soil layers if their root systems reach that depth.

Grasses in the Gramineae family and some herb species rely on wind for pollination (anemophily). However, wind velocity near the ground is typically slower than higher up due to friction between the ground and airflow. Thus, anemophily is effective only when wind speeds near the ground reach a certain threshold. The widespread emergence of herb varieties during the Tertiary Period and the later appearance of grasses indicate that winds and rainfall were common natural phenomena at that time, occurring on a significant scale across the land.

Overview

Earth's development has lasted about 4.6 billion years, dating back to pre-Cambrian times. We have divided time into convenient eras:

Cenozoic Era	The present to 65 million years ago
Mesozoic Era	65 to 250 million years ago
Paleozoic Era	250 to 570 million years ago
Pre-Cambrian Era	570 million to 4.6 billion years ago

These eras, the associated epochs within each era, and their boundaries are not rigidly fixed. Geologic and fossil discoveries in different countries reveal comparatively different times when certain life forms prevailed. However, it can be postulated that the essential natural phenomenon of wind was not always present and more likely to have appeared in the more recent two eras.

To understand this phenomenon, if we consider the life that originated in the sea and its movement to the land, a primary environmental condition is vast stretches of shallows as a transitional environment between the sea and the land. From the viewpoint of soil mechanics, the soil layer from its early beginnings under the water would most likely be relatively loose in the Paleozoic Era, and the bond strength between the root system of vegetation and the soil layer would be weak. Plants growing in shallows or marshlands could not have withstood the significant force exerted on their bases by the wind.

Thus, for trees with large crowns and shallow root systems present in the Paleozoic Era, a primary pre-condition would be the lack of strong winds for the stability of the pteridophyte forests, which were present during that long evolutionary process. Not until the completion of the evolutionary process, when some species of sealife were transferred by natural selection from a sea-based to a land-based existence, freed from the sea to become an entirely land-life system, was their structure able to change. This change took place in the Mesozoic Era, and subsequently, the evolutionary process of the land-life system entered a new epoch that is much more magnificent than the sea-life system.

Scientists believe that gentle winds generally began to appear in the Mesozoic Era and gradually developed. Trees began to evolve with deeper root systems, enabling them to withstand the wind forces developing with different scales and patterns in the Cenozoic Eras.

The described process above concerns the period when the natural wind phenomenon first occurred over a long geologic period and can be traced by identifying fossil remains as the evolving course of plant life.

ANOTHER VIEWPOINT: THE NATURAL CONTRIBUTING FACTORS

We turn now to another viewpoint to understand this phenomenon, the natural contributing factors for forming the planetary wind system.

The Change in Global Temperature over Millions of Years

As described above, the origin of winds on our planet is limited to the geologic period after the landing of life from the sea about 400 million years ago. The question of whether the natural phenomenon of winds occurred before that period or not is inconsequential because life existed then only in the sea.

According to research in paleontology, which is significant in the study of stratified rocks, the remains of plants and animals allow for time marking of various events and species. Scientists believe that the atmospheric temperature from the Paleozoic Era through the Cenozoic Era was comparatively high except for two relatively short geologic periods, the Great Ice Ages in the Carboniferous Period-Permian Period (about 290 million years ago) and in the Quaternary Period (since 2.5 million years ago).

Nonetheless, the Great Ice Age of the Carboniferous Period is generally accepted to have had no expansive ice sheets in the Arctic region, and pteridophyte plants at that time still existed in some middle and high latitudes in the South. In the Paleozoic Era, a hot and humid climate prevailed worldwide, favoring the evolutionary process in which the pteridophyte proceeded from the small slim plants of psilophytes, gradually evolving into pteridophyte forests. Today, the pteridophyte still prefers to grow in the shallows in a tropical climate.

What were the natural conditions in the Mesozoic Era? The discoveries of dinosaur fossils in high latitudes reveal that the atmospheric temperature in those regions was similarly high. Scientists think the climate in the Mesozoic Era was even more hot and humid than in the Paleozoic Era, and the difference in the atmospheric temperature in different parts of the world and at various times of the year was insignificant. Therefore, we can infer that the planetary wind system in these two eras was weak, if it existed at all, because of the slight difference in the atmospheric temperature between the high and low latitudes. In contrast, a wide temperature difference is a factor in forming today's planetary wind system. Currently, the yearly average temperature in the south Polar Regions is about—13°F (-25°C), the north 0°F (-18°C), and the equator 77–82°F (25–28°C).

The global temperature reached its highest level about 100 million years ago in the Cretaceous Period, according to geohistory, and then began gradually falling in the middle and high latitudes during the Cenozoic

Period, lasting until the onset of the Ice Age in the Quaternary Period. Along with the falling of the global temperature, the expansive ice sheets in the south and north Polar Regions were gradually formed before the Quaternary Period. Therefore, the conditions involved in creating the planetary wind system were understandably formed gradually with the falling of the global temperature in the Cenozoic Era (Figure 1–4).

Today, under the effects of the Sun heating the Earth's surface, winds blow in most parts of the world in different time periods. Two factors, namely, the revolution of the Earth following an elliptical path around the Sun once each year and the rotation of the Earth turning once upon its inclined axis (the obliquity of the ecliptic, or 23°26') in 24 hours, are the root cause of the four seasons, which are opposite in the Northern and Southern Hemispheres in the same month of a year.

These factors above cause the six trade wind zones, the seven atmospheric pressure belts, and the subtropical high known as the planetary system of winds.

Factors Related to the Earth's Surface

The distribution of sea and land, geographical features, and the condition of ground cover gave variety to the distribution of atmospheric pressure around the world in different seasons, resulting in various patterns and wind scales. Scientists believe that about 200 million years ago, there was only one interconnected landmass, Pangea, surrounded by the primitive sea, Panthalassa. After Pangea formed, it began to divide into continents through continental drift and seafloor spreading during the Mesozoic and Cenozoic Eras. Gradually, the complex distribution of sea and land and the intricate geographical features we see today were formed by the collision of crustal plates due to continental drift, a process known as orogeny. This includes notable events like the Alpine orogeny and the Himalayan orogeny.

So, Pangea likely had much fewer mountains before continental drift than what exists today. With this generally flat land surface, the distribution of the atmospheric pressure on the Earth's surface could not possibly have been as complex as today. If this is correct, then the atmospheric circulation of air in the Mesozoic and Paleozoic Eras would not have been nearly so active as today, lacking in the relevant natural contributing factors necessary to create strong winds (see Chapter I.1, The Timely Change in Ancient Land).

In short, the natural geographic factors before continental drift could not have influenced the formation of wind phenomena as they do today. It wasn't until the Cenozoic Era that the contributing factors for wind gradually developed. By this time, continental drift had largely settled (though it never completely ceased), and the current distribution of sea and land and the Earth's geographical features had taken shape. The Earth's surface began to resemble its present-day form during the Eocene and Oligocene Epochs. These changes in geographical features significantly intensified atmospheric circulation on both local and global scales. As temperatures steadily decreased and Greenland and the Antarctic continent drifted to their current positions, expansive ice sheets began to form at the poles (see Chapter II.3, The defining of the Antarctic and South American Continents).

Ultimately, the important natural phenomenon of wind developed, with various patterns such as monsoons, sea and land breezes, valley breezes, and foehn winds, and was classified on a scale from 1 (light breeze) to 12 (hurricane strength) in different seasons. Although air movement undoubtedly occurred in earlier eras, the establishment of wind patterns, particularly the planetary wind system, can be argued to have gradually begun in the Cenozoic Era.

The planetary wind system triggers a series of natural changes around the world. The system brings about rainfall seasonally, accelerating the weathering process and forming much of the soil layer. Wind, rain, and soil accelerate the evolutionary process of plants, resulting in the appearance of herbs, including the grass family, which contributes to the evolution of herbivorous animals. A form of wild wheat emerged with horses and oxen finally appearing in the Quaternary Period of the Cenozoic Era. Thus, the needs of early humans were met, allowing them to develop early agriculture, one of the bases of civilization. The planetary wind system also contributed to sea voyages, strongly accelerating the development of civilization.

During the Earth's development, water circulation worldwide, especially rain, was an ideal form to provide fresh water for life on the land, and it was much more effective than the shallows or marshland of the Paleozoic and Mesozoic Eras. However, the first formation of the planetary wind system was a gradual, complex process, which matured in the late Cenozoic Era with the appearance of the expansive ice sheets of the

Polar Regions. The planetary wind system resulted in global atmospheric circulation, affecting water vapor and heat worldwide, and thus brought about rainfall on the land (see Figure 2-11 in Chapter II.1). Of course, there may have been some locally strong winds infrequently occurring in the Paleozoic and Mesozoic Eras. Still, they could not possibly have had the same effect as the planetary wind system and, therefore, had little significance in the evolution of life.

The planetary wind system, which includes the six trade wind zones, the seven atmospheric pressure belts, and the subtropical high, is not the inevitable outcome of the development of our planet. Its formation is a complex process involving several natural contributing factors. Only with all the natural contributing factors can the planetary wind system emerge from the Earth's development.

As argued above, the first occurrence of the natural phenomenon of wind derives from two main factors: the evolutionary process of plant life and natural contributing factors. As mentioned at the beginning, winds, as a common, natural global phenomenon, play a crucial role in the evolutionary process of life and the development of human civilization. We may say that without such a natural phenomenon as wind, civilization would not exist as we know it. Air movement, consequential precipitation, and evaporation are essential to life itself and the development of civilization.

If the planetary wind system had originated 400 million years ago during the Paleozoic Era, when life first emerged on land, could the evolutionary process of plant life from small, slim psilophytes to angiosperms have occurred? Based on the arguments above, this seems unlikely. This also highlights the significant timing of when the planetary wind system fully developed on the Earth, underscoring the thought-provoking rationality behind the planet's development.

Now, let's review. Winds leave no trace in the geologic stratum, so there seems no other way to research the subject but to examine the evolutionary features of living organisms in each geologic period. Evolutionary features always depend on the changing conditions of the environment through which we can find valuable information for researching this abstract subject.

Winds, including strong winds, are a natural phenomenon that could occur at any geologic period, but the planetary wind system is a different story. It could not have appeared without specific conditions, which need a long natural development process to establish.

The planetary wind system is regular and global, but other types of wind systems that may have occurred in the Paleozoic, Mesozoic, and Cenozoic Eras are irregular and local and would have little influence on evolutionary processes. So, we may say that the emergence of the global planetary wind systems is a major natural arrangement.

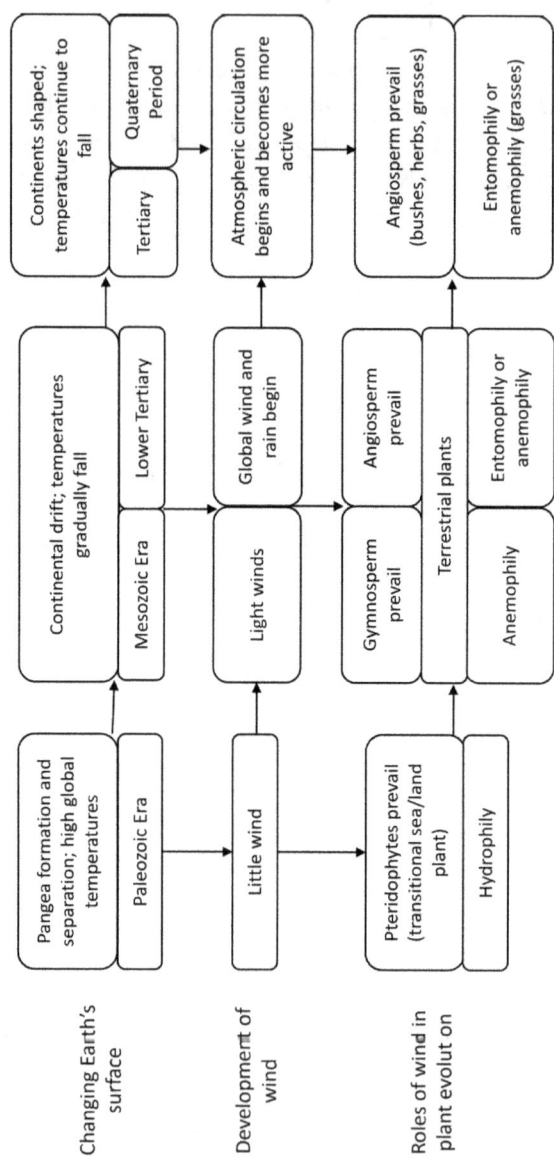

The Development/Roles of Wind on the Earth

THE RATIONALITY OF THE EARTH'S DEVELOPMENT IN TIME

ANCIENT SEAWATER

The world's freshwater resources are principally developed through precipitation, whether rainfall or melting snow. Rainwater on the ground becomes runoff discharging into streams, rivers, and lakes or passes into the ground, becoming shallow groundwater, which can contribute to and maintain river flow to avoid evaporating during dry seasons. This is an ideal way to support all land-living organisms on our planet. Yet how is rain formed?

Many natural factors are involved in the formation of rain, of which winds are essential. Without wind, warm or cold air masses couldn't move, preventing the creation and dispersion of rain. As argued in the previous section, there would not have been as much wind in the Paleozoic Era as today. If so, the probability of rain in the Paleozoic Era would have been much less than now. Long-duration, heavy rainfall would have been unlikely, and rain would have been confined to convective rain, which is brief and limited in area. This raises the question of how the vast stretches of shallows and marshlands were maintained to the benefit of the Pteridophyta plants that could have evolved from small and slim psilotosida plants to vast stretches of pteridophyte forests lasting about 200 million years, not hundreds or thousands of years.

Any body of water on the ground will decrease in volume through evaporation and outflow if there are no compensating replacement sources. What, then, was the compensating water source to maintain the vast stretches of shallows and marshlands in the Paleozoic Era? The sea is the most likely and only possible alternative source that could have supplemented the vast stretches of marshlands and shallows in that era, suggesting that shallows, marshlands, and sea were substantially contained within the one water system,* although they appeared physically separated. This supposition is significant to what follows in this section.

*Coal layers formed from the sediment of dead forests in shallows and marshlands during the Paleozoic Era, specifically in the Carboniferous and Permian periods, around 300 million years ago. Some scientists believe that the characteristics of coal layers (such as forks and wedges within a single layer) were influenced by the rise and fall of sea levels, which affected the water levels in the shallows and marshlands during those geological periods. This indicates that the sea, shallows, and marshlands were part of a unified water system at that time.

As to the question of how the shallows and the marshlands formed, scientists think that their appearance was mainly due to a series of geological processes known as the Caledonian movement, which occurred in the early Paleozoic Era, and the subsequent Hercynian movement of the late Paleozoic Era. We can imagine that some areas of the sea first became shallow coastal or neritic zones, then shoal water zones, followed by shallows that may have separated from the sea to become marshland and eventually dry land. This complex geological process likely followed this course, spanning millions of years. We refer to this process as regression, while its opposite is transgression (see Chapter I.1, The Timely Appearance of Life on Land). These two processes could have determined the water levels of the shallows and marshlands. The sea has always played a significant role in the development of human civilization.

The question arises: when did seawater begin to become saline? Was seawater salty from the beginning, or was there a particular stage in its progressive development when this occurred? What is the relationship between the seawater salting process and biological evolution? The following explanation attempts to answer the question from an evolutionary perspective.

Water is fundamental for life. There are two types of life on our planet: freshwater-supported life (FSL) and seawater-supported life (SSL). These two categories of life, whether animal or plant, are distinct because they do not share a common water environment. The most significant difference is that marine life, constantly surrounded by water, does not need to develop the tissues and organs to conserve water that land-based life forms have. For instance, seaweeds have a root system mainly for anchoring, whereas trees have both a root system and a vascular system for anchoring and collecting water necessary for growth. Seaweeds and kelps absorb water and nutrients through their bodies and roots, which also anchor them.

Typically, whether animal or plant, organisms can only survive in one type of environment—either freshwater or saltwater, but not both. However, there are some exceptions among fish species.

All life, whether FSL or SSL, is generally accepted to have begun in the same water environment: the sea. Thus, a new question arises. Was the seawater salty before the appearance of FSL? There are only two possible solutions to this question.

Solution One: The seawater was initially salty.

Solution Two: The seawater was initially fresh or had minor salinity.

THE RATIONALITY OF THE EARTH'S DEVELOPMENT IN TIME

Before any determination of the two possible solutions, a look back at seawater formation could contribute to exploring this question. People have thought that the sea-forming process was completed about 1.8 billion years ago. However, the shape and location of seas around the globe differ from age to age. Around 45 million years ago, the most recent significant geological event—the rise of the massive Himalayan Mountain ranges—shaped the seas into their current form and location. It's thought that the sources of seawater were from the Earth, probably mainly from volcanic eruptions and magmatic activities, usually with a large amount of steam. The steam condensed to liquid water, accumulated on the Earth's surface, and gradually became today's seas and oceans. Data collected by satellites indicates that the Earth also received water that originated from space. Scientists, therefore, consider there must be a considerable proportion of seawater that came from space over the past millions of years.

Concerning the changing salinity of seawater, modern science provides little to no proof of this geologic transformation that occurred before the Cenozoic era. Paleo-marine geology merely touches on the geologic events in the Quaternary, Tertiary, and Cretaceous Periods. Researching the changing salinity of seawater before the Cenozoic Era is difficult due to present technological constraints. Given time and future technical advancements, perhaps we will come to understand this unknown phenomenon.

Now, let's explore two solutions to when life emerged on land, starting with the Paleozoic Era.

The study of biological evolution tells us that environmental change is necessary in the evolutionary process. As the environment changes, some species, through the evolutionary processes of gene mutation, natural selection, and reproductive isolation, may become new species that adapt to the changed conditions. Otherwise, natural selection may eliminate them. The rate of change in the natural environment is so gradual that organisms can continue to exist as they adapt. The evolution process is measured in terms of millions of years. This study tells us the way for sea life to move or adapt to land-based life is as follows:

The evolutionary process: Life in the sea evolved into life on land	
Environment	The sea
Life forms	Primitive sea life
Time (Phase I)	Before 400 million years ago

Environment	Shallows and marshland
Life forms	Pteridophyta (hydrophily), amphibians, fish
Time (Phase II)	From about 400 to 250 million years ago

The shallows and marshland represent a transitional environment between the sea (phase I) and the land (phase II) left by the sea through a combination of crustal movement and sedimentation, or regression. By way of example, about 500 million years ago, the greatest part of China was once covered by seawater and became land through gradual geological processes in the Tertiary Period. During this prolonged development period, vast expanses of shallows and marshland were once a feature of China.

Environment	The land
Life forms	Gymnospermae, reptiles (dinosaurs)
Time (Phase III)	From about 250 to 65 million years ago

If Solution One were correct, it would require a lengthy process in which the salty waters left by the sea gradually became fresh during the transitional environment (Phase II). In this scenario, all living organisms (Pteridophyta, amphibians, etc.) in such an environment would have gradually evolved from SSLs to FSLs. However, no widely accepted geological evidence supports a process where salty shallows transformed into freshwater. Therefore, we can conclude that Solution One is doubtful.

Solution Two is consistent with evolutionary processes. Before reaching a certain stage in natural development, seawater had been fresh for an extended period. Around 400 million years ago, geological processes and atmospheric changes began a new stage in biological evolution. During this period, "transitional life" emerged in vast stretches of shallows and marshland. Over time, through further evolution, some species, such as pre-gymnosperms, gradually evolved into new species entirely independent of the sea, becoming land-dwelling organisms.

Some may argue that several fish species living in salty seawater at river mouths moved upstream into the freshwater and gradually evolved into freshwater-supported species. If so, the evolutionary process would have had nothing to do with the change in environment. Furthermore, there would have had to be many large rivers lasting millions of years on the ancient land. Yet, we have argued that there were only light winds in the Paleozoic Era, and therefore, rivers would have been few, if any, that might have been caused by crustal movement rather than heavy rainfall.

The "animal landing schedule" and the "plant landing schedule" are almost parallel, as shown in the above chart, which also indicates that the two processes of evolution result from the same cause, environmental changes, rather than any other individual cause.

The ocean is the largest stagnant body of water on the Earth, enabling it to accumulate more chemical elements and retain them in the seawater or the seabed. The sources of the chemical elements that make seawater salty can be roughly classified into two categories:

One: The undersea lithosphere. Lighter materials continue to move upward through the lithosphere due to gravitational differentiation. Volatile constituents or volatile components (comprised of water and dissolved gases, typically involved in magmatic activities) are released from large active faults and volcanoes (volcanic gases contain trace amounts of chlorine elements). In addition, hydrothermal fluid (or hot seawater) is expelled through hydrothermal activity from oceanic ridge zones and other active seafloor areas. This process primarily introduces chemical elements of metals into seawater.

Two: Rivers. Rivers transport large amounts of weathered solids from the land into the sea, introducing chemical elements such as sodium (Na), magnesium (Mg), and calcium (Ca) into seawater, contributing to its salinity. This process is widely accepted today.

Compared with the first category (the source of salt), rivers have played a more significant role in seawater becoming salty. Perhaps the next question would be, when did rivers begin to appear in most parts of the land? Although not discussed in detail, we believe rivers formed due to heavy rainfall gradually developed mainly in the Cenozoic Era as the planetary wind system developed. In the Paleozoic Era (before 400 million years ago), the ancient land area was relatively small, so large rivers were difficult to form. Further, living organisms in the sea played a role in adjusting, but not creating, the amounts of chemical elements introduced to seawater by these two sources.

Building on the arguments above, we can imagine that the percentage of seawater salinity increased throughout the process that lasted millions of years. The distribution of the level of salinity throughout the ocean varies. The saline percentage of seawater (S percent) around the seafloor, such as large active faults, active volcanoes, oceanic ridges, etc., was greater than any other space distant from such features before the Mesozoic Era.

Correspondingly, the increasing rate of seawater salinity was once slow but became relatively rapid since the appearance of massive rivers. Eventually, the salinity of the seawater became roughly homogeneous. On the ocean's surface, salinity varies between about 3.2 percent to about 3.6 percent. In the Red Sea, it is more than 4.0 percent and less than 1.0 percent in the Baltic Sea (except for underground artesian water, which can be saline, fresh water on the land generally has a salinity in the range of 0.1 percent).

Here are some examples to illustrate the idea:

- Mangroves (angiosperms) have developed complex physiological adaptations to high salinity in seawater. However, there is no evidence that plants (pteridophytes) in the Paleozoic Era possessed similar physiological traits.
- Dinosaurs living in forested, wet, and coastal habitats suggest that seawater may have been freshwater or had very low salinity.
- Numerous coalfields worldwide were formed in the vast shallow seas of the Carboniferous Period, indicating that the shallows and marshlands where pteridophyte forests grew were connected to these shallow seas.

If seawater had always been fresh instead of salty, humans would not only miss out on the abundant mineral resources in seawater but also face adverse effects from atmospheric circulations that would be unfavorable to the development of civilization. This is due to the slight differences in physical properties between freshwater and salty water. Therefore, Nature had two seemingly incompatible missions to accomplish: first, to create the initial living organisms in the sea and allow some to evolve into FSLs on land, and second, to make seawater salty. Remarkably, seawater became salty in a way that supported both the evolutionary process and the development of human civilization.

THE RATIONALITY OF THE EARTH'S DEVELOPMENT IN TIME

Nature gradually created the first living organisms in the sea and, through the evolutionary process, enabled some species to move onto dry land and survive independently of the sea environment. Following this, Nature began to increase the salinity of seawater, a process that accelerated since the Cenozoic Era due to the massive appearance of rivers. This process allowed sea life to continue evolving, culminating in a rich and crucial legacy for humanity. We believe this is the process by which seawater became salty. The evolutionary process of life "moving" from a salt-water environment to existing in a freshwater-dependent environment other than by the course of events outlined here is difficult to imagine.

Pay close attention: As the largest body of water on the Earth's surface, the ocean accumulated its salinity gradually over time. Chlorine likely began slowly accumulating in seawater through volcanic eruptions (HCl). However, this process accelerated significantly with the massive increase in precipitation and the formation of rivers on land (introducing Na, Mg, and Cl). Given the relatively low abundance of chlorine on the ocean floor (including areas with active volcanoes) and in the ground, the high chlorine content in seawater remains a puzzling mystery with many questions yet to be answered (see Chapter I.1, The Timely Process of Seawater Becoming Salty).

So, you may find that continental drifting, orogeny, the sustainable temperature drop in the Cenozoic Era, precipitation, and the massive appearance of rivers are a series of natural processes that eventually make seawater salty. Before these natural processes, there was hardly any primary source of saltiness in seawater. This is what modern science suggests.

Next, we turn to another interesting and important question: the composition of inorganic salts in seawater.

In the process of seawater becoming salty, proper timing and the right kinds of salt are required. Of all the elements in seawater, sodium chloride content is 70 percent on average, and magnesium chloride is 14 percent. A large amount of chlorine and sodium in seawater is required not only by living organisms (particularly humans) but also by the development of civilization. If seawater were made up of other elements, the global environment would be changed accordingly, and it would undoubtedly be adverse to human subsistence. No matter how we explain the current condition of salt in seawater, ocean salinity is a perfect natural selection, favorable to the global environment, human subsistence, and the development of civilization.

Now, let's compare the content of elements in seawater with the abundance of elements in the crust (or in rocks). An interesting question arises: Why are Cl, Na, and Mg the plentiful elements in seawater rather than other elements? There seems to be no answer to that question, but any substitute for these three elements in seawater would certainly change the global environment (climate) and may be adverse to life. Table salt, or sodium chloride (composed of chloride ions and sodium ions in seawater), is essential for animals and a food preservative. People have a great demand for salt for both subsistence and civilization. No other element could substitute.

The average concentration of the main elements in seawater includes Cl (19.10 g/L), Na (10.62), Mg (1.29), S (0.904), Ca (0.412), K (0.399), Br, C, Sr, B, Si, and F, in descending order of abundance.

The abundance of elements in the Earth's crust or rocks is O (46.95), Si (27.88), Al (8.13), Fe (5.17), Ca (3.65), Na (2.78), K (2.58), Mg (2.06), Cl (0.05).

At first sight, seawater seems to have many elements, but the most plentiful are Cl, Na, and Mg. Notably, the abundance of chlorine on the ocean floor (active volcanoes, etc.) and in the ground is very low. There are some explanations, but we would call it a natural arrangement instead.

The process of ocean water salinity, in terms of time, element make-up, and volume, is far from simple!

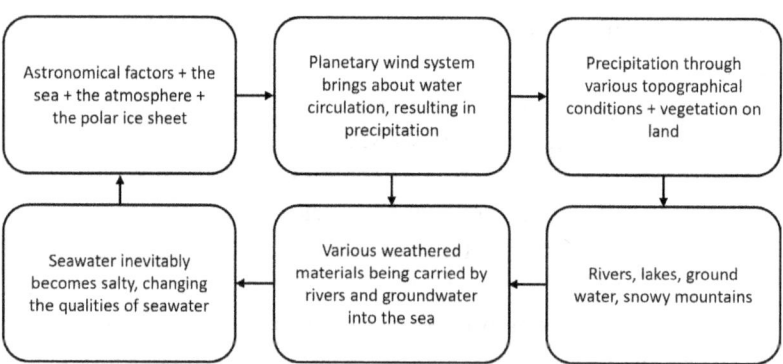

The Cycle of Seawater Becoming Salty

Notes:

1. Rivers are the primary source of seawater salinity.
2. The appearance of the planetary wind system is the key to the massive appearance of rivers.
3. We can easily imagine that today's global environment would be much different if seawater were fresh rather than salty. Furthermore, the main chemical elements in seawater are Na-sodium, Cl-chlorine, and some Mg-magnesium in optimum amounts. If seawater contained chemical elements other than Cl, Na, and Mg, or if their amounts were different, the global environment would be dramatically different and no longer suitable for human life. This cycle seems unlikely because seawater becoming salty must have been a complex, stochastic process. And that may represent the reasonability of the natural world (see Chapter III, Water (H_2O)).

THE ILLUSTRATION OF NATURAL DEVELOPMENT CHAINS

Nature appears to be a rational, logical, and efficient network governed by numerous natural factors. While modern scientific knowledge allows us to imagine this network extending back into the distant past, forming a natural developmental sequence across time and space, humans have yet to fully map such an extensive network. Here, we outline a chain of natural development based on earlier discussions—time characteristics, the planetary wind system, and seawater salinity. If you disagree with this proposed chain of natural development, you are welcome to suggest an alternative that could meet the conditions necessary for the emergence of civilization.

The charts below illustrate natural development from a simple to a complex process involving an increasing number of natural factors over time. At the core of natural development lies the movement of tectonic plates, which ultimately facilitated the rise of civilization. The rationality behind plate movement remains one of the greatest mysteries of natural development, perhaps representing "the Reason that manifests itself in nature." Astronomical factors also play a crucial role, emphasizing the interconnectedness of earthly phenomena within the wider universe—a pivotal theme in this book.

THE RATIONAL UNIVERSE EVOLVING FOR HUMANS

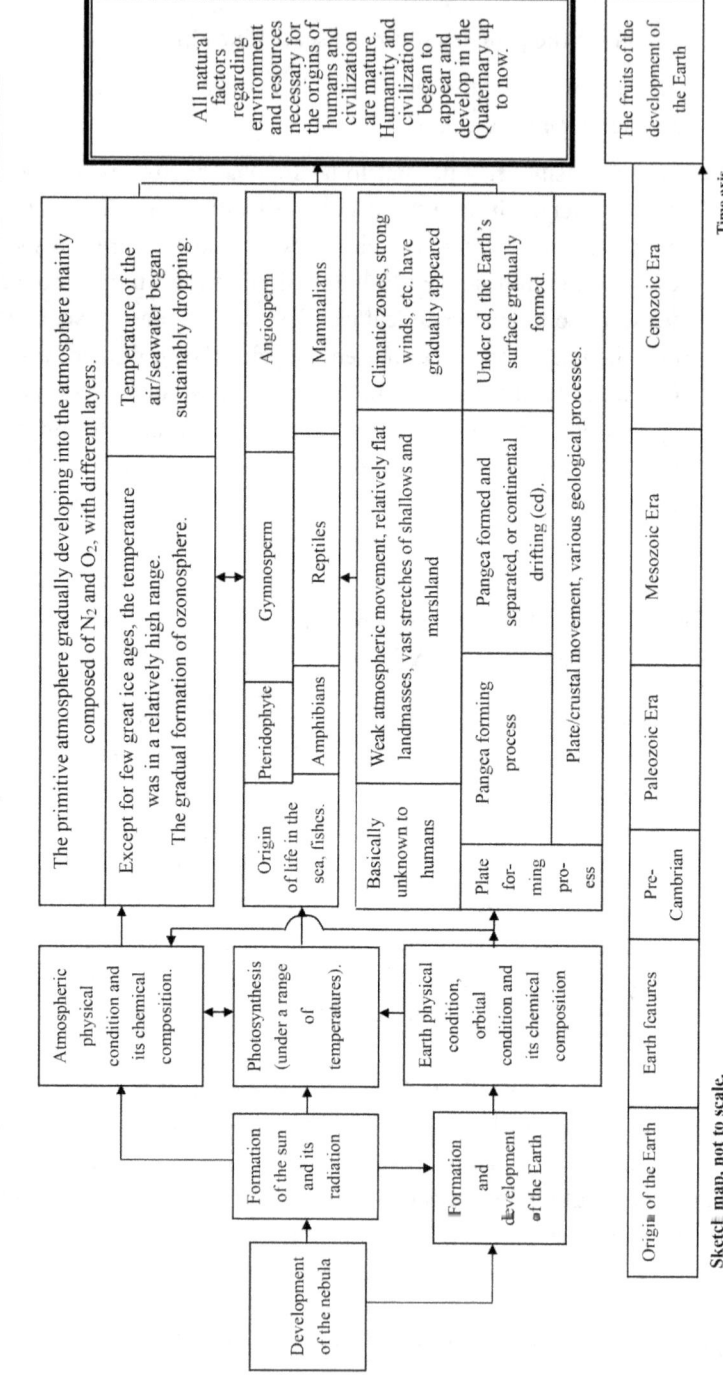

The Natural Chain 1 (A stochastic natural development process)

Sketch map, not to scale.

THE RATIONALITY OF THE EARTH'S DEVELOPMENT IN TIME

The Natural Chain 2 (Under certain of astronomical conditions, a complex but progressive natural process)

4600	2500	570	250	65	0.01(10^6a)
Archean	Proterozoic Era	Paleozoic Era	Mesozoic Era	Cenozoic Era	

Plate forming process → plate/crustal movement (volcanism, seismic activities, etc.)
Mineralization (inorganic/organic mineral forming processes)

The development of the atmosphere/the sea (including being layered, ozonosphere, etc.)

The temperature began dropping
Continental drift; the Earth's surface gradually formed

Pangea forming process, Pangea formed and then began to separate, or continental drifting.

Planetary wind system, gradually formed

Primitive global environment, relatively simple, weak atmospheric movement
(Atmospheric movement becoming more and more complex as continental drifting)

Strong winds, heavy rain; and long rivers becoming commonplace.

Strong winds, and heavy rains perhaps appeared but rare. Relatively flat landmasses, vast stretches of shallows and marshland.
(Topographic conditions becoming more and more complex as continental drifting)

Salinization accelerated; becoming salty seawater.

Fresh water or with minor salinity
(Seawater becoming more and more salty as rivers increased)

The prevalence of angiosperm (wheat, etc.)

The prevalence of gymnosperm (terrestrial plants)

Pteridophyte (a transitional plant from the sea to the land)

Landing

The prevalence of mammalians (horses, etc.)

The prevalence of reptiles (terrestrial animals)

Amphibians (transitional animals from the sea to the land)

Landing

The origin of humanity and civilization

Time axis

Little is known to humans

The origin of life in the sea, and evolved in the sea.

developing tendency

Dotted line refers to sketch map, not to scale.

The Natural Chain 3 (Plate movement is a primary factor)

Under the astronomical conditions → four seasons, day and night, climatic zones, etc

The natural terrestrial environment suitable for habitation

Plains, mountains, plateaus, rivers, lakes, groundwater, etc.

Weathered rocks, diverse in size and quality, including stones, sand, soil layers, etc.

Metamorphic rocks ↔ Sedimentary rocks

Magmatic rocks (Igneous rocks)

Landmasses with various landforms

The shaped surface of the Earth

The sea (Rivers) — The primitive sea

The atmosphere (Life) — The primitive atmosphere

Atmosphere circulation, water circulation and weathering process

Plate/crustal movement (volcanism, seism, metamorphism, magmatism, sedimentation, mineralization, etc. (What is the power to drive the movement?)

Does the inexhaustible heat energy inside the Earth for the power of plates' movement mainly come from the neutrinos of solar radiation?* We don't know.

---

*Zhang, Guowen. "Can Solar Neutrinos Heat the Earth?" *HANS Preprints*, May 15, 2020. https://pdf.hanspub.org/hanspreprints20200100000_41404959.pdf.

The Earth's Great Geological Events

Era	Period	Epoch		mya	Plates	Environment	Evolution
Cenozoic Era	Quaternary	Holocene		0.01	Pangea separated into a few continents, continental drifting began	Temperature of the atmosphere and seawater began to drop	Angiosperm, mammalians
		Pleistocene		2.47			
	Tertiary	Upper	Pliocene	5.2			
			Miocene	23.3			
			Oligocene	35.4			
		Lower	Eocene	56.5			
			Palaeocene	65			
Mesozoic	Cretaceous			135	↑	Global temperature was in a relatively high range except for two great ice ages	Gymnosperm, reptiles
	Jurassic			208			
	Triassic			250			
Paleozoic Era	Permian			290	Pangea was being formed		Life landing Pteridophyte
	Carboniferous			362			
	Devonian			409			
	Silurian			439			
	Ordovician			510			Life was born in the sea 3.5 bya. All living things existed in the sea.
	Cambrian			570			
	Precambrian			800	↑		
				4.6 bya	Plates were being formed ↑	Little is known	↑

✦ Great ice age
💥 Great extinction
☆ Cambrian explosion

bya/mya = billion/million years ago

All the numbers in the blocks refer to the beginning of the period. All those numbers vary slightly but remain roughly the same between different reference materials.

Chapter II.1

The Rationality of the Earth System
The Composition of the Earth System

After reading the previous chapters and as you read this chapter, you will further understand the rationality of the Earth's environment. This environment is composed of major and minor components, each with its own complex formation process. By the Quaternary period, all these components had merged to form the global environment necessary for the birth and development of human beings. Thus, Earth appears to be a mysterious planet that has developed over time and space to support the eventual emergence of humans and civilizations.

The Earth System is comprised of numerous natural factors, which can be summarized into two categories: environmental and resource. Humans and civilization were born and developed due to these natural factors.

We call them *essential factors*.

This important fact represents the nature of planet Earth. Unfortunately, people tend to take this fact for granted with little understanding of the complexity of the Earth System. The following explanation illustrates the importance of these factors, stressing the roles, functions, and meaning of the factors in the development of the Earth, resulting eventually in the appearance of humanity and civilization.

The formation processes of every essential factor are complex and rigid, even mysterious, and their occurrence in the universe is probably rare. The research for this is ongoing.

ASTRONOMICAL FACTORS

The Appropriate Distance, the Appropriate Size

If there were a slight change in the distance between the Earth and the Sun or in the size of the Earth, would such a change matter? Would the slightest change have a serious impact on the Earth System?

The temperature at the Earth's surface has the power to change the existence of everything animate and inanimate. The solar radiation at the Earth's surface is so strong that any change, either in the Earth's distance from the Sun or in the size of the Earth, could severely impact the temperatures on the Earth's surface, resulting in a dire fate for our planet.

This factor is demonstrated by the principle that the higher the solar altitude, the stronger the solar radiation intensity (Figure 2–1). It is responsible for the four seasons, with the polar regions and the equator forming the coldest and warmest parts of the world, respectively. The temperature difference between these regions can reach 212°F (100°C) or more, essential for forming the planetary wind system. Furthermore, in the temperate zone, the temperature difference between summer and winter can range from 120–40°F (50–60°C), which is suitable for the local biosphere.

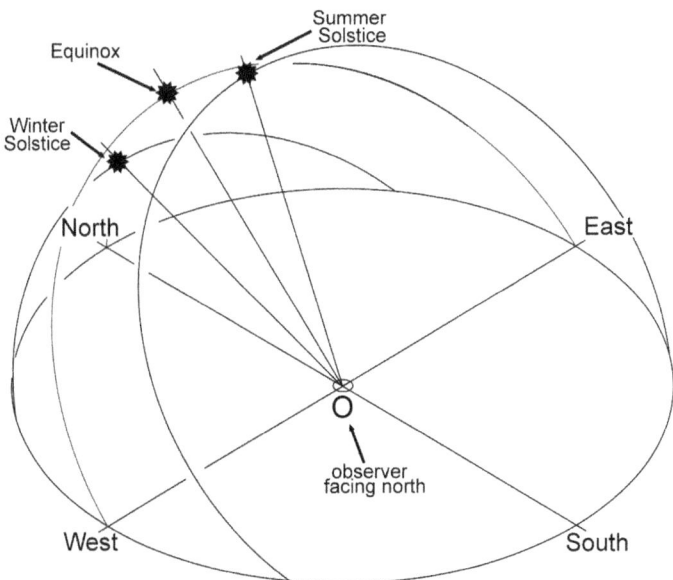

Figure 2–1: The relation between solar altitude and solar radiation intensity (Source: S.Wetzel, CC BY 4.0, labels updated).

Everyone has experienced the strikingly different feeling between being in the sun and being in the shade. The ground temperature in the sun and that in the shade can differ by 50–86°F (10–30°C). An unclothed person's bearable temperature ranges from about 63–86°F (17–30°C) with a relative humidity of less than 70 percent. People are sensitive to temperature changes and feel more uncomfortable as temperatures deviate from the bearable range. A slight change in the distance between the Earth and the Sun or in the size of the Earth would have a devastating impact on our ability to adapt and survive.

According to the theory of Milankovitch cycles, the changes in temperature between the interglacial and glacial periods in the Quaternary Period were caused by slight changes in the three orbit-motion elements of the Earth. Even a slight change can significantly impact the global environment (Figure 2–2).

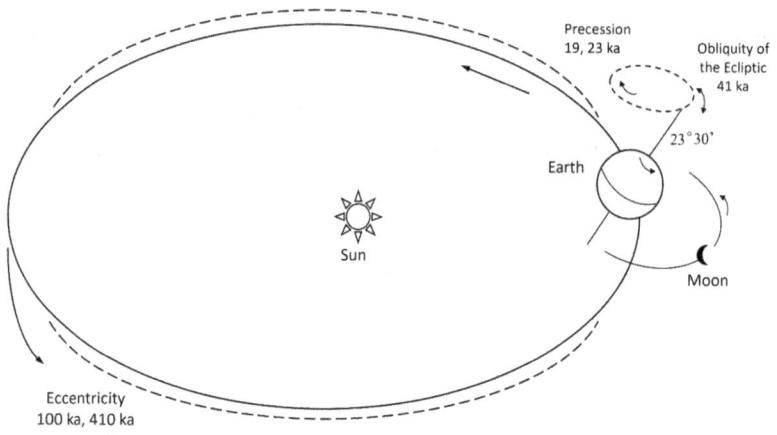

Figure 2–2: The theory of Milankovitch cycles. The three elements of Earth movement (ka=1000 years), though the distance from the Earth to the Sun is unchanged (not to scale) (Source: Benpei Liu, et al. *Geohistory*. 3rd ed. Beijing: Geology Press, 1996.).

A suitable temperature range on the Earth's surface is critical for the global biosphere to flourish. Any deviation in the Earth's motion would cause a change in temperatures at the Earth's surface, which would undoubtedly impact human life. Therefore, the Earth's distance and size are essential factors influencing the environment, particularly temperature, which is needed to sustain a life-sustaining biosphere.

Aside from temperature, there is a subtle relationship between the Earth's size and its atmosphere, oceans, structure, and evolutionary process. If the Earth were larger or smaller, such differences would undoubtedly alter its physical properties (such as gravity and atmospheric pressure), which would have significant adverse effects on the global environment and the existence of all life (see Chapter VI, Earth—A Perfect Planet for Forming the Human Brain).

The Sun, Moon, and Earth belong to a nonlinear system. Any change in distance or size would break the harmony essential for life on Earth.

Four Seasons in a Year, 24 Hours in a Day

You may enjoy the changing seasons, but have you considered the subtle yet significant relationship between the four seasons and the origin of civilization? Without the seasons, modern times as we know them would not exist.

The four seasons, along with day and night, are essential for people to live and develop on this planet. According to Kepler's law, the distance from the Earth to the Sun will result in a certain revolution period, one year, which, with the obliquity of the ecliptic, makes the global environment change periodically, resulting in the four seasons (Figure 2–3, 2–4).

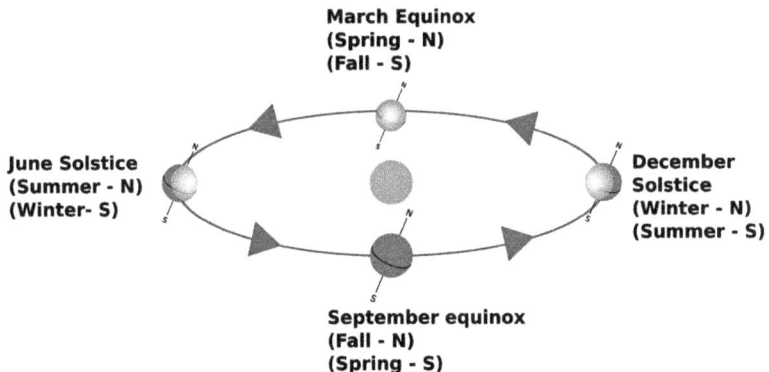

Figure 2–3: The revolution of the Earth with its inclined axis (Source: Public domain).

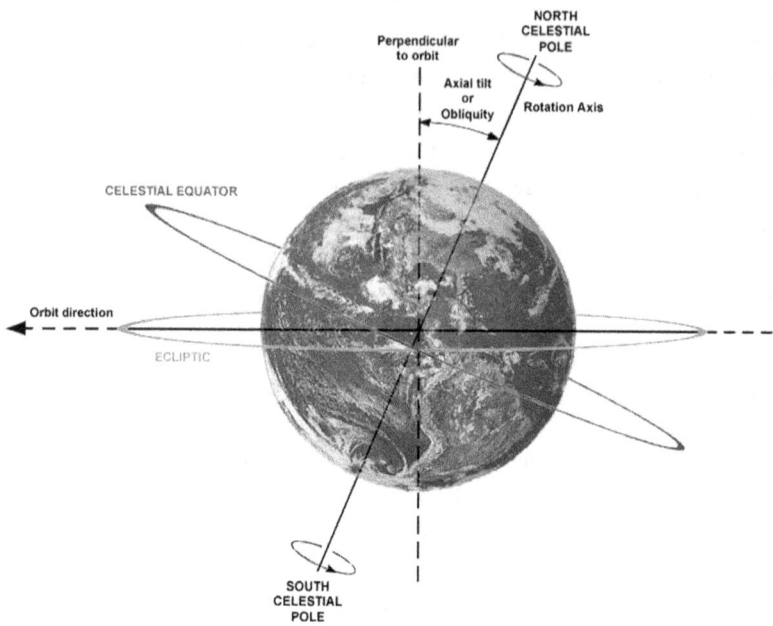

Figure 2–4: The obliquity of the ecliptic (axial tilt) and the equinox—Autumn Equinox/Spring Equinox (daytime equal to nighttime) (Source: Dennis Nilsson, CC BY-SA 3.0).

The orbit-motion relationship between the Earth and the Moon works so well that the resulting motion periods are perfect for living and working on Earth. These periods are the three months that make one season, about thirty days each month, twelve months, 365 days in a year, and the twenty-four hours that make one day. The complex, nonlinear system produced by the Sun, Earth, and Moon is a great miracle, critical for the evolution of life and our counting of the passage of time. For example, determining specific dates would be impossible if there were no months.

The repetition of the four seasons spurred early humans who lived many thousands of years ago to naturally conform their behavior to patterns that enabled them to survive and flourish. The seasonal change from summer to winter would cause them to seek bodily cover and burn wood and dry grass for warmth. This seasonal change also necessitated the ability to control fire. In spring, plant life awakens to resume growth, flowering and producing fruit and seeds. Early humans would eventually learn how to cultivate some species of plants effectively.

THE RATIONALITY OF THE EARTH SYSTEM

Like the conductor of an orchestra, the four seasons control climate and life on Earth. In response, human activity varies periodically, enabling people to develop civilization. On the one hand, nature's seasons repeatedly and progressively spurred early humans to adapt, while on the other hand, nature produces everything needed for humans to discover and use. This subtle course of development may be summarized as follows:

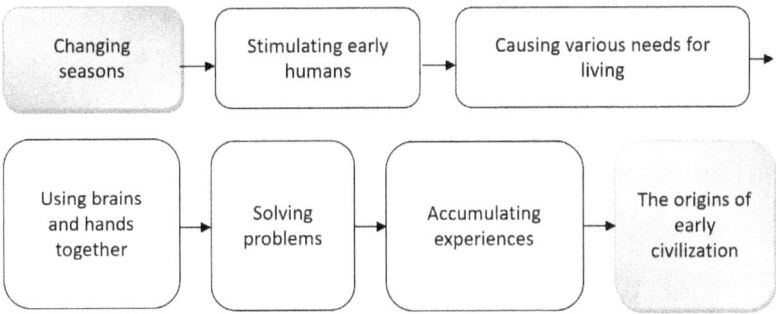

This course may be why early civilization first appeared in the temperate regions rather than the torrid or frigid zone. As a matter of fact, the New Stone Age began in the North Temperate Zone, the area of most significant human activity in the world. Four seasons also make our planet's landscapes more attractive and cause changes in our neuroendocrine system secretions, influencing emotion and creative power.

In addition to the Earth's revolution with the obliquity of the ecliptic, factors such as altitude, latitude, atmosphere, and the sea also influence different climatic and seasonal features. However, it is primarily the obliquity of the ecliptic that creates the four seasons and climatic zones (Figure 2–5). This tilt is a simple yet ingenious mechanism that causes the temperatures on the Earth's surface to change periodically, gently, and gradually throughout the year. Its current value is ideally suited for life. Any alteration would significantly impact the Earth's surface temperatures (see The Appropriate Distance, The Appropriate Size, above). Can you think of an alternative solution that would have the same effect?

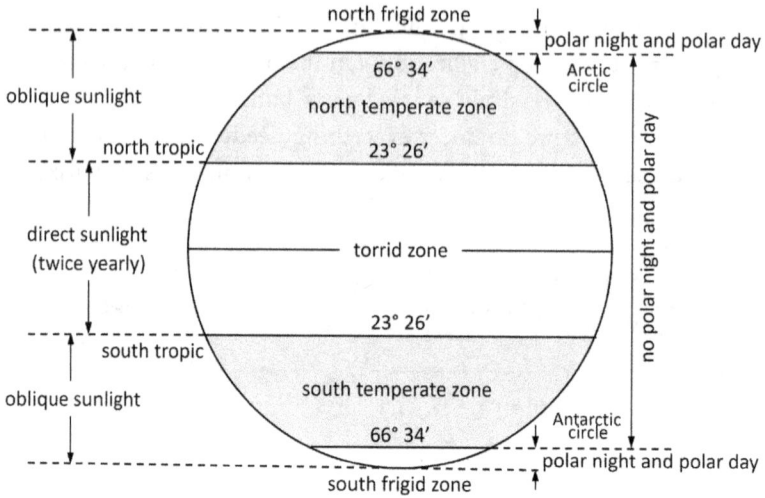

Figure 2–5: Global climate zones, based on the obliquity of the ecliptic (Source: The author's book in Chinese, published 2016).

Now, let's consider the Earth's rotation from the perspective of civilization's development. The Earth's daily rotation matches well with the Earth's revolution around the Sun and is crucial for life. For thousands of years, the duration of a day has been approximately twenty-four hours, regulating all activities through day and night cycles. A day significantly longer or shorter than twenty-four hours would drastically impact the environment and civilization.

Gravity is another natural factor influencing evolutionary processes on Earth. The average height of people today is largely determined by gravity. Given our current biological structure, it seems unlikely that humans would evolve to sleep for twenty hours a day if the length of a day were forty-eight hours. It is hard to imagine the effects on animals, humans, and plants if a day's duration differed from the current twenty-four hours. The twenty-four-hour day is ideally suited to life on Earth.

Furthermore, the spiritual world fuels our creative activities. The day and night cycle is ideally suited to our lifestyle and work habits, aligning well with our physiological rhythms, including our metabolism, digestion, excretion cycles (which range from seventeen to twenty-nine hours), and our need for eight hours of sleep. However, this specific cycle does not appear to be as essential for animals and plants. Notably, the day cycle and the physiological period of metabolism, in a sense, are independent of each other. The Sun, Earth, Moon, and humans' size and physiological systems match so well that the symbiosis supports our subsistence on our planet.

The Earth's rotation also contributes to the formation of the planetary wind system, consisting of the six trade wind zones and the seven atmospheric pressure belts (Figure 2-6), moving within a range as seasons change. To a certain extent, the Earth's rotation also contributes to forming its geomagnetic field, which is necessary for the subsistence of life and human civilization.

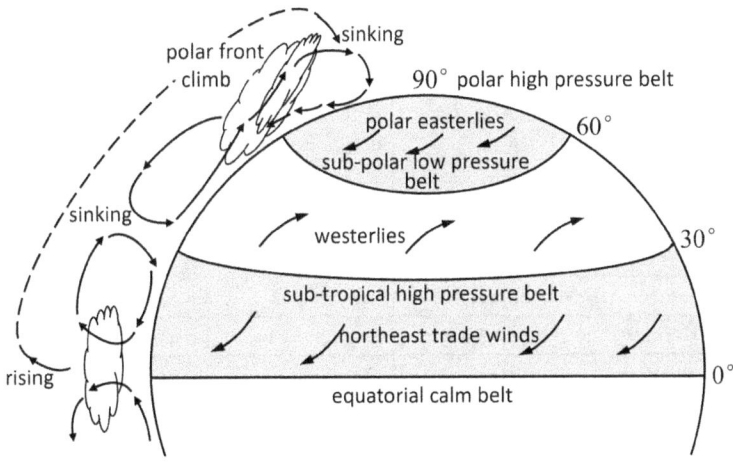

Figure 2-6: The planetary wind system (Northern Hemisphere). The global and regular wind system is based on the obliquity of the ecliptic and the rotation. Air flows from the equator, where hot air rises, to the polar regions, where cold air sinks. The Earth's rotation causes this flow to divide into three sections (Source: The author's book in Chinese, published 2016).

Suppose the Earth's rotation was too slow, resulting in a significantly longer day. Could those two important natural missions, the wind system and the geomagnetic field, have been established? Could life and civilization remain?

The Earth's rotation, its revolution, and the obliquity of the ecliptic are three basic astronomical features independent of each other that form the global environment. Here, we would like to call them natural arrangements created by nature to fulfill their natural missions.

Keep in mind that natural arrangement and natural missions are two significant concepts with broad philosophical meanings in this book that are helpful for us to comprehend Nature.

Our Mysterious Moon

Everyone enjoys the beauty of the Moon and its moonlight, but do you understand the Moon's necessity for life on Earth? Though nature has produced a perfect satellite for the Earth, the Moon is a mystery that has puzzled scientists (see Chapter I.1, The Timely Birth of the Moon). The Moon is an extraordinary satellite, the largest one relative to the size of its planet in the solar system.

Planet	Radius (km/mi)	Radius of largest moon (km/mi)	Moon/planet radius ratio
Earth	6378/3963	1738/1079	0.272
Mars	3395/2109	11.1/6	0.003
Jupiter	71,400/44,365	2631/1634	0.037
Saturn	60,000/37,282	2575/1600	0.043
Uranus	25,900/16,093	789/490	0.030
Neptune	24,750/15,378	1350/838	0.055

Is it necessary for the Moon to be so big? Absolutely necessary!

The Moon's presence is an essential factor stabilizing the movement of the Earth, probably including the obliquity of the ecliptic. This is because the Earth is not a homogeneous, round ball, and the Moon, Earth, and Sun are in one nonlinear system, interacting with one another to maintain its whole stability. Any change in its movement would undoubtedly influence the stability of the entire system.

The Moon also causes the Earth's tides, which influence the exchange of materials and heat between different areas within bodies of water, particularly relatively shallow ones. Therefore, the tidal zone is always the busiest area for living organisms. Tides would be much weaker without the Moon's presence (Figure 2–7). All bodies of water would have been stagnant for billions of years, adverse to the subsistence of life. Scientific evidence suggests tides slow the Earth's rotation and gradually cause the Moon's orbit to recede.

THE RATIONALITY OF THE EARTH SYSTEM

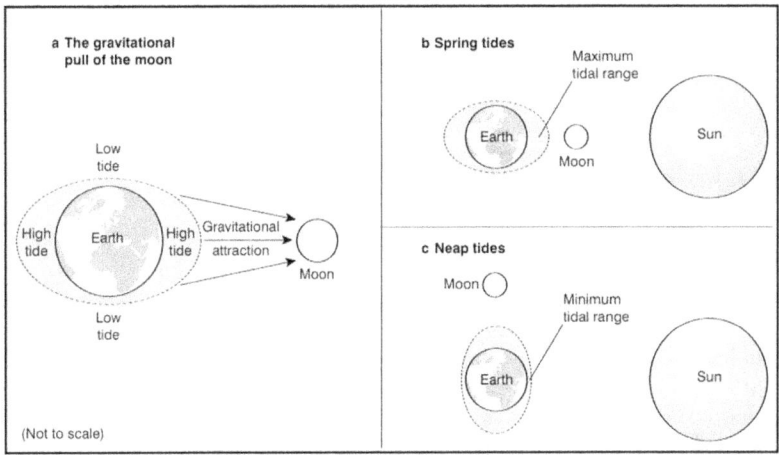

Figure 2-7: The formation of tides. The spring tide occurs when the Sun, Moon, and Earth align (gravity overlaid). The neap tide occurs when the Sun, Moon, and Earth form a right angle (gravity offset) (Source: Waugh, David. *Geography: An Integrated Approach. United Kingdom: Nelson, 1995.*).

One might ask why the tides nearer to the Moon are not much stronger than those on the other side (Figure 2-7). In other words, why is the tidal shape of water an ellipse rather than like a water drop pointing to the Moon? This complex, dynamic problem involves universal gravitation, the weight of the Earth's water, the Earth's rotation, etc. Any change in the distance or the magnitude of the Sun, Moon, and Earth would affect the size of tides on the Earth's surface, seriously impacting life. We cannot undervalue it!

Of greater interest is the fact that the ratio of the distance between the Sun and the Earth to the distance between the Sun and the Moon is roughly equal to the ratio of the radius of the Sun to the radius of the Moon. As a result, during a solar eclipse, the Sun is precisely and fully covered by the Moon's shadow, providing the best opportunity for us to make observations of the Sun.

Relativity was first proved in a solar eclipse. We would miss this opportunity if the Moon were larger or smaller or if the distance from the Earth to the Moon was further or nearer.

The Moon has other impacts on civilization, but they are not as rigid as its size. For example, one year consists of twelve months, and one month is one revolution of the Moon around the Earth, leading to the creation of the lunar calendar, a convenient method of dividing time and living according to lunar cycles (Figure 2-8).

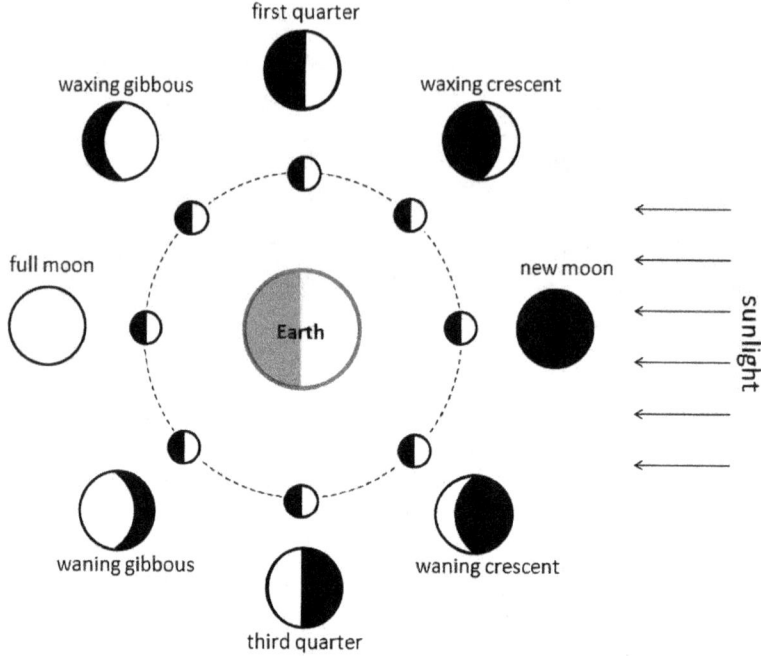

Figure 2-8: The phases of the Moon. Its revolution period is equal to its rotation period, or 27.32 days, forming the basis of the lunar calendar (Source: Andonee, CC BY-SA 4.0).

The Moon is a frontier we desire to reach, explore, and mine its rare minerals. The Moon makes the night beautiful, inspiring our souls. The Moon has many mysteries, and we have a long way to go to investigate its wonders (see Chapter I.1, The Timely Birth of the Moon).

Do you think our planet could survive the loss of the Moon? Could the Moon be larger or smaller, further or nearer?

The Mysterious Solar System and the Galaxy

Our solar system is located in the "life zone" of the Milky Way galaxy and may be the most complex one, a topic we will explore in greater detail in Chapter VI.

THE RATIONALITY OF THE EARTH SYSTEM

A question naturally arises. Is our solar system's complexity an essential condition for the appearance of humans and civilization on Earth? Theories abound, but how a simple star system could produce the complex Earth System, resulting in the birth of humans and civilization, is difficult to fathom. Such a result is arguably the most extraordinary natural event.

Regarding the complexity of the solar system, the following factors may be favorable to the formation of the Earth System.

First, the movement of the Earth is subject to the solar system. Our solar system is an immense dynamic system with various celestial bodies, including large and small planets, gaseous and solid planets, comets, and satellites. As such, its gravitational field might affect the movement of the Earth.

Second, our solar system has maintained a dynamic balance since its formation, establishing a stable system. The solar system is a complex, nonlinear system, also known as a chaos system in physics, and it is important to note that it is highly sensitive to its initial conditions.

Third, the complexity and stability of the solar system contribute to its suitability for the birth of humanity and civilization. That human life and civilization were born in a simple and unstable solar system is improbable.

Lastly, comets consisting of water molecules may have contributed significantly to the presence of life on Earth. Oxygen and hydrogen, under certain conditions, can be combined into water molecules. How did the Earth obtain and store its tremendous quantity of water? Some scientists believe significant amounts of water and possibly the first organic matter, the "seeds of life" on our planet, were brought by a comet through an impact on the Earth.

We know little about comets, the most mysterious celestial bodies in the system. What is the role comets play in our solar system? Without comets, what would the solar system be like?

How were comets formed? Their origins seem different from those of planets. Why are their orbits so different from other celestial bodies? Some orbits, such as Hale-Bopp, C/1995 O1, are nearly 90° crossed with the Earth's orbit. Is the existence of comets necessary for keeping the dynamic balance of the solar system? Considering the system as a whole, how do the strange motion behaviors of comets conform to the theory of Relativity (curved space)? Might numerous comets be necessary for a solar system to be conducive to the birth of humans and civilization?

Some scientists believe the immense gravitational field of Jupiter helps the Earth largely avoid possible collisions with comets and meteors, which would be potentially fatal for life on our planet. Such a collision may have produced the Moon and led to dinosaurs' extinction.

We do not know whether these theories are true, but they could not be possible without the complexity of the solar system. Civilization could not occur on our planet if Earth were the only one in the solar system or if Jupiter was much smaller. The development of the Earth could not be free from the influence of all other celestial bodies in the solar system. Fortunately, such influences seem to be favorable to the Earth.

The solar system is far from simple, but the ideas above must be subject to further scientific inquiry and research.

There are many other solar systems in our galaxy, but all the systems we have discovered are simple in structure and could not produce civilization.

Today, there is growing interest among scientists in researching the relationship between the cosmic environment and the Earth's development. Some factors outside the solar system might contribute to some natural events on our planet. For example, the galaxy possesses an immense gravitational field and strong radiation. Scientists reasonably imagine the solar system passing by the galactic pericenter and the apocenter might respectively affect the global climate (the occurrence of Ice Age) and, consequently, the evolutionary process. A supernova outburst might have resulted in the appearance of humans.

Many statements aim to prove the relationship between the events on the Earth and factors in the universe. Scientifically, every idea has its grounds. Considering the gravitational field of the galaxy, natural events that occurred on the Earth could not be independent of the impact of the cosmic environment, although scientists have not generally accepted some statements (see Chapter VI).

A Discussion

Comets could hold the key to unlocking many of the mysteries surrounding our solar system. These icy bodies, originating from the far reaches of space, offer valuable clues about the formation and evolution of planets, moons, and other celestial objects. By studying their composition and

THE RATIONALITY OF THE EARTH SYSTEM

behavior, scientists may gain a deeper understanding of the early conditions that shaped our solar system.

> Perhaps some of the most exciting discoveries with comet Hyakutake were the detections of many new molecules in the coma using radio telescopes and infra-red telescopes. Among the molecules discovered include a large suite of organic compounds such as methanol (CH_3OH), methyl cyanide (CH_3CN), hydrogen cyanide (HCN), formaldehyde (H_2CO), methane (CH_4), ethanol and ethane (C_2H_6).[1]

> The primary goal of the [NASA] Stardust mission was to collect samples of a comet and return them to Earth for laboratory analysis... One of the most unexpected was the 2009 discovery of the amino acid glycine by a team of scientists from the Goddard Space Flight center.[2]

> All terrestrial organisms depend on nucleic acids (RNA and DNA), which use pyrimidine and purine nucleobases to encode genetic information... Our results demonstrate that the purines detected in meteorites are consistent with products of ammonium cyanide chemistry, which provides a plausible mechanism for their synthesis in the asteroid parent bodies, and strongly supports an extraterrestrial origin.[3]

> The ESA-Giotto mission (1986) found that cometary nuclei are extremely dark objects... The reason for such a low albedo is a cover of the surface by dark complex organic compounds. Solar heating drives off volatile compounds leaving behind heavy long-chain organics that tend to be very dark, like tar or crude oil.[4]

> One of the major discoveries from the analysis of the comet samples was finding particles rich in organic matter. "Comets are believed to have brought water and organic matter to the early Earth, and it is important to understand the nature of these materials because they are necessary ingredients for the origin of life," said Lindsay Keller, NASA scientist at JSC and Stardust co-investigator. "One of the first analyses we obtained on the samples showed abundant hydrocarbons in many of the particles."[5]

1. Meech, *1997 Apparition*
2. Brownlee, *Stardust*
3. Callahan, et al., *Carbonaceous meteorites*
4. Hanslmeier, *Habitability*, 91
5. NASA, *Comet Stardust Findings*

These findings raise questions about understanding the role of comets in the solar system. Could comets have formed with or attracted substantial organic matter and retained it ("rich in organic matter")? The organic substances mentioned might encompass complex molecular compounds, although methane, a simple organic substance found widely in the solar system, could have formed independently of life.

Given that some chemical changes in molecular structure may occur in the unique and variable environment of comets, this would provide conditions for the possible formation of a great deal of organic matter. The formation of comets must be a complex process that is little understood. Yet, "Subsequent analyses revealed that some of the organic matter formed in the cold cloud of dust and gas that was the precursor to the solar system."[6]

Why are the Moon, Mars, and even Venus not rich in organic matter? As their neighbor, logically, Earth would be unlikely to have any organic matter on its early surface. Are comets, then, the primary source of organic matter for the Earth? If so, our planet was a favorable environment to receive those organic materials for further development.

Modern science may not yet be able to answer these questions, but the potential answers will be of profound significance in comprehending the nature of the wondrous universe.

GEOLOGICAL FACTORS

Earth's Structure

Earth's structure is the basis of its function, so what kind of Earth structure can give the Earth such a magical function? This question will be explored in the next chapter.

The Geomagnetic Field

The existence of the geomagnetic field leads us again to consider the thoughtfulness of our planet's design. The geomagnetic fields of Venus, Mars, and Mercury are much weaker than that of Earth, although Venus is similar in magnitude.

---

6. NASA, *Comet Stardust Findings*

The geomagnetic field is like an umbrella shielding particle flows from the Sun and outer space, protecting all life on Earth (Figure 2-9), and using a compass depends on it. Without a compass, seafaring navigation would have remained as it was in the early days, and world history would have unfolded differently.

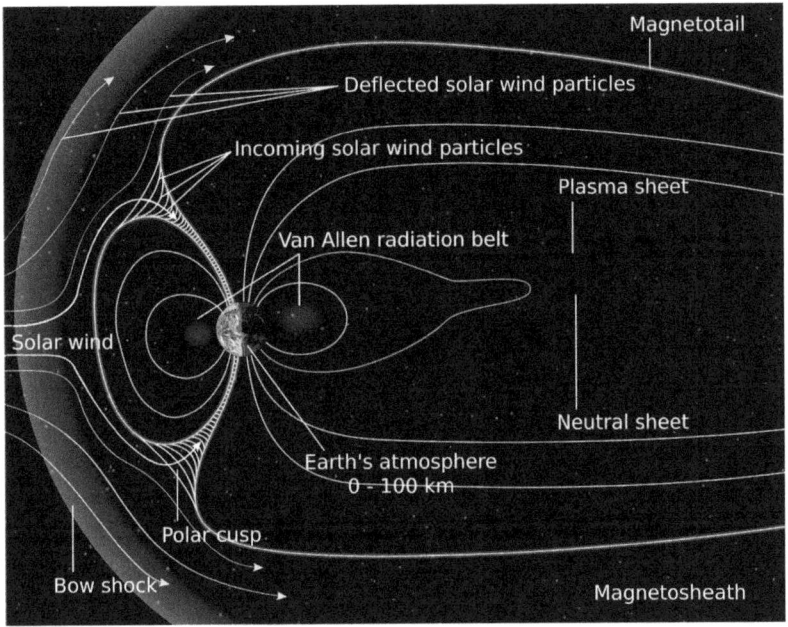

Figure 2-9: The geomagnetic field (Source: NASA, CC BY-SA 4.0).

Like other essential factors of the Earth, the formation of the geomagnetic field must have been a complex and precise process. Its existence depends on the planet's magnitude with its rotation speed as well as on its structure. Our planet is lucky. Its complex structure cannot be overstated.

FACTORS RELATED TO EARTH'S SURFACE

The Sea

Do you agree that a planet without plenty of ocean water could not produce human civilization? If you do not, you may not completely understand the role of the ocean and the complexity of the origin of civilization on our planet, although you may enjoy its magnificent scenery.

Water covers three-quarters of the Earth's surface. In that sense, Earth is a water planet. The earliest life was born in the sea, a wealth of many species and minerals.

The atmospheric circulation and ocean currents in the deep and on the surface jointly regulate global temperatures, moisture, and precipitation worldwide. Precipitation on land is formed mainly by ocean evaporation circulating through the atmosphere over the continents. The phycophyta in the ocean produce most of the oxygen in the air and absorb a significant amount of carbon dioxide from the air, which can also be directly dissolved by seawater. The origin of life, global temperatures, the content of oxygen and carbon dioxide in the air, and global precipitation are essential factors for human life and the development of civilization. The functions of the ocean and all the essential factors require a certain amount of seawater in terms of both area and volume.

If the oceans were smaller, they would be unable to fulfill their natural missions, and the Earth would not have had the chance to develop into an optimal planet for civilization, even with all other necessary factors in place. Therefore, if the extent of the Earth's water-covered area were significantly altered, whether larger or smaller, it would profoundly affect the global environment. Moreover, the functions of the oceans are intricately influenced by the salinity and temperature of seawater. The Earth's oceans, which humans have relied upon for millennia, are immense, and their salinity and temperature have stabilized over time (see Chapter I.2, Ancient Seawater).

Although modern science has not been able to quantitatively determine the area and volume required by the oceans to meet the needs of its various functions, the ocean is an important source of the Earth's environment, which cannot be stressed too much.

Mountains

It is difficult for most people to imagine the Earth's surface being uniformly flat. If the surface were uniform, our beautiful natural world would not exist. If you disagree with this notion, you might not fully appreciate the importance of mountains.

THE RATIONALITY OF THE EARTH SYSTEM

If oceans are the sources of water vapor, then mountainous landmasses serve as the consumers. Mountains, through precipitation, act as "water towers" and sources of rivers. Without mountains formed by plate movement, water would remain in isolated, stagnant pockets of various sizes. Watersheds create river basins; in other words, precipitation in mountainous regions gives rise to rivers. Therefore, mountains are always the birthplace of rivers. Without flowing waterways, the land would be much drier (see Water, Precipitation, and Rivers, below, Figure 2–10, and Figure 2–11 below).

> Figure 2–10 (below): Xinjiang in China is an arid area with little rainfall each year, but many snow-covered mountains provide necessary water resources for life. The pictures below show us the importance of the snowy mountains, formed by plate movement about 45 million years ago, or Himalaya orogeny.

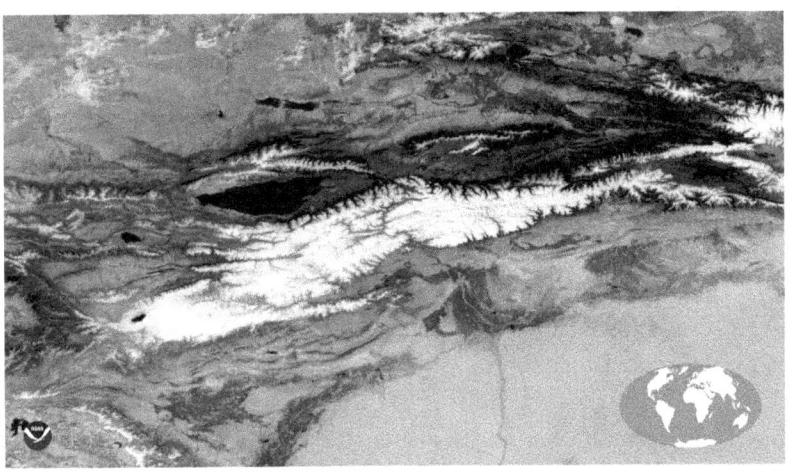

> a. The Tain-Shan Mountains provide the Karez irrigation system with plenty of water from snowmelt (Source: NOAA-20, public domain).

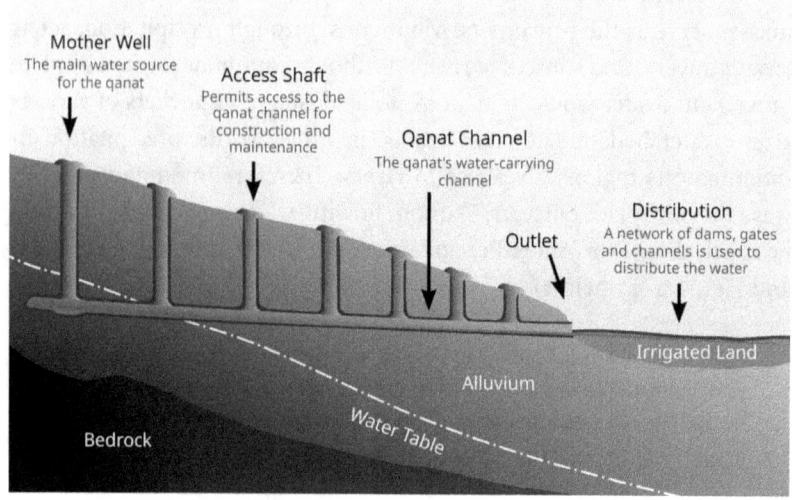

b. The Karez (Qanat) irrigation system was first used over 2,000 years ago. It has about 1700 channels and a total length of over 3100 mi (about 5000 km) (Source: Samuel Bailey, CC BY 3.0).

Mountains have many other natural missions. The existence of mountains promotes the variety of climates in different regions and, correspondingly, the evolutionary processes of life. Mountains are rich in species.

High-elevation mountains with snow and glaciers are essential water resources in the natural world. Notably, most snow-capped mountains are located inland, even in deserts where rainfall is much lower than in coastal areas. For instance, the Tian Shan mountain range of the Euro-Asian continent, stretching about 1550 mi (2500 km), is situated inland and serves as the source of many major and minor rivers. These snow-covered peaks provide ample water resources for life in arid regions, which is vital for the surrounding desert areas. How thoughtful! Without snow runoff from mountains, there would be little life in the expanse of the inland continent.

Mountainous areas tend to be rich in mineral deposits necessary for human life and the development of civilization (see Factors Related to Civilization, below). If the land were flat everywhere, we would not enjoy flowing rivers, let alone life and civilization as we know them.

Land

It's indisputable that human existence is grounded on land. Alluvial plains worldwide, rich in water resources, soil, plants, and animals, provide the ideal conditions for the birth of civilization. Earth is indeed the mother of humanity. While this statement stands alone, this book, as a work of natural philosophy, digs deeper into the significance of the distribution of land and sea, the land environment, and land resources in human development to highlight nature's inherent rationality. For more details, see Chapter II.3.

Water, Precipitation, and Rivers

Water is a necessity for life. But how does nature provide water molecules everywhere on the land, effectively supporting the subsistence of life? This is indeed a big problem for nature to solve.

The structure of the water molecule H_2O is simple but has many important roles to play in the Earth System, for which nothing can substitute. Thus, we may well call water a God molecule.

Water, however, cannot fulfill its natural missions without the vital form of precipitation. Nature uses this (simplified) process (Figure 2–11) in which precipitation results in water resources on land:

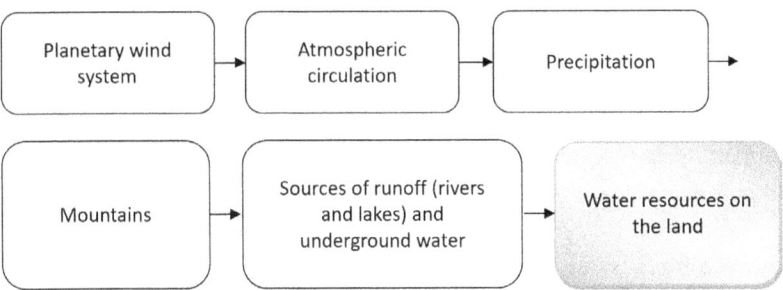

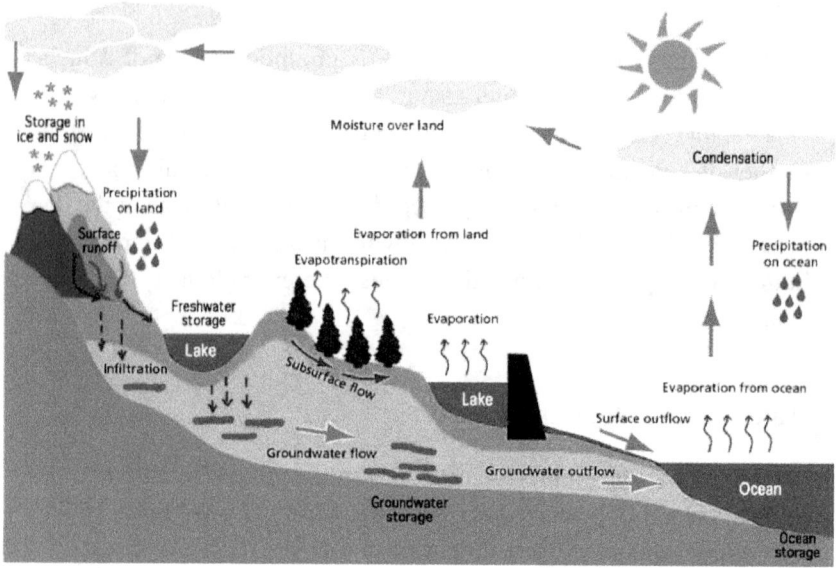

Figure 2-11: Water circulation between ocean and continent. A river or lake's main function is collecting rainwater (Source: Sunsono8, CC BY-SA 4.0).

This is a complex and subtle process in detail. For example, how raindrops form is still a problem for scientists to research and understand.

Underground water through capillary tubes (a quality of water molecules) in the soil supports life on the land during dry seasons. Rivers, through flooding, produce many expanses of fertile alluvial plains, supporting all living organisms on the land. Rivers, lakes, and underground water cause the land to become an animate world, a fairyland of life. Precipitation also accelerates the weathering processes, producing soil layers almost everywhere worldwide.

Precipitation, rivers, and underground water are the best ways to provide fresh water for life on land. In the development of the Earth, the massive occurrence of precipitation might not have appeared until the late Cenozoic Era. Before then, precipitation may have been scant because of the weakness of atmospheric circulation. The presence of life before the Cenozoic Era depended on shallows and marshland, and there were few organisms on land before about 400 million years ago.

Precipitation is an ideal way to provide the land with water resources. Do you understand the complexity of the formation of rain? Isn't it surprising? Can you design another way to make water resources available on land besides precipitation? Can you recognize the complexity of precipitation in detail?

The Atmosphere

As mentioned above, the formation of water resources on the land is a complex natural process. The following will tell you that the formation of the atmosphere is complex and magical. The atmosphere has a layered structure, and every layer has its own natural mission.

Any change in an atmospheric layer would result in a change in its function. There are many problems about the formation of the atmosphere being researched by scientists.

Looking at our planet from space, you will find it covered with a thick layer of gases under which we live. We think ourselves incomparably powerful, but you cannot live longer than a moment if you leave the atmosphere. The atmosphere is essential for life (Figure 2–12).

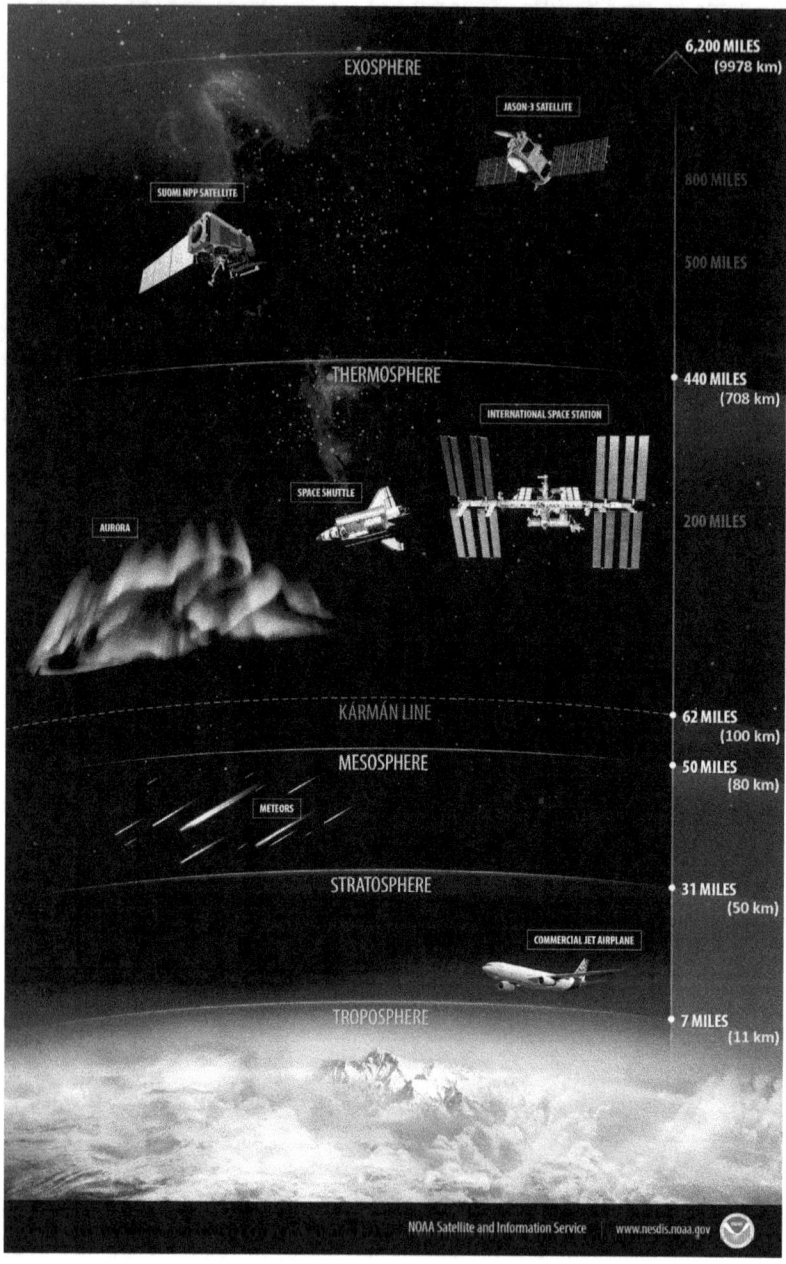

Figure 2-12: The layered atmosphere, with its many functions, is necessary for humans and civilization (not to scale) (Source: NOAA, public domain).

Every layer of the atmosphere has its own natural mission, but the troposphere is the most important one for all life on the Earth's surface. It is believed that the troposphere could not have been formed without the layers above it. How was the layered atmosphere formed during the past millions of years? Modern science knows little about that question. Its formation must have been a complex natural process, lasting for millions of years, involving many factors such as the gases that make up the atmosphere, solar radiation, terrestrial radiation, atmospheric radiation, atmospheric absorption, ground absorption, etc. Furthermore, the content of every gas in the atmosphere is an independent variable, always in gradual change. So, we must be aware that the formation of the atmosphere is far from simple!

The atmosphere plays several essential roles in shaping the global environment, such as facilitating circulation, enabling precipitation, and regulating heat and humidity. Remarkably, the precise proportion of gases that make up the air—oxygen, nitrogen, a small amount of carbon dioxide, and traces of methane and other inert gases—is necessary for life. These proportions have gradually formed over millions of years. Scientists believe that the Earth's temperature is significantly influenced by the content of these trace gases, particularly carbon dioxide, which contributes to the greenhouse effect. Today, even a slight increase (0.0x percent) in carbon dioxide levels has impacted global temperatures, threatening the existence of life on land. How sensitive this balance is!

About 21 percent oxygen content, an ideal value, is principally required by all living organisms, is essential to the two qualities of fire and profoundly affects the evolutionary process of human intelligence. The two qualities of fire are its universal usefulness to humans and its availability (see The Mystery of Fire, below). The remaining component of air is mainly nitrogen, an important gas. Plants must absorb a significant amount of nitrogen, an inactive chemical element, making combustion safe.

The following table shows how a change in the air's density influences life.

The Theoretical Relation Between Altitude and Oxygen Content Percentage			
Altitude above sea level (m/ft)	Atmospheric pressure (millibars)	Oxygen Content Percentage	Boiling Point (°C/°F)
7000/22,965	420	9.75	77/170.6
6000/19,685	481	11.35	80/176.0
5000/16,404	549	12.95	84/183.2
4000/13,123	624	14.55	87/188.6
3000/9842	707	16.15	90/194.0
0	1013.25	20.95	100/212.0

The atmospheric pressure at sea level, known as standard atmospheric pressure, provides optimal conditions for comfortable breathing and easy combustion. At this pressure, water boils at 212°F (100°C), enabling food to be cooked, most germs in the water to be killed, and harmful substances in some raw vegetables, like certain types of beans, to be neutralized. Also, evaporation from the ocean surface at sea level generates sufficient moisture for global precipitation, sustaining numerous rivers and lakes.

The density of air gradually decreases with the increase in altitude. The physiological reaction to altitude begins in humans at about 9800 ft (3000 m). Here, ultraviolet rays cause serious harm, and the boiling point lowers, causing difficulties in daily life. Accordingly, we can easily see how our atmosphere, including its mass, composition, and gas proportions, is an optimal design for life.

The following exercise aims to deepen our understanding of the rationality of the atmosphere. The amount of oxygen and carbon dioxide in the air is so optimum for life that any change in their proportion would be detrimental, so much so that the current proportions have likely remained stable for millions of years. If any change in the amount of oxygen and carbon dioxide occurred, what would happen? The following results are what we can logically imagine.

Any significant decrease in the oxygen content of ancient air (below 21 percent) would have led to dire consequences: survival would have been difficult for all life, the combustion rate would have slowed, soil health would have deteriorated, and the evolutionary process from primates to humans could not have been completed. Conversely, an increase in oxygen content (above 21 percent) would have accelerated metabolic and physiological processes, leading to faster growth and larger sizes of all living organisms while also increasing the combustion rate.

Likewise, any effective increase in carbon dioxide content in the air would have led to global warming, as we see today. Any decrease would weaken the greenhouse effect; the Earth would have become a frozen planet.

Such changes would have been unsustainable for the Earth System. Therefore, oxygen and carbon dioxide levels must remain stable, and the proportions of other gases are accordingly defined, ensuring that all gases in the air are precisely balanced (see Chapter III, The Mysterious Abundance of Elements on the Table).

Let's review. The average global surface temperature of the Earth varies depending on location, season, and other factors, ranging well below freezing in polar regions to much higher temperatures in equatorial regions. This range of temperatures, known as air temperature, feels familiar to us because our bodies are predominantly composed of water. However, achieving these temperatures on such a minuscule spot in the vast universe is anything but straightforward; it requires a precise interaction of numerous natural factors. Can you identify the primary factors involved?

Remember, the formation of the atmosphere is also a stochastic natural process. But how was the exact proportion of the different gases formed over billions of years? Don't you think this natural process is magical (see Chapter I.1, The Timely Appearance of the Atmosphere)?

The Soil

Soil is a critical component of the natural world. Soil is the only material on the Earth that plants can firmly root and grow in. A little soil can support a huge tree. Living organisms contain about 60 chemical elements, all from the soil. Without extensive soil, Earth would lose its value to support life, even if all other essential factors are present.

Plants and animals display diverse colors, smells, and organic substances, all fundamentally sustained by soil. Have you ever considered the remarkable roles soil plays? Besides its essential support for life, soil also serves as a crucial link between the inorganic world and human civilization (Figure 2–13).

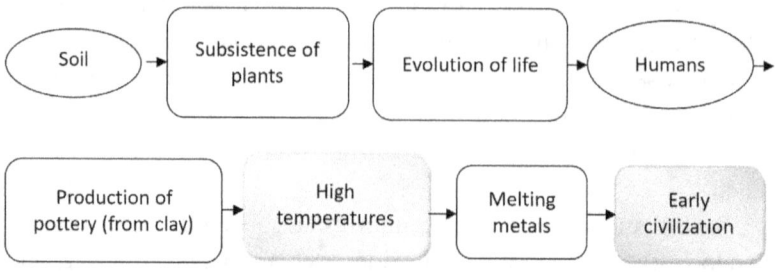

Figure 2-13: The miracle of soil—a little soil supports the growth of a huge tree (Source: Marathon CC BY 2.0).

Archaeologists have surmised that firing pottery was the only way for early humans to have learned how to master high temperatures, opening the gate to the most essential material in the development of civilization—metal.

Soil is an excellent outcome of the teamwork between astronomical, geological, atmospheric, biological, and erosion factors, which compose a complex soil-forming process. The most valuable substances in the soil are clay minerals, essential elements in forming the soil that provide qualities such as water retention and sintering that are necessary for plant growth and pottery-making. These clay minerals are mainly kaoline, illite, and montmorillonite. Each is in the form of a crystalline structure in microscopic view, with a complex molecular formula. For example, the formula of kaoline contains Al-Aluminum, Si-silicon, and OH, or $Al_4[Si_4 O_{10}](OH)_8$ (see Chapter III, Soil and Sand).

The clay mineral formation process is complex, consisting of a series of geological processes: magmatism, metamorphism, sedimentation, and erosion. Scientists are still researching this process. Aren't you amazed that the Earth produces so many clay minerals all over the world (see Chapter III, The Mysterious Abundance of Elements on the Table)? Soil layers worldwide could not possibly form without a tremendous quantity of clay minerals. Without the soil, could civilization arise?

Having learned these soil qualities, don't you think nature creating so much soil with a tremendous quantity of clay minerals on the Earth is magical?

Wetlands

People may think wetlands, including estuarine wetlands (seawater wetlands), seem to be an "unnecessary" part of the natural world. On the contrary, wetlands are an essential condition in evolutionary processes. From minute psilophytes that existed about 400 million years ago, the first plant on land, to pteridophyta forests, to dinosaurs and gymnospermae forests, none of these could exist without shallows and marshland.

From the viewpoint of evolution, wetlands are a treasure trove of species, including terrestrial and aquatic animals and plants, as well as birds. If all wetlands disappeared from the land, many species would lose their optimal environment, and water resources would be diminished, probably affecting the global climate. Wetlands can purify harmful substances, so they are also called the kidneys of the Earth. Wetlands today are an essential component of the Earth System with many natural missions. We cannot undervalue them.

Today, wetlands are formed by special geographic and climatic conditions entirely different from those in the Paleozoic and Mesozoic Eras.

Deserts

Deserts seem to be the most desolate and unwelcoming areas of the Earth. However, deserts are inevitable natural phenomena resulting from the relationship between the planetary wind system and geographic features. For example, one of the factors generally accepted as forming the Sahara is the consistent subtropical high. However, some interesting questions have puzzled scientists. Where did such a tremendous quantity of sand come from? How and when was the sand produced?

Typically, sand grains are formed through weathering processes, often involving substantial water, primarily from precipitation. It's intriguing to ponder that desert sands might not originate locally. Deserts could instead be created by the deposition of sand produced elsewhere and transported by winds. Before the development of the planetary wind system, there might have been few, if any, deserts.

Is the existence of deserts essential? That Nature does nothing that has no role in the world is understandable. From the viewpoint of evolution, specific environmental conditions are needed to form life. Thus, the desert is like a school where life learns to survive adverse natural conditions, creating a unique ecological system. An example is the *populous euphratica*, a diversiform-leaved poplar, which lives in arid and semi-arid regions (Figure 2–14).

Figure 2–14: The *Populus euphratica* subsists in the desert, and the root system is more than 65 ft (20 m) deep (Source: lwtt93, CC BY-SA 2.0).

People may find sandy ocean beaches beautiful and relaxing, but the desert warns us to tread carefully lest we lose our lives. Today, some deserts are expanding due to destructive human activities. Note that the quality of desert sand is far less than that of sea and river sand, and there is little use value.

The idea of terraforming, which is transforming land and climate to generate a more hospitable environment, has been speculated by science fiction writers and scientists. Is terraforming all the deserts in the world into new green land a good idea? What would happen to the global climate? The fact is we do not know how such a transformation would affect the Earth, producing a planet-wide garden or spiraling our atmosphere out of control into catastrophic weather conditions.

Deserts represent a relatively small portion of the Earth's landmass, but they may play an essential role in the synergy of the Earth's climate. This is an important topic of ongoing research.

BIOLOGICAL FACTORS

Forests and Grasslands

Undoubtedly, forests and grasslands are essential components of the Earth System. But do you agree that any planet without them could not produce humans and civilization? Let's look at their importance.

Adapting to different climatic conditions, forests and grasslands are similar ecological systems widely distributed on land. Both have many essential roles in the natural world, such as releasing oxygen and absorbing carbon dioxide, maintaining the proportion of the gases that make up the air. Both can adjust the local temperature and moisture, and they affect the volume of regional precipitation.

However, what we are particularly interested in is not only their roles but also their natural missions over millions of years. Both are also a treasure trove of species from the viewpoint of evolution, giving rise to some species crucial for developing civilization, such as primitive horses, oxen, wild wheat, wild rice, wild cotton, silkworm, etc.

Among the vast numbers of species found in forests and grasslands, the highlight is the early primate from which humans evolved. Hands and fingers possess essential functions in human life and are one factor promoting the human brain's evolutionary process. Forests provided favorable conditions for the completion of the evolutionary process of the primate's fingers.

As an omnivore, the primate particularly enjoys fruits as its food. Primates use fingers instead of nails (unlike cats) to grasp trunks and branches and break some kinds of nuts. Subsequently, the primitive fingers gradually evolved into hands. During this evolutionary process, fruits played an important role, enticing primates to live in trees. Here, we may say the evolution of plants promoted the evolution of animals, and the development of hands could not have taken place without the help of forests.

Notably, the human hand has a thumb with an opposite direction of movement from the fingers and is shorter and nimbler than the non-human primate hand (Figure 2-15). The distribution of the five fingers decidedly influenced the function of our hands, which are more agile than primate hands, indicating an extensive process for primate hands to evolve into human hands.

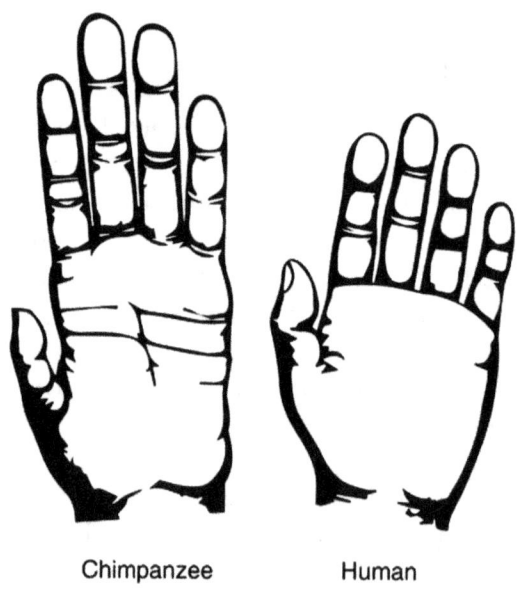

Chimpanzee   Human

Figure 2-15: Human hand versus chimpanzee hand (Source: Almécija, Jeroen B. Smaers & William L. Jungers, CC by SA 4.0).

As time passed, primates migrated to grasslands. They adapted to bipedal locomotion as forests gradually receded and savannas (grasslands interspersed with trees) became dominant in the GRV around 3–4 million years ago, marking the evolution toward early human ancestors. Thus, both forests and grasslands played crucial roles in the development of primates into early humans (Savanna Hypothesis).

Regarding forests and grasslands, what is surprising are the following two points:

1. Herbaceous plants, particularly grasses, have root systems with a strong regeneration capacity at the top, differing from woody plants. This capacity makes grasslands major providers, ceaselessly and generously offering food for herbivorous animals to enjoy.

2. Before humans became dominant, woody plants were widespread, offering abundant wood materials crucial for the early development of early civilization, which lasted for thousands of years. Early humans would have faced significant challenges in starting their civilization without access to these essential natural resources.

How thoughtful these two natural arrangements are! Nature has always provided, allowing humanity to progress step by step. In this regard, the tropical rainforest is significant. The number of species in tropical rainforests is more than half the total species worldwide, although its area is only about seven percent of the land. However, forests and grasslands require a large enough area to fulfill their natural missions. Destruction of significant regions of both is the problem we are facing today.

Shrubberies

Do you consider shrubbery as insignificant as wetlands in the natural world? Nature doesn't create anything without a natural mission. Given the diverse climatic conditions on the Earth, shrubs play a vital role as a complement to forests and grasslands or as a transition between the two. They serve functions similar to forests and grasslands but also contribute significantly to human civilization. Many fruit trees, beautiful flowering shrubs that enhance our spiritual life, and various useful plants originated from shrubs. Imagining a world without them is challenging. If apple, peach, pear, and other fruit trees were tall like forest giants, harvesting their fruits would be inconvenient, making them less enjoyable for us. Gardens lacking a variety of both large and small flowers would seem dull without shrubs.

The Biosphere

The biosphere is the most valuable treasure in the Earth System, consisting of three kingdoms: producers (plants), consumers (animals), and decomposers (microorganisms, some insects, and some small animals, etc.). A reasonable and subtle system developing from the earliest days, the biosphere is critical for human existence as well as the development of civilization.

Solar radiation is the primary energy supporting the function of the biosphere. All living organisms are interdependent, forming a whole system, and all life is dependent on the biosphere (see Chapter III, The formation of global biospheres).

Therefore, we can naturally imagine that life on a planet implies not only an individual being but also an immense system. The formation of a biosphere may be a primary condition for a planet to support the development of civilization in the universe.

FACTORS RELATED TO CIVILIZATION

To form a civilization on a planet in the universe is far from simple! Many natural factors are unnecessary for the subsistence of living organisms but critical for the birth and development of civilization. We may call them civilization factors. The two such factors are various mineral deposits and the two natural primary substance chains (see below). Even if all the essential factors stated above are present but lack the civilization factors, could civilization be born? We would say, "It could not!" If you disagree, consider the factors below.

Minerals and Deposits

Scientists generally agree that chemical elements are formed in the first appearance of stars in the universe, further developing into supernova explosions, scattering all the chemical elements into space. But minerals are mainly formed by geological processes on a solid planet.

With a few exceptions like sodium, living organisms generally cannot directly absorb minerals. For humans, minerals such as iron, copper, sulfur, and phosphorus are crucial elemental sources derived from mineral deposits.

Earth is a fortunate place, not only possessing various minerals sufficient for people to use but also producing all the minerals in large quantities for us to mine to develop our civilization. However, the formation of various minerals from chemical elements and their subsequent aggregation into mineral deposits results from intricate geological processes.

We have identified approximately 3000 distinct ore minerals to date. Among these, only twenty to thirty types are categorized as rock-forming minerals, predominantly non-metallic, and widely distributed in the Earth's crust. However, several dozen ore minerals are crucial for advancing civilization, yet these are less abundant. Fortunately, various geological processes have concentrated these valuable minerals into different types of primarily metal-bearing deposits. Mineral deposits enable people to locate and extract substantial quantities of ore minerals through mining. If these minerals were dispersed in minute amounts rather than concentrated in significant deposits, their extraction in usable quantities would be challenging, if not impossible.

An Interesting Fact Regarding the Depth of Mineral Deposits

All mineral deposits result from a series of geological processes, such as magmatism, metamorphism, or sedimentation, often two or three processes occurring at different times in the same location. Further, most ore mineral deposits and all organic mineral deposits have been formed deep down in sedimentary strata over millions of years. However, a fascinating fact is that many of the mines we exploit are near the ground's surface. Most metallic mineral deposits are less than 2300 ft (700 m) deep in the ground, some excavated using open-pit mining methods.

The phenomenon of deposits being formed in deep layers but found in shallow layers is significant for people who could not excavate sufficient quantities of coal and other ore minerals such as iron without modern technology, particularly in ancient times. Crustal movement and weathering are responsible for deeply formed mineral deposits near the ground surface. The upheaval of the crust causes the deeply formed coal deposits in sedimentary rock to rise nearer to ground level, and the deposits become thin in the shallow layer due to weathering.

Considering the complexity and subtlety of the geological processes that transform chemical elements into minerals, which are then concentrated into deep deposits or nearer to the Earth's surface for human use, wouldn't you agree that the Earth behaves like a master magician of sorts, orchestrating these essential substance chains of energy and metals?

Even more interesting are the arguments below.

Utilization of the Two Primary Substance Chains

There are two primary substance chains in the natural world: energy and metal. The chart below shows the natural energy chain serving civilization:

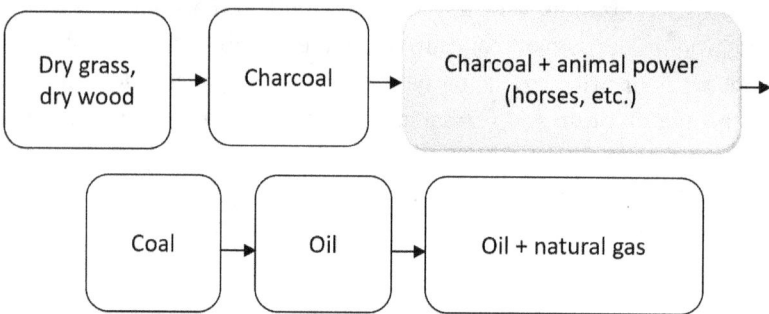

There are many sources of natural energy, such as waterpower, wind energy, etc., but their utilization must be done by machines. The link in the chain mainly refers to those that can be used without relying on machines. In ancient times, energy sources such as coal and oil could be utilized for burning. Early humans lacked the technology to make machines needed for other energy sources.

And this chart shows the natural metal chain serving civilization:

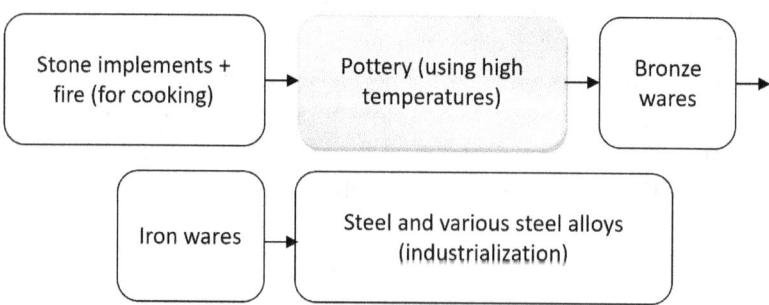

In a sense, these two chains symbolize the trajectory of human progress, where each component has served as a cornerstone in its time and remains relevant today. For instance, grass and wood remain crucial fuel sources in certain remote regions today. The pivotal role of animal power in civilization cannot be overstated. It facilitated essential tasks such as farming, communication, and transportation during early times, without which humans would have faced formidable challenges.

The various species useful for work far exceed humans in physical power, endurance, and speed. These animals have functional traits in body and temperament (Figure 2-16). For example, the horse, originating from a primitive horse the size of a fox about 50 million years ago, is an ideal herbivore with a temperament conducive to communication with us and a physical form with a capacity for strength, movement, and durability. In addition, riding a horse increases a person's combat ability far greater than fighting on foot. Horses played a vital role in the development of early civilization. No other animal could substitute for the role of the horse.

Figure 2-16: Section of a Chinese Eastern Han Dynasty (25-220 AD) fresco from a tomb in Luoyang, China. Horses had many uses in ancient times. The horse's appearance in the evolutionary process is indeed one of the natural miracles! (Source: Public domain).

Before the invention of motorized vehicles and trains about 100 years ago, horses, donkeys, camels, oxen, etc., were provided by nature for human use for several thousand years. All these animals' significant contributions to the development of civilization in early times can never be undervalued! How thoughtful nature is!

As for the second primary substance chain, the melting point of metal with impurities is considerably lower than that of metal in its pure state. So, early humans could refine copper ores and iron ores with charcoal because they contain many impurities.

The links in the two primary substance chains in the natural world seem to be entirely unrelated to each other, randomly distributed. Yet all the links in either chain seem to be naturally arranged in such a way from being easily obtained but of fewer uses, such as dry leaves on the ground, to being difficult to obtain but of more uses, such as oil deep underground. Thus, early humans could discover and use them from simple to complex over the last few thousand years.

The advancement of civilization is closely tied to the use of primary substances within each chain, interlinked throughout its long development. Our capabilities have advanced as we have mastered each link and moved on to uncover the following intriguing link in the chain, driven by human curiosity. This progression follows a clever natural logic. As humanity's abilities have grown, so too have the challenges, with each subsequent link being more difficult to discover and solve than the previous one. This story of civilization's development is like climbing a ladder, one step at a time. If any link within these two chains had been missing, humanity could not have progressed beyond the early stages of civilization to achieve what we have today.

Since humans gradually and naturally learned how to use these two natural chains, step by step, humanity's intelligence has improved. The two primary substance chains represent the same way people living in different parts of the world each developed their early civilizations. This significant fact indicates that the development of civilization seems to be purposely arranged by nature rather than simply by human instinct. Without the appeal from nature, human interest could not step on the "rung" toward the development of civilization. Don't you agree?

The utilization of the substances in the chains is also helpful to the evolution of intelligence resulting from the teamwork between animate and inanimate factors, and each one has its own complex development process, such as clay (the two qualities of fire are examined in the next section).

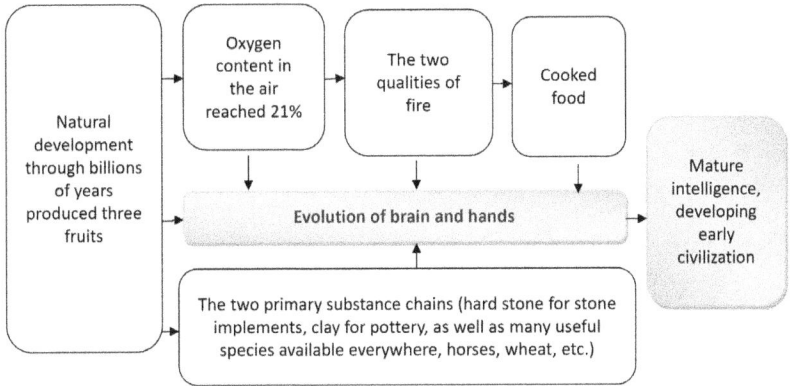

Therefore, we may say the two chains are civilization factors rather than factors of the emergence of life. However, the processes of forming these two natural chains are complex. They can be traced back to the beginning of our planet, including the Earth's development and life's evolution (see Chapter I.1, The Timely Formation of Two Primary Substance Chains).

Why are there only a few kinds of metals that function as links in the primary substance chain in metal? The metallic elements of copper (Cu), tin (Sn), lead (Pb), zinc (Zn), as well as iron (Fe) are easier to extract from their mineral deposits and smelt by ancient peoples (from about 6000 years ago), and they are commonly found in the natural world. Many other metallic elements are in the periodic table but are relatively difficult to extract from their mineral deposits and smelt. Most of them are obtained today through electrolysis, which was impossible in ancient times. Nevertheless, it is important to recognize that all metallic elements hold significant value. Every element in the periodic table plays a role in the development of civilization.

Aside from various metals, nature also provides many useful, non-metal materials, primarily wood, types of cement, rubber, glass, etc. Modern civilization could not have developed without rubber or types of cement. For civilization to develop, how thoughtful the series of natural arrangements is!

There seems to be a special relationship between nature and humanity. On the one hand, nature, through evolution, gives humans intelligence and intense curiosity. On the other hand, nature produces everything useful for people to discover and use to meet our needs for subsistence and development. In this way, humanity develops from the earliest days into the remote future.

So, we might well say that humans could not possibly have begun to develop civilization without thoughtful help from the two primary substance chains.

The Mystery of Fire

Obviously, without the phenomenon of fire, people could not use the two primary substance chains discussed above. Fire is the foundation of the roles the two chains play.

The first fire ancient people used was undoubtedly caused by nature. Of all the natural causes that ignite a fire, only lightning strikes occur with enough regularity and across a wide distribution in the world to enable early humans to progressively experiment with the control of fire once ignition has occurred. Accordingly, lightning strikes undoubtedly played an irreplaceable role in the early utilization of fire, which lasted about 500,000 years. This is the significance of lightning strikes, which are produced by the planetary wind system and would seem to represent a natural arrangement for the origin of civilization.

Since early humans became aware of the usefulness of fire, the world has changed dramatically. Life had evolved without using fire for about 3.8 billion years before the appearance of early humans. The drilling of wood to make fire began just a few tens of thousands of years ago.

However, the real significance of fire consists of two qualities: its universal usefulness and its ready availability, which are critical to the development of civilization. Without these two qualities, fire would have no value for humans, and without the existence of people, fire would have little meaning or value. The wonder of fire in the natural world is inextricably linked to humanity and the remarkable evolution of human progress toward civilized development. Consequently, fire is an essential requirement for civilization to exist and progress.

Fire could be loosely referred to as the father of civilization. In the natural world, everything animate and inanimate "fears" fire, yet the advancement of civilization absolutely depends on it. Fire seems to exist specifically to enable the development of civilization.

Why does fire possess these qualities of usefulness and availability? The question is thought-provoking. The formation of the two qualities is subject to a few natural factors, mainly fuels and oxygen in the air. The first fuel early humans used was dry grass and wood, and then later, coal.

With the advancement of science and technology, people began using oil and natural gas as significant fuels within the last century. Chemically, the primary element involved in the combustion of firewood or coal is carbon. In contrast, oil and natural gas combustion consists of both carbon and hydrogen. These two elements are particularly notable for being highly flammable and possessing high heat energy within the periodic table. The other remaining combustible elements are inorganic salts and are relatively inert in the natural world. Thus, these qualities are rooted in and determined by the patterns of the microstructure of matter itself.

Coal originated from the wood of trees that lived millions of years ago. Oil and natural gas originated from ancient decayed animals and plants, particularly those that lived in water. In other words, they all come from organic matter. We may well say that fire could not have occurred without life-forming substances. No life, no fire. In the natural world, there are only two kinds of substances: organic matter and inorganic matter. The former is used to form life and fuel, while the latter is used to form our planet. Inorganic matter cannot be ignited, or the world would be consumed by fire! This is how both life and the Earth coexist.

Air and life-forming substances are essential for the two qualities of fire. In other words, the "design" of air and life-forming substances is based on the two qualities of fire.

Without the two natural substance chains and fire, could civilization occur on a planet? Civilization could not occur on a planet without this significant foundation!

Fire can destroy everything except water molecules, which have a simple yet stable chemical structure. This stability with high specific heat capacity makes water highly effective for extinguishing fires, serving as the only unbeatable opponent of fire. Without this unique property of water, the world would face complete destruction from fire hazards. Fortunately, through precipitation, our planet has ample water to combat any potential fire threat.

Furthermore, both water and fire hold fundamental roles in the natural world: life originated from water, while civilization emerged from fire. The relationship between water and fire is a profound yet complex concept for both life and civilization within the Earth System. This balance is ultimately resolved through various stochastic natural processes.

From the viewpoint of the subsistence of life and the development of civilization in the Earth System, do you realize the proposition's complexity and significance? Doesn't nature surprise you (see Chapter III, Understanding the Element Carbon)?

The following is, in a sense, the relationship between chemical elements, atmosphere, life, organic matter, fire, and civilization. Water also exists in this chart (atmosphere and life). Indeed, we could not design another way to realize the natural process from elements to civilization. This natural process, we must understand, can only occur with each essential factor in precise availability, such as the proportion of the gases that make up the air.

Thinking over the logic below, perhaps you will realize that the process from elements to human civilization is far from simple!

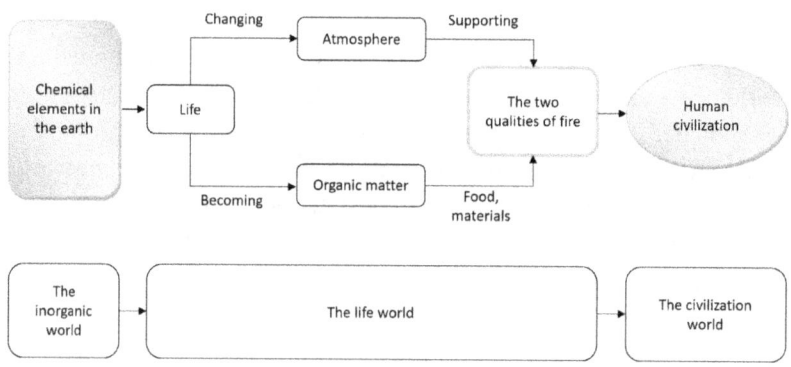

The progression from chemical elements to life to fire and to civilization: a wonderful natural arrangement!

Do you have any other solution that could transform the inorganic world into a world with civilization? After understanding this chapter, can you comprehend the existence of the Reason (rationality) in nature?

THE DEFINITION OF THE EARTH SYSTEM

Finally, let's define the concept of Earth System in this book. The Earth System is defined as follows:

THE RATIONALITY OF THE EARTH SYSTEM

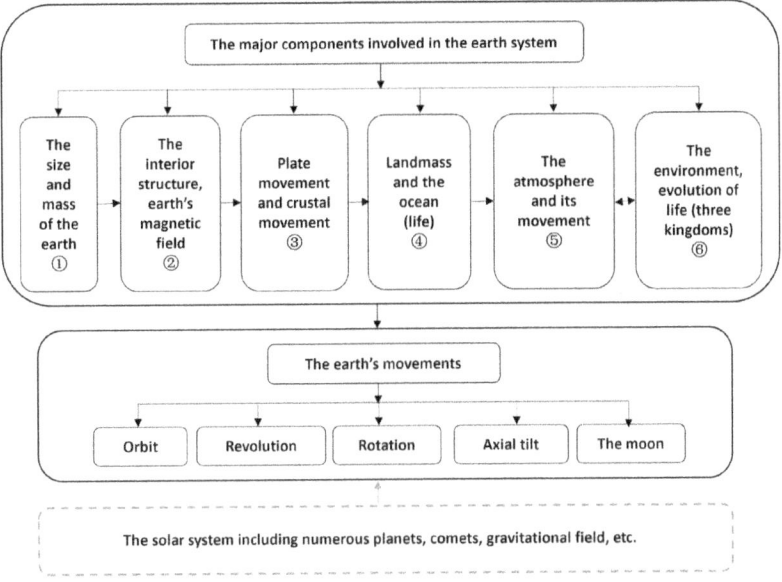

The Earth System in this book comprises "the major components involved in the Earth System" plus "the Earth's movements," making up a dynamic system. With a specific size and interior structure of the Earth (item 1, 2), the components of the system from items 3 to 6 have been slowly developing up to the present. Still, the changing atmosphere is also highly influenced by the biosphere.

The formation and development of every item diagramed above (such as the obliquity of the ecliptic, Earth rotation, the size of the Earth, and the plates) is based on a particular motion law, subject to certain conditions, and determined by chance. Therefore, it would seem every movement and component above is randomly selected by nature among numerous potential solutions, and all the items have been assembled into our Earth System. Fortunately, along the time axis, all the items above function well with each other, resulting in the ceaseless and progressive evolution of life.

Perhaps the greatest puzzle in the solar system is whether there is a relationship between the formation of the Earth System and the formation of the solar system (including the Sun and all other celestial bodies). In other words, could the Earth System have been formed without the given complexity of the solar system? Is the solar system's complexity required to form the Earth System? Today, we cannot provide more evidence, including long-term accurate astronomical observations, to answer this question. However, some scientists have advanced relevant hypotheses, such as the Theia Impact (a comet strike that may have formed the Moon). However, it is logical to believe that other celestial bodies influenced the formation and development of the Earth's movements and components. Our planet was born in a complex environment with numerous celestial bodies, all within a shared gravitational field. If the solar system had initially lacked these other celestial bodies, could Earth have developed into the system we know today?

Overall, we can compare the Earth System to our own bodies, where each component has a specific role and requires precise timing during its development. Any abnormal change, whether in an element of the system or in the cosmic environment, impacts the global environment, much like how any abnormal change in a part of our body or our surroundings affects our health.

The rationality of the Earth System explored in this chapter forms the foundation for the emergence of the human world and civilization. The question of whether the solar system exists specifically for the birth of the Earth will be explored further in the final chapter.

Chapter II.2

The Rationality of the Earth System
The Structure of the Earth System

Evolution took 4.6 billion years to develop chemical elements into life, humans, and the civilized world. Earth's structure governs its function. With the help of astronomical conditions, this structure can cause the surface of the planet to change constantly, and these changes lead to the continuous evolution of life and, eventually, the emergence of humans and civilization. According to the principles of natural philosophy, the Earth's structure, function, and astronomical conditions must exist in a perfect combination of extreme complexity in time and space. Still, modern science knows only a little about this dynamic.

Do you agree the Earth itself is like an automatic machine for life's evolution? Every factor influencing the Earth System, such as the atmosphere and the sea, is rooted in that machine. A planet without a complex structure could not produce a civilization.

    An essential function of this structure is to make plate movement possible and favorable to the eventual appearance of humans and civilization (Figure 2–17).

EARTH STRUCTURE

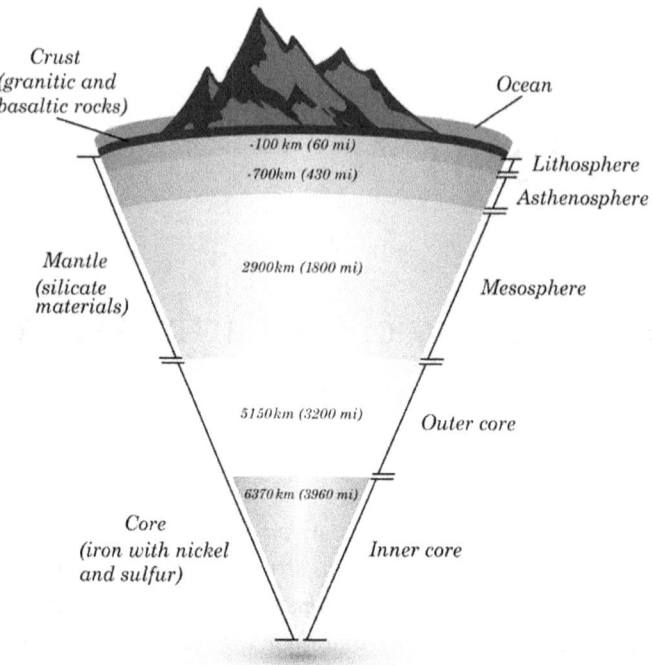

Figure 2-17: Profile segment of the Earth's structure (Source: Steven Heath CC BY 4.0). Note: All the numbers refer to the depth from the Earth's surface. Among them, the depth of the lithosphere and asthenosphere has regional differences (e.g., ocean vs. continent) and remains controversial.

EARTH'S CORE

Earth's core has a radius of about 2,156 mi (3,470 km) and is mainly composed of iron, nickel (Ni), and traces of other elements. Note that nickel is more magnetic than iron, so adding nickel to the core is conducive to the Earth's function, but the abundance of nickel is too low to add much to the core. Structurally, the core consists of the inner solid core (center sphere) and the outer liquid core with a slow flow. There is also a transition layer between those two. It's not hard to see why the temperature and pressure inside the core are so high.

The creation of our planet's massive magnetic field remains a key focus of scientific study. It's believed that the core's size and element composition are crucial; without these, the magnetic field wouldn't exist. While Earth, Mercury, Venus, and Mars share solid material compositions, their formation likely parallels. However, Earth's magnetic field is significantly stronger than other planets, vital in safeguarding life.

THE MANTLE AND THE CRUST

Earth's component above the core is the mantle, divided into the upper and lower mantle layers. The density of the upper mantle is less than that of the lower mantle, owing to different pressures. Structurally, the lower mantle is solid, and the upper mantle may be partially molten, but the whole is "plastic." The outermost layer is the crust, the main element of which is silicon. The upper part of the upper mantle, along with the crust, makes up the lithosphere (Figure 2–17), which "floats" on the asthenosphere. The asthenosphere is relatively soft, located in the upper mantle.

Note that the abundance of silicon is second only to oxygen in the Earth's crust. Its common form is silica (SiO_2), which is very hard and is not only a necessary but also an ideal element to compose the crust.

THE PLATE AND ITS MATERIALS

The lithosphere that "floats" on the asthenosphere is not wholly connected but divided into many pieces called plates (like a cracked eggshell). The plates are constantly moving, although very, very slowly. The forces driving these vast plates in perpetual motion remain a significant scientific problem. We know that plate movement somehow contributes to the Earth machine, and its "purpose" seems to create a variety of landforms (natural geographic environments) on the Earth's surface.

What are the plates made of? The plates forming the Earth's outer layer, known as the lithosphere, consist primarily of lighter elements like silicon, aluminum, and magnesium. This composition allows the lithosphere to essentially "float" on the asthenosphere. Despite its lightness, the lithosphere possesses some durability, capable of bending to form Fold Mountains without shattering into fragments. These plate movements wouldn't be as impactful if the surface rock layers were prone to breaking when bent or squeezed (see Chapter III, The Distribution of Element Abundance in Earth's Layers and Their Properties).

The image below (Figure 2-18) shows a small rock with many curved inner layers, likely shaped under high temperatures and pressures, unlike an eggshell, which remains unchanged even when cooked. This natural occurrence holds significant importance.

Figure 2-18: The curved layers of rock (folded rock) (see Figure 1-8) (Source: The American Museum Journal, c.1900-18, public domain).

The final chapter will explore the correlation between the Earth's size and gravity, exploring their rational connection.

HOW TO PROCESS THE EARTH'S SURFACE

As discussed above, according to modern plate tectonics theory, the lithosphere floating on the asthenosphere is divided into many plates (large plates can also contain small plates). Some plates on the surface are land, some are oceans, and some have both land and oceans (Figure 2–19).

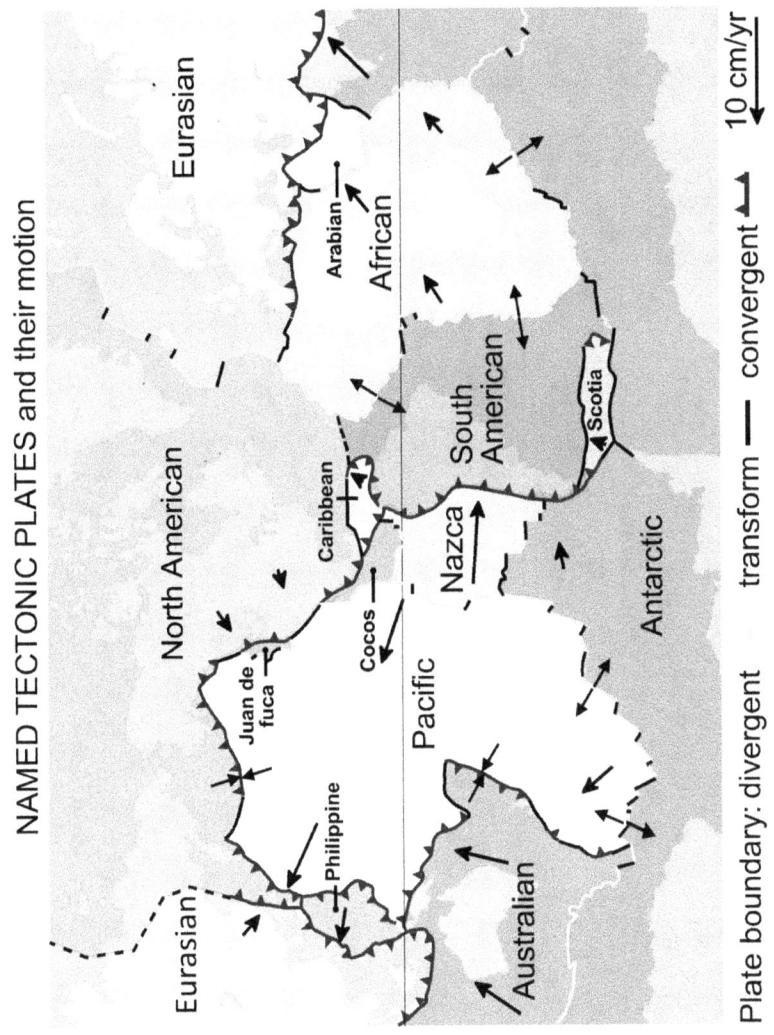

THE RATIONAL UNIVERSE EVOLVING FOR HUMANS

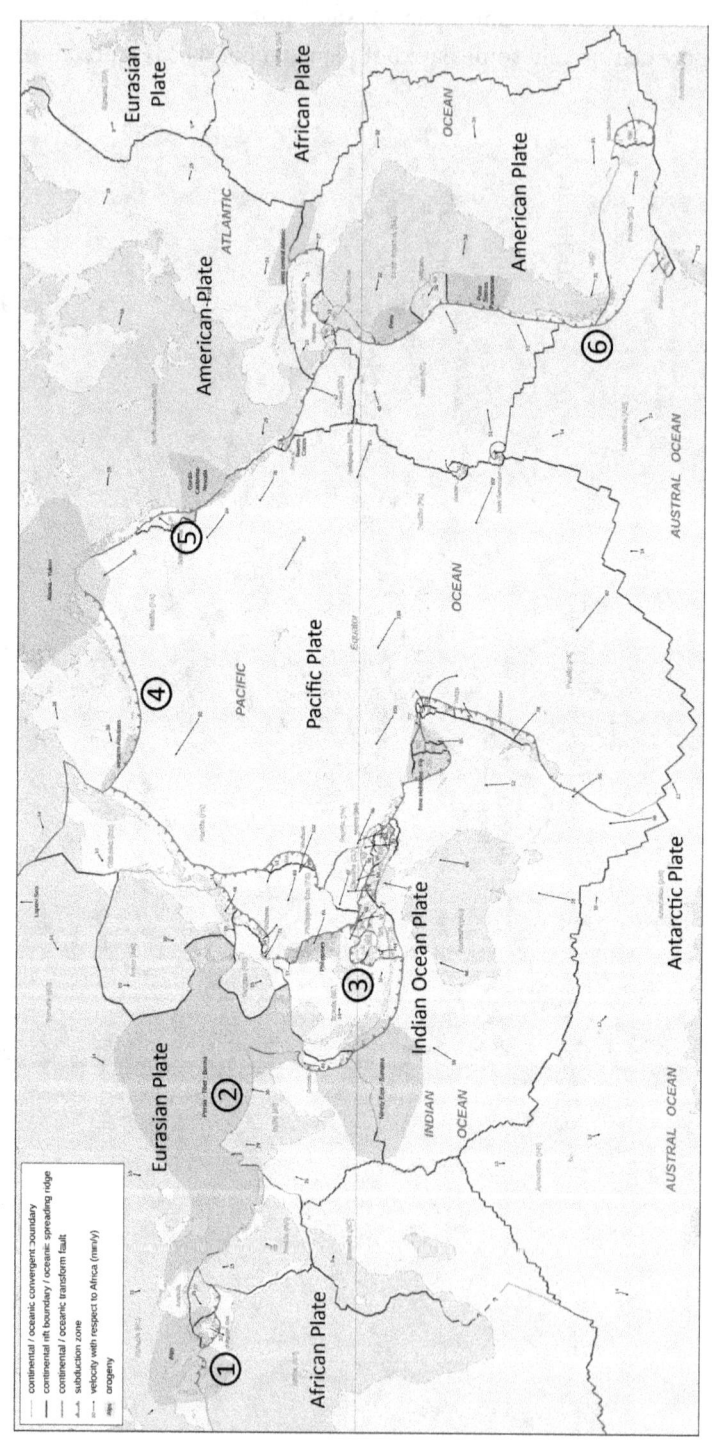

Figure 2-19: Global plate boundary distribution. The plate distribution shown above is a conceptual representation. Plate distribution maps vary somewhat across different sources. (Source: [Top] Hughrance, CC BY 4.0, plate boundary arrows added, [Bottom] Eric Gaba, CC BY-SA 2.5, labels added).

Plate boundaries can be broadly classified into two categories: divergent boundaries, also known as growth boundaries or ocean ridges, and convergent boundaries, also known as subduction boundaries. At divergent boundaries, oceanic plates crack and magma rises from beneath the surface, spreading outwards (seafloor spreading). Simultaneously, the edges of the oceanic plate are pushed towards the continental plate, where they subduct due to their lower position, forming subduction boundaries. This process leads to the creation of various surface features and islands (Figure 2-20).

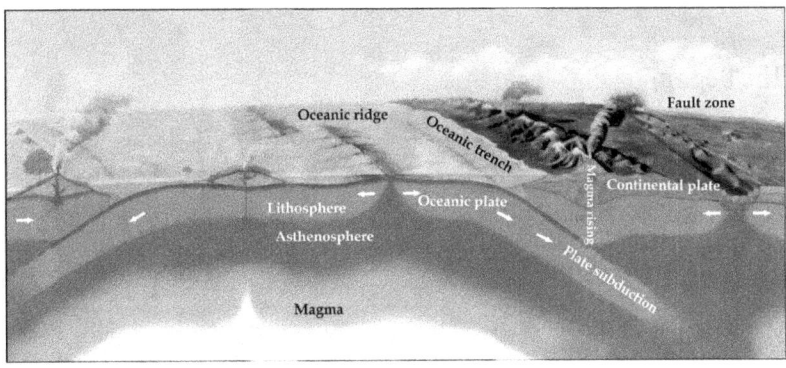

Figure 2-20: Profile of a typical collision between an oceanic plate and a continental plate (Source: USGS, public domain).

The boundaries where continental plates collide are a different type of subduction boundary (collision boundary), where mountains rise from the rock layers of the Earth's surface as the two continents push against each other. For example, the Himalayas are called Fold Mountains (see Figure 2-41 in Chapter II.3).

Mountains collect precipitation, forming streams through their terrain to create rivers. These rivers, shaped by diverse geological processes, carve out large and small alluvial plains, providing an optimal environment for human subsistence.

From the above, we can also see how reasonable the Earth's structure is, from the thickness of each layer to the elements found and the function of each layer to plate movements and formation of various terrains on the surface to promote the evolution and development of life. Describing Earth as a machine or an intricately crafted design isn't an exaggeration.

Hence, Earth adheres to a singular path in its structure, materials, and movement. Let's now dive into understanding the mechanics of this system and the corresponding results.

THE FORMATION OF PLATES AND THE MOVEMENT OF PLATES

Over billions of years, the plates formed, and life on Earth evolved from its inception to the emergence of humans within the context of the plates' movement.

The Beginning of the Earth

What the Earth looked like in the first place 4.6 billion years ago is poorly understood from a scientific perspective. Initially, the world was likely a hot celestial body, and the mantle had not cooled. But as the Earth's materials moved, tremendous amounts of water molecules evaporated into the primitive atmosphere, and they couldn't escape because of the Earth's gravity. Water molecules accumulated and condensed into liquid water as the Earth's surface cooled and the first oceans began, accelerating the rate of surface hardening. But where the immense volume of water came from remains a mystery.

The Formation of Plates

Some sections of the mantle condensed first into a solid, or the continental nucleus, which formed between 2.5 billion and 3.85 billion years ago. As the continental nucleus expanded, the plates formed gradually from 2500 million to 800 million years ago. Some of these plates had land on them, and some had oceans. Land, however, was continually expanding in area. Life started in the primitive oceans an estimated 3.5 billion to 3.8 billion years ago and has existed in the oceans ever since.

The Uncertainty of Plate Movement

Once established, the Earth machine had many possibilities for development. In other words, the results of plate movement were uncertain and driven by chance. For example, there could have been different ocean sizes, continental distributions, landforms, etc., and these different surface shapes would undoubtedly have affected the fate of life's evolution. Fortunately, throughout the Earth's development, plate movements at every stage created a corresponding shape of the Earth's surface perfectly suitable for life's evolving needs, which is what we will explore next.

Plate Movement Leading to the Evolution of Life on Land

The global plates, formed about 800 million years ago, have remained in constant motion. Landmasses gradually converged, forming Pangaea, a vast ancient continent, about 200 million years ago. Before that, about 400 million years ago, the oceans began to recede, and correspondingly, the land expanded in area, which is called regression.

During this geological process, life gradually expanded to the land (life landing), starting a new evolutionary process. In this initial stage (400 million to 250 million years ago), life transitioned from exclusively water to land life, also called semi-land life. At this stage, the environments on land, similar to the marshlands of today, adapted to this transitional stage of life.

By the time Pangaea formed, transitional life (from sea to land) had evolved into land life (Figure 2–21). Since then, with the fragmentation of Pangaea, new oceans and continents have emerged, and various landforms have formed through plate interactions, including continental drift and the spreading of the seafloor. Land life evolved as the environment on land developed. These different environments led to the evolution of life, including that of humans. It's a highly complex process, and this is just a brief introduction (see Chapter I.1, The Timely Appearance of Life on Land, The Timely Change in Ancient Land).

Figure 2-21: As the sea receded and the land expanded (regression), life gradually evolved from water-living to land-living. This process is the landing of life (Source: Public domain).

There are some questions in this process that science has yet to answer. For example, regression allowed life to land; marshlands allowed water life to evolve into land life. But why did plate movement form such different environmental conditions necessary for the evolution of life? It is a stochastic process that will be further explored from a natural philosophical point of view in Chapter IV.

In addition to the shapes of the Earth's surface, the terrestrial environment also includes the atmosphere and temperature, which are essential factors influencing the evolutionary process of life, yet they are still rooted in plate movement.

The Formation of Today's Land Distribution

The current land-sea distribution took shape as the African and Indian Ocean plates moved northward, converging with the Eurasian continent roughly 40 to 50 million years ago. Approximately two million years ago, today's configuration and major landforms, such as high mountains and great rivers, became recognizable. A detailed exploration of this topic will follow in the next chapter.

Chapter II.3

The Rationality of the Earth System
Who Predetermined Our Human World?

Previously, we explored the rationale behind the Earth System and structure, examining it from a spatial and physics perspective. In this chapter, our focus shifts towards comprehending the rationale behind the distribution of land and sea over time, considering the development of human civilization. You may never think that the current distribution of land and sea should hold such amazing secrets. We will devote a lot of space to examining the complex and logical process of forming the global distribution of land and sea. It is a more complicated topic that warrants extensive discussion. Here, we'll observe how nature created diverse environmental conditions and resources tailored to the needs and traits of human existence and development. While this concept may seem far-fetched, everything unfolds under specific conditions. Without these conditions, humanity wouldn't have reached its current state. Within this chapter lies the possibility of understanding "the Reason that manifests itself in nature." Human intelligence is remarkable, yet it might not surpass or replicate the achievements of nature, such as the intricacies of the Mediterranean, as argued in this chapter, for instance.

INTRODUCTION TO ANCIENT CIVILIZATIONS

So-called ancient civilization refers to the civilization that originated long before the current era. At that time, humanity was still in the middle of the Neolithic Age when people mainly used pottery, bronze wares were few, and iron wares were in their beginning stages. Humans then had little ability to withstand natural disasters. Although irrigation was available, they benefitted mainly from nature and were subject to its fury, such as drought, flood, etc.

The development of human civilization is continuous. Although ancient civilization was simple, it was the beginning of the development of today's civilization. Without a beginning, there is nothing.

Human intelligence relies on both the physiological structure of the human brain and the accumulation of knowledge. The human brain reached maturity over 10,000 years ago. However, the buildup of knowledge is a slow, continuous process, starting with the accumulation of practical experiences, evolving over thousands of years. Only in the past few hundred years has this gradually transitioned into theoretical innovations with numerical symbols and experimentation. Broadly speaking, nature has facilitated this transformation.

In this chapter, we will determine the importance of Eurasia to the origin of human civilization, the importance of global land and sea distribution to the development of human civilization, and how nature created many geographical environmental conditions to help human social progress and civilization develop.

The Eurasian continent is the earliest birthplace of human civilization. In ancient times, the civilization on the Eurasian continent was always ahead of the rest of the world and promoted the overall development of human civilization. The following are the earliest civilizations that developed on the Eurasian continent.

Ancient Mesopotamian Civilization (Sumerian Civilization)

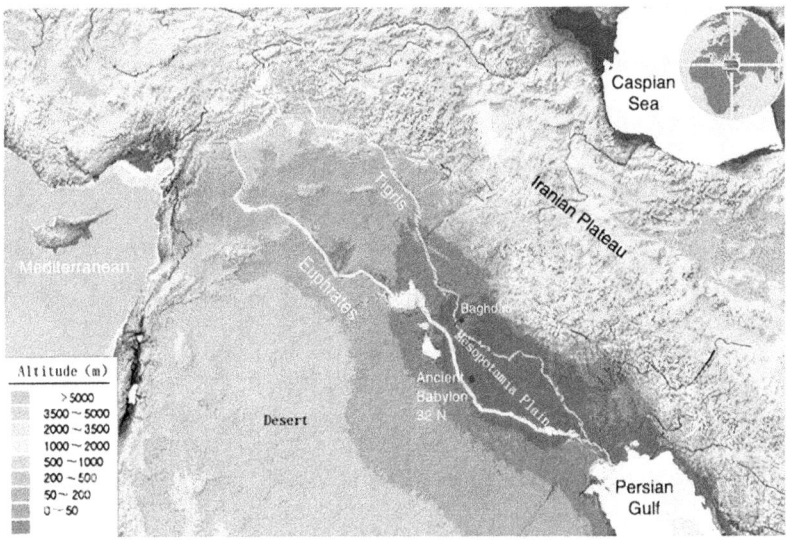

Figure 2-22: Terrain of Mesopotamia (Source: Public domain).

Around the 32°N latitude in present-day Iraq, the Mesopotamian civilization, also known as the two rivers culture, thrived on the Mesopotamian alluvial plain, shaped by the Tigris and Euphrates rivers. These rivers converged in the south, forming the Shatt Al-Arab and its delta. Stretching from northwest to southeast, the Mesopotamian alluvial plain resembles a crescent and is thus called the "fertile crescent." In ancient times, the region was well irrigated with canals, resulting in highly fertile lands. Some speculate it might have been the site of the Bible's Garden of Eden.

Compared with other ancient civilizations, Mesopotamian civilization was small in size and population but considered the most advanced early civilization. From 3500BC to 500BC, the Mesopotamian civilization created many of the "firsts" in the world:

- The earliest written language, cuneiform ("cuneus" and "forma" in Latin), appearing between 5,000 and 6,000 years ago.
- The first epic in the world, *The Epic of Gilgamesh*.
- The first legal codes, the Code of Ur-Nammu and the Code of Hammurabi, written on clay tablets.

- The earliest agricultural almanac, The Peasant's Lunar Calendar, helping determine optimal times for planting, harvesting, and various farming activities.
- The first library, the Library of Ashurbanipal, with about 24,000 clay tablets, built by King Ashurbanipal.
- The first marvel of architecture and engineering, the Hanging Gardens of Babylon.

These are just a few of the wonders of ancient Mesopotamian civilization. Many more are not listed here due to space constraints.

Ancient Egyptian Civilization (Ancient Egypt)

Although not part of Eurasia (Africa and Europe are only human divisions), Egypt is geographically similar and closely connected with Sumer ("fertile crescent").

Ancient Egypt spans from the Nile's first waterfall in Africa to its delta region. Situated north of present-day Cairo, the delta covers an expanse of approximately 9300 mi^2 (24,000 km^2), serving as the location for numerous pyramids.

Egypt is located mainly between the Tropic of Cancer and 30°N, always under the control of subtropical high pressure, and has a hot climate with little precipitation. Therefore, it did not have the conditions for the emergence of early civilization. Fortunately, the Nile, which runs through its land, brings abundant water and fertile land on both sides. The Mediterranean Sea influences the climate in northern Egypt, which is relatively mild and humid. These climatic factors allowed for the birth of ancient Egyptian civilization.

Egyptian civilization lasted from about 6000 years ago to the seventh century AD when foreign nations conquered native Egypt.

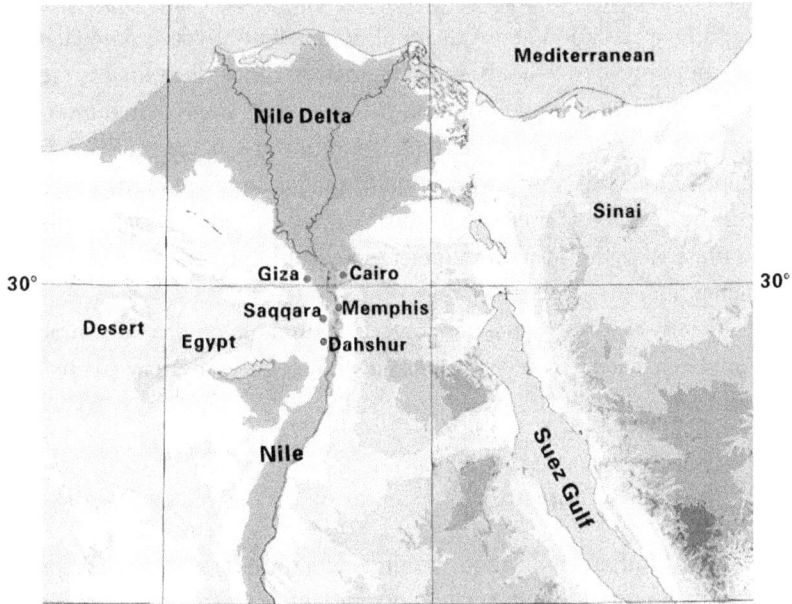

Figure 2-23: Terrain of ancient Egypt. There are many pyramids in Giza, but most Egyptian rulers lived in the old kingdom capital of Memphis (Source: Public domain).

The Ancient Egyptian civilization is widely recognized, and while we won't dive into a comprehensive introduction here, we'll focus on the ancient Egyptian pyramids.

Egypt has more than 80 to 90 pyramids varying in size. Their construction occurred between the twenty-seventh and twenty-second centuries BC, constituting the Pyramid Period—approximately 500 years. Notably, three of the most famous pyramids stand in the Giza Highlands, southwest of present-day Cairo. They were Khufu, Khafra, and Menkaura. These remarkable pyramids are celebrated as one of the Seven Wonders of the World.

There are two points to consider:

1. The pyramids in Egypt must have been constructed by the Egyptians at that time, who likely lived in the region for generations since the Neolithic Age. They would have possessed deep familiarity with the area's geography and adapted to its local climate over time.

2. The precision and construction methods behind the pyramids pose a challenge that modern science finds difficult to explain fully. These

structures exhibit remarkable geometric alignments and dimensions, prompting speculation about the depth of ancient Egyptian knowledge. While they appeared to use basic mathematical concepts needed to create such precise features, we don't know how much they relied on numerical concepts versus the level of sophistication in our current mathematical concepts. Moreover, the construction of the pyramids remains puzzling due to the absence of metal tools during that time. Even today, replicating a massive pyramid with modern tools would present a considerable challenge.

As early as more than 4,000 years ago, humans could create miracles that modern science could not explain, showing the mystery (secret) of the human brain.

Ancient Indian Civilization

The ancient Indian civilization usually refers to India's current footprint as well as the entire South Asian Subcontinent. Like the ancient civilizations in Mesopotamia (the two rivers) and Egypt (the Nile), ancient Indian civilization originated in the valleys of great rivers, the Indus Valley and the Ganges Valley.

Now, let's first talk about the Indus civilization.

The Indus River in present-day Pakistan spans approximately between 24° and 36°N latitude. Originating in the Himalayas, it stretches nearly 1,864 mi (3,000 km), primarily flowing through Pakistan from north to south before reaching the Arabian Gulf. The Indus Basin has subtropic conditions and four seasons.

Although not considered long, the Indus River is significant to the origin and development of ancient Indus civilization and the natural environment of today's Pakistan. Not discovered until about 1921–22, the culture is also referred to as the Indus Valley Civilization because of its concentration in the Indus Basin. Its most significant archaeological sites are at Harappa and Mohenjo-Daro, which were occupied later than Mesopotamia and ancient Egypt.

In terms of geographical conditions, the plateau to the west and the north, deserts to the east, and the ocean to the south, much like the geographical conditions of the Mesopotamian alluvial plain, limited the overall range of the Indus Valley Civilization (Figure 2-24).

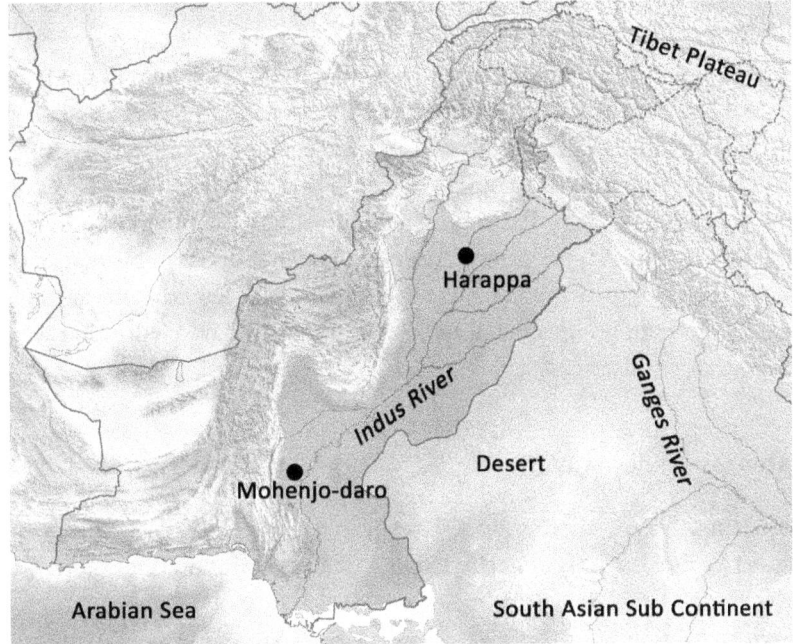

Figure 2-24: Terrain of the Indus River, Harappa, and Mohenjo-Daro in the Indus Valley of present-day Pakistan (Source: Carport, CC BY-SA 3.0, labels added).

Leading up to 2600 BC and 1500 BC, the Indus Valley Civilization flourished, giving rise to cities like Harappa and Mohenjo-Daro. A striking feature of this civilization is its advanced urban development, which is remarkably ahead of its time. Their cities boasted exceptional street layouts, sewage systems, and well-designed housing structures. The interconnectedness between these cities fostered vibrant commerce.

What particularly stands out is that over 4,000 years ago, while brick and wood construction was reserved for palaces elsewhere, the average home in the Indus Valley cities was built with these materials, featuring comprehensive indoor and outdoor drainage systems along with courtyard wells. Their elaborate underground sewage networks are impressive even to seasoned archaeologists, showcasing an urban civilization's astonishing advancement. Moreover, the Indus civilization excelled in agriculture, demonstrating expertise in irrigation methods, granary storage, and cultivating various crops and livestock.

Another remarkable achievement of the ancient Indian civilization of the South Asian Subcontinent was the emergence of numerical symbols. Humans have known about numbers for thousands of years, but writing them down in symbols is a relatively recent and complex development. The Punjab region of the South Asian Subcontinent was probably where humans first attempted to use symbols to express numbers. Then, after hundreds of years of advancement, people began using the ten-based number system in the ninth century. The introduction of the concept of zero and its representation using a symbol marks a profoundly significant and groundbreaking moment in human history.

In addition to the achievements mentioned above, there are too many others to go into detail here. The ancient Indian civilization was complex and advanced.

Ancient Chinese Civilization

Ancient Chinese civilization first appeared in the middle latitudes of eastern Eurasia and is unparalleled in scope and duration (Figure 2-25).

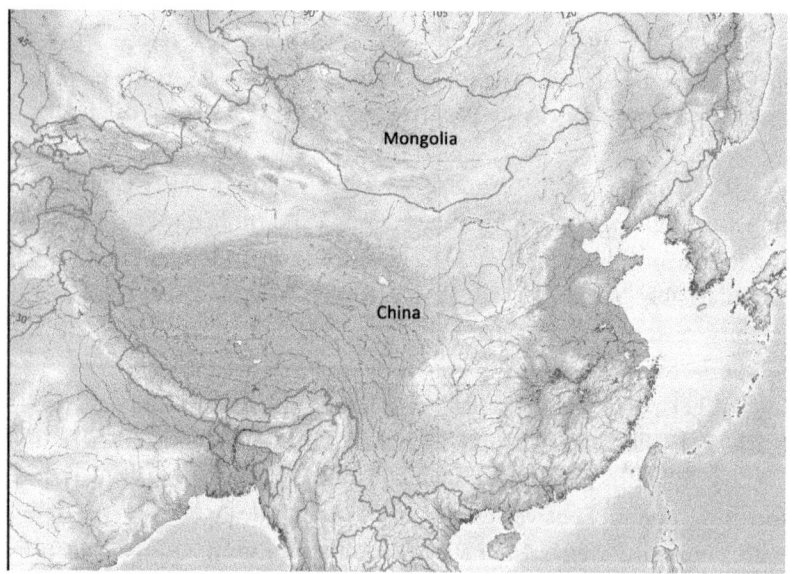

Figure 2-25: Terrain of China (Source: Ksium, CC BY-SA 3.0, labels added).

Bordering the ancient Chinese civilization to the east and south is the sea, to the southwest and west is the Tibet Plateau (Qingzang Plateau, known as the roof of the world), to the northwest is the desert, and to the north are desert and the Mongolian Plateau. Surrounded by these harsh geographical environments are alluvial plains of varying sizes formed mainly by two great rivers, the Yellow River and Changjiang River (Yangtze). This region boasts fertile soil, ample rainfall, numerous rivers, and abundant natural resources. It provided the ideal conditions for the emergence and flourishing of one of the world's most populous ancient civilizations. The people who have lived here for generations are the Han Chinese.

When comparing the ancient Chinese civilization with the three previously discussed civilizations, we see what they have in common. Each developed on subtropical alluvial plains formed by great rivers rising from massive mountain ranges. Keep this feature in mind as we continue.

Despite being constrained by surrounding deserts and plateaus, China's geographical environment profoundly influences its culture and economy, marked by a self-sufficient natural economy that had endured for thousands of years. This environmental backdrop also significantly shapes the national character, impacting Chinese history. For millennia, the densely populated Han nationality resisted assimilation by foreign groups. In contrast, the Hans successfully assimilated some foreign invaders and incorporated the diverse cultures they brought with them.

Ancient China had many noteworthy cultural achievements, such as the Great Wall, the Grand Canal, giant stone-carved Buddha statues, the Terra-cotta Warriors, etc. The difficulty of excavating mountains and carving Buddha statues is unimaginable without explosives and metal tools. For example, at 232.94 ft (71 m) tall, the creation of the Leshan Giant Buddha is a marvel.

While the accomplishments mentioned above are noteworthy, the following four inventions significantly impacted the world: the compass, papermaking, printing, and gunpowder.

The Chinese knew about the magnetism of magnetic iron ore about 2000 years ago and later used it to make the compass. Paper typically denotes the use of plant fibers processed into thin, resilient sheets. In this context, China possessed papermaking technology around 2000 years ago. In the seventh century, the Chinese invented printing (although movable type didn't appear until the eleventh century) and explosives, which were introduced to Europe in the twelfth and thirteenth centuries. The invention of paper in China is probably related to the region's abundance of bamboo and silk, which are the raw materials used to make paper.

Ancient Mediterranean Civilization

Ancient Mediterranean civilization refers to the civilizations that occurred on the Italian and Greek peninsulas.

Greece stands out as a unique country whose ancient civilization left a profound and widespread influence on the world. Although the commencement of Greece's ancient civilization on the Eurasian continent was relatively late, it reached its pinnacle over 2,000 years ago. Greece boasted a robust city-state system, introduced the world to the first democracy (Athenian democracy), and produced renowned philosophers like Socrates, Plato, and Aristotle, whose ideas, especially Aristotle's logic, significantly shaped subsequent generations. The Greeks also contributed mythology, the Olympics, distinctive architecture, and lifelike sculptures. Today, the Global Olympic Games have become a symbol of modern civilization and are held every four years. With a keen interest in nature, deep philosophical thinking, and the development of scientific concepts like atoms and Euclidean geometry, the ancient Greeks became a vital source of modern civilization. Their civilization also played a crucial role in influencing the Renaissance.

Another contributor to modern civilization is the ancient Roman civilization, which flourished over 2,000 years ago. While the ancient Greeks influenced their Roman counterparts, the Romans were known for their practicality in contrast to the abstract thinking of the Greeks. Ancient Rome left a lasting impact on subsequent generations, particularly in architecture and sculpture. Notably, the military strength of ancient Rome played a pivotal role, fostering the exchange of civilizations through warfare. In the fourth century, Christianity became the official state religion of the Roman Empire, eventually leading to the establishment of the Vatican, a force with significant global influence.

What Do Ancient Civilizations Imply?

What is the historical and geographical significance of the emergence of ancient civilizations on the Eurasian continent? The answer to this question is necessary before understanding who predetermined the modern world. Only by studying the question can you peek into the secrets of nature. What follows is a discussion about the question from different points of view.

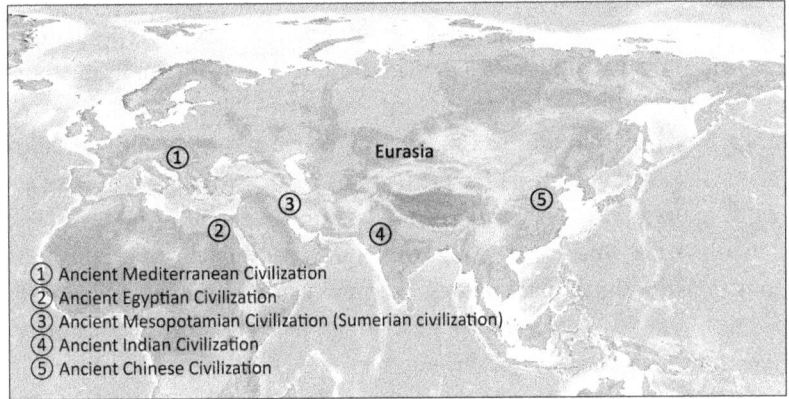

Figure 2-26: In addition to the sites marked above, there are many other early human civilizations in Eurasia, such as Arabia and Persia. Eurasia is unique because it is the only continent with numerous early civilizations. During the same period, other continents were relatively undeveloped (Source: Giorgi Balakhadze, CC BY-SA 4.0, labels added).

Civilization's Official Arrival

The rise of ancient civilizations signifies the formal emergence of human civilization. A key indicator of this shift is the transition from merely utilizing natural materials, such as stone tools, to creating items previously absent in nature—starting with pottery and advancing to bronzes and beyond. Concurrently, primitive forms of writing surfaced, and various tools were used for writing. Humans progressed from the hunter-gatherer stage to engaging in cultivation, animal husbandry, and crafting increasingly complicated and practical tools, thus marking a significant turning point in the trajectory of civilization.

A Significant Foundation for Modern Civilization

Ancient civilizations serve as crucial starting points for the emergence of modern civilization. Essential elements like logical thinking, numerical symbols, papermaking, printing, and the compass—all vital for the birth of modern civilization—originated in ancient civilizations. Without these foundations, the subsequent development of civilization would not have been possible. This underscores a fundamental principle: the evolution of human civilization is a continuous process of accumulation.

From the Paleolithic Age to the Neolithic Age, from the Bronze Age to the Iron Age, and onward to early modern civilization and today's modern civilization, there has been progressive and successive development. Despite the demise of ancient countries or dynasties, a diverse array of material achievements such as pyramids, canals, grottoes, stone statues, murals, and more, along with cultural and intellectual contributions in astronomy, mathematics, geography, meteorology, lunar calendar, historical records, and various writings, persist in the long historical river of human civilization's development.

Witnessing events like the pole vault and triple jump highlights the importance of athletes having a run-up before takeoff. Similarly, the ancient civilizations can be likened to a preparatory run-up process essential for modern civilization's takeoff.

Eurasian Continent: A Well-Designed Progression

In the development of human civilization, the Eurasian Continent appears to have been thoughtfully designed (see The Logical Step 3, below). Except for the atomic world, all processes are, strictly speaking, continuous, implying that evolutionary stages are interconnected without exception. Consequently, human civilization must undergo a natural and continuous evolutionary process, starting from its origin, progressing through development, and culminating in the emergence of modern civilization.

In the evolution of civilization, the Eurasian Continent functions as an expansive platform. The subsequent chapters will explore in detail the role played by its geographical conditions in the origin and birth of civilization. For now, we will focus on the key highlights:

Most of early civilization is in the temperate zone. The history of human development indicates that cold and tropical regions are not the birthplaces of civilization. Eurasia happens to be the world's largest temperate continent.

The long, continuous process from ancient to modern civilization requires many environmental conditions. Diverse natural environments form the foundation upon which ancient and modern civilizations gradually emerged on the same continent.

THE RATIONALITY OF THE EARTH SYSTEM

Exchanging materials and ideas is crucial to developing civilization. Interactions among ancient peoples, whether through trade or war, primarily occurred on land, and the expansive Eurasian Continent offered favorable conditions for exchanges between different civilizations. At the western end of Eurasia lies the Mediterranean Sea, a crossroads between ancient Egypt, ancient Europe, and ancient Asia. It served as a convenient and broad platform for widespread exchanges among them. Essential elements from various ancient civilizations, such as ancient Chinese contributions like paper and printing, ancient Indian numerical symbols, and other pivotal "tools" for the birth of modern science, converged. Stimulated by the Renaissance, modern civilization emerged through the diligent efforts of scientists. In this process, Arab merchants in the heart of the Eurasian Continent played a vital role in transmitting information (see Chapter II.3, Geographic Conditions Assist Human Development).

Human civilization would not have reached its present state if the Eurasian continent had separated into two regions. The ongoing process of civilization's development could not have unfolded on an isolated island, as we acknowledge that everything occurs within specific conditions.

The existence of modern science is contingent on social progress. Its birth occurred during the Renaissance, with the Mediterranean Sea providing a distinctive natural backdrop. This uniqueness becomes apparent when considering that, during the Renaissance, the rest of the world remained dormant, emphasizing the significance of the Mediterranean's presence.

Throughout human history, only Eurasia has undergone a continuous evolutionary process from early civilization to modern civilization. It appears unlikely that such a continuous evolution could have occurred anywhere else but in Eurasia.

Eurasia is also the birthplace of the world's major religions (especially in the eastern region of the Mediterranean), and all influential cultures originated from there.

If Africa is the cradle of humanity, then Eurasia is the cradle and stage of human development. Given their locations and natural environments, these two continents go hand in hand (!) for the origin and development of human beings.

Given these significant points, which continent do you think could replace Eurasia? None! Eurasia seems to have been designed especially for the birth and development of human civilization (see Chapter II.3, The Ancient Civilization Sites on the Collision Boundary).

The profound significance of the Earth's land-sea distribution—mysterious and intriguing!—will be explored in greater detail below. Our discussion on the ancient civilizations of other continents will continue in the following chapters.

A VERY LOGICAL DISTRIBUTION OF LAND AND SEA

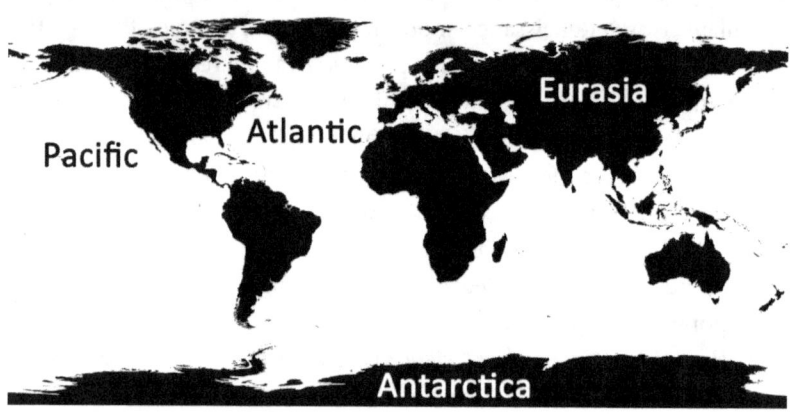

Figure 2-27: Global land-sea distribution (Source: Ebrahim, CC AT-SA 4.0, labels added).

While everyone is acquainted with the world map, have you ever noted how the global distribution of land and sea profoundly influences human development and progress? The following examines this topic more deeply.

Have you ever wondered how human civilization began? You might think the question is straightforward or that progression is inevitable because human beings are intelligent. If this were so, people worldwide would simultaneously progress from early to modern civilization. History, however, demonstrates otherwise! The birth and development of human civilization is a complex and subtle process that has occurred unevenly worldwide. The reason is that *having human intelligence and making the most of it are two different things*, and *everything happens conditionally*.

Let us explore this profound question just in the context of land-sea distribution. Consider the idea that nature operates as a clever designer, where land and sea function as an extensive set of tools, and human intelligence serves as the raw material to be processed. Let's examine how nature, through the global distribution of land and sea, has transformed human intelligence into civilization. Civilization, in turn, represents the value of human existence.

The Logical Step 1: The Birth of Human Beings

The evolution of life depends on environmental change, so the transition from early mammals to primates to early humans to modern humans is a complex process requiring certain ecological conditions.

Although primates appeared in some parts of the world about 50 million years ago, anthropologists generally believe upright-walking primates originated only in the GRV about 3–4 million years ago. At that time, the wide valley possessed special environmental conditions: high altitudes with low latitudes (where appropriate temperatures formed for the subsistence of life), desirable climates, plenty of water resources, fertile soil layer (containing a lot of volcanic ash), as well as a large stretch of forests with plenty of fruits year-round (see Chapter II.1, Forests and Grasslands). Later, an expanse of grasslands (savanna) prevailed, replacing the forest. This transition and other natural conditions promoted the evolutionary process that led primates to evolve into early human ancestors. Modern molecular genetics supports the theory that humans originated from Africa.

While savannas thrive on other continents like South America and Australia, and it's logical to imagine their evolution from expansive forests millions of years ago, scientists have discovered limited evidence of human origins in these regions. The notable absence of such evidence leads us to conclude that human evolution is a complex and subtle natural process, relying on various natural conditions such as climate, geography, plants, animals, and water resources. This insight emphasizes the significance of the GRV. In the GRV, and nowhere else, these particular natural conditions acted as the geographic "womb" for humanity to be born. Thus, the question arises: Without the specific environmental conditions of the GRV, could modern people have come into existence?

Furthermore, it's worth highlighting that the GRV (Figure 2–28) is adjacent to the Red Sea. Marine organisms in the Red Sea contain significantly higher concentrations of lecithin (DHA) and iodine than terrestrial organisms. Since primates primarily rely on fruits and the meat of some herbivores, they likely lacked sufficient DHA and iodine, which are essential for the evolution of the brain with intelligence. Therefore, the eastern edge of the GRV, with its proximity to the Red Sea, plays a significant role in the evolution of primates.

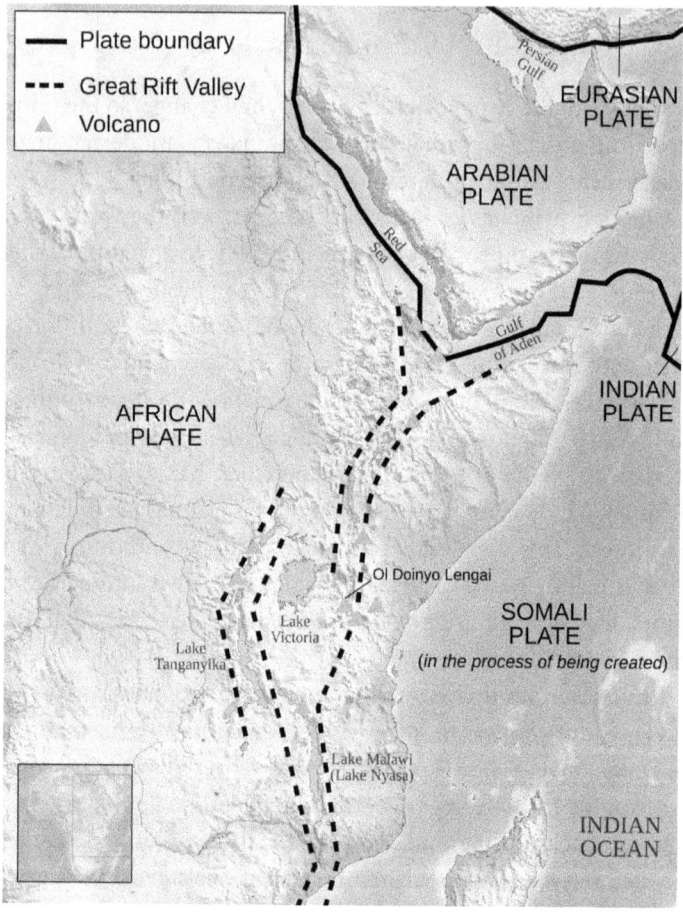

Figure 2-28: The Great Rift Valley's maximum depth reaches about 6600 ft (2000 m), and its width is well over 6 mi (10 km) (Source: Sémhur, CC BY-SA 4.0).

The origin of humans in time and space has long been a subject of research, but the significant role of the GRV is widely acknowledged. It is unnecessary to go deeper into its significance; suffice it to recognize the importance of the GRV to the origin of humanity.

Based on the above, we can see the GRV as a "starting point" for the global distribution of land and sea. So then, how do you "design" the position and area of the other landmasses on the Earth? Such a design should be mainly to address the needs of human survival and development. If this notion seems absurd, I invite you to continue reading (see The Logical Step 5, below).

The Logical Step 2: Out of Africa—The Birth of Civilization in Eurasia

Humans, known for their intelligence, are naturally inclined to seek environments conducive to survival. Early humans, unwilling to limit themselves to a single region indefinitely, inevitably ventured beyond Africa.

Interestingly, the expansive Mediterranean serves as a barrier between the Eurasian continent and Africa. However, nature establishes a connection at the northeast corner of Africa through the Suez land bridge, a topic we will explore later in this chapter. This location provided a gateway for bipedal creatures to access a broader and more habitable world—the Eurasian Continent. Subsequently, human beings embarked on a new journey of survival.

The historical progression of human development reveals that, in the millennia following the migration of early humans into Eurasia, numerous human civilizations outpaced those in Africa and other parts of the world from approximately 10,000 to 2,000 years ago. The accelerated progression vividly illustrates that the shift from the African to the Eurasian continent marks a significant leap in human developmental history. Eurasia took the lead in entering both the Bronze Age and the Iron Age, preceding the rest of the world. The transition from the Ceramic Age (Pottery Age) to the Bronze Age and then the Iron Age collectively signifies the early stages of civilization, bringing humanity closer to the emergence of modern civilization.

Therefore, the Eurasian continent stands as the exclusive cradle of the world's earliest human civilizations, as determined earlier. The collaboration of Africa and Eurasia forms a synergistic partnership crucial for the birth and development of humanity, an essential natural arrangement.

The Logical Step 3: The Birth of Modern Civilization

Modern civilization's origins are more complex than those of early civilizations. Early civilizations had multiple origins across different parts of the world, while modern science emerged in a single location. This distinction may be attributed to the latter necessitating much more stringent environmental conditions than the former.

Early civilization's development depended mainly on manual (physical) innovation. In contrast, the development of modern civilization relies predominantly on mental innovation (many theoretical innovations), a considerable shift in the history of human development, allowing the rapid growth of human intelligence. Consequently, compared to early civilizations, the birth of modern civilizations or science occurred under much more complex conditions.

Following their migration to Eurasia, humans gradually transformed from small settlements scattered across the continent several thousand years ago to establishing larger villages and cities throughout ancient Eurasia. However, the early civilizations in Asia and other parts of the world appeared to remain in the Iron Age for an extended period, lacking significant new turning points. The question arises: How could the world transition into modern civilization? Could this transition have been automatic? Looking at human history, we say the answer must be no! The answer lies in the European continent (Figure 2–29).

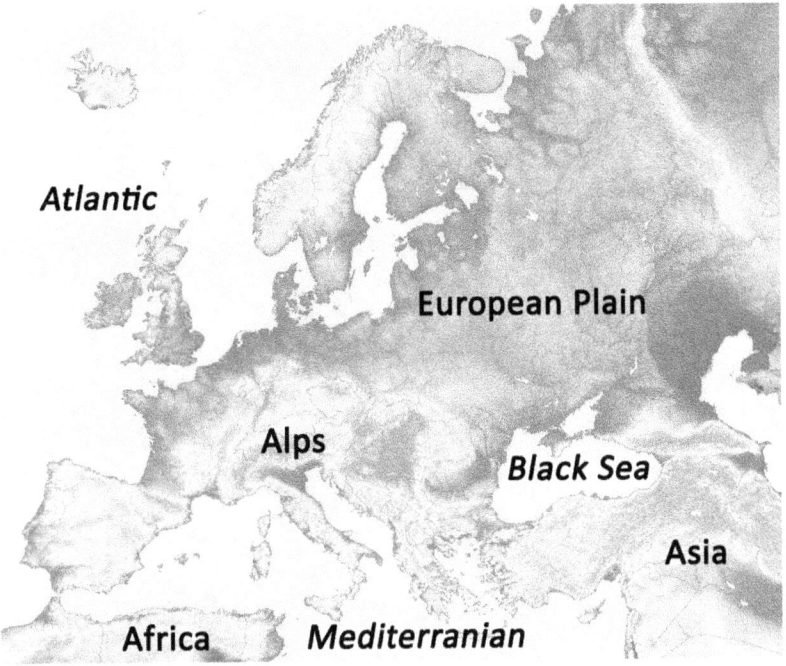

Figure 2-29: The geographical map of Europe (Source: San Jose, CC BY-SA 3.0, labels added).

The African continent logically determines the position of the European continent, as it is expected to be near the continent where humans originated. Consequently, the European continent is positioned at high latitudes, which might suggest a cold climate. However, this is not the case. Due to global thermohaline circulation, the continent enjoys a warm and humid environment. Scientists estimate that one cycle of this circulation takes about 1,000 years.

Furthermore, Western Europe is also affected by the strong North Atlantic warm current and southwest winds from the Atlantic Ocean. All these result in elevated winter and reduced summer temperatures and more precipitation in the region (Figure 2-30). Regarding topography, continental Europe exhibits a relatively temperate landscape dominated by plains. The plateaus thrive above 200 m above sea level, and hills and mountains account for 40 percent of the European area. There are many rivers, lakes, and wetlands. In addition, Europe has a long, winding coastline with many natural harbors.

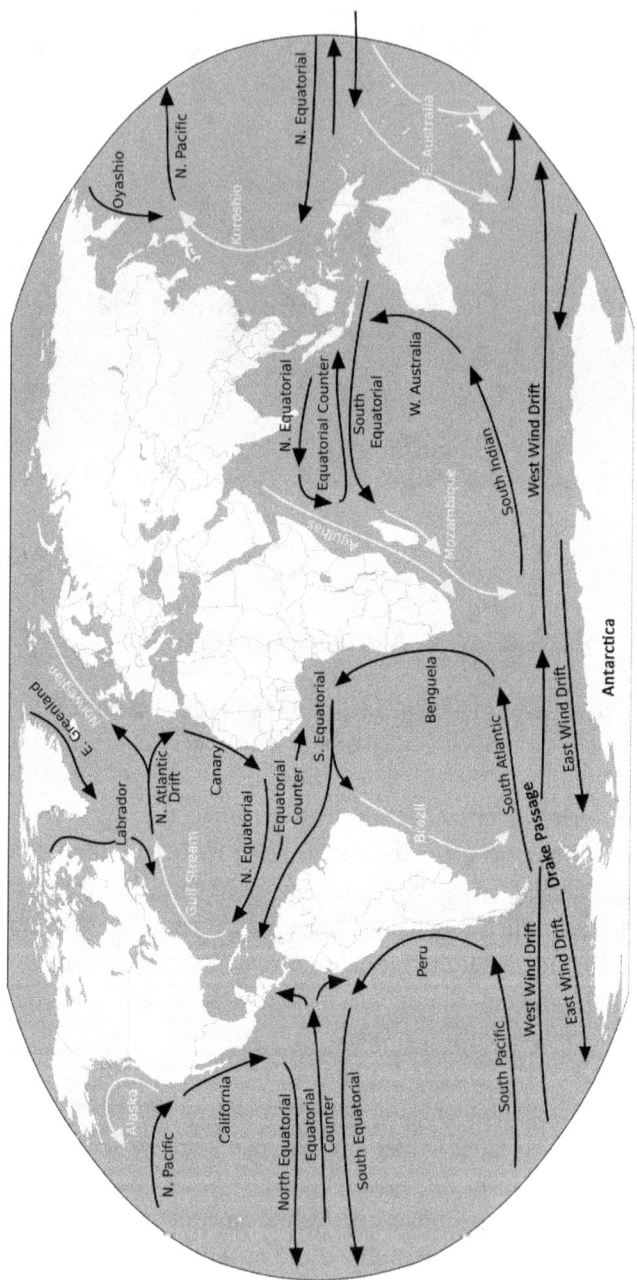

Figure 2-30: The world's ocean surface. Warm currents dominate the equator and Northern Hemisphere, while cold currents dominate the Southern Hemisphere (Source: Public domain).

THE RATIONALITY OF THE EARTH SYSTEM

The south of Europe borders the Mediterranean, the largest inland sea in the world, historically serving as a vast and convenient platform for trade and cultural exchanges between Africa, Asia, and Europe. The Mediterranean played an irreplaceable role in the emergence of modern civilization, including advancements in modern science and technology.

In the Renaissance period (fourteenth to sixteenth century in Europe), three societal developments emerged around the Mediterranean, contributing significantly to the overall advancement of science.

The first societal development involved the utilization of symbols in natural science. Arabic numerals, originating in India, along with the introduction of the concept of zero, were brought to Europe by Arabic merchants. These early numerals transitioned into our modern numerals during the fifteenth and sixteenth centuries. Imagine the difficulty of mathematical calculations without numeric characters. In ancient China, solving a simple equation required an entire page, even though some mathematical achievements predated those in Europe by over a thousand years. Modern science could not have developed without applying symbols, including Latin letters and Greek letters.

The second societal development involved the emergence of experimentation as a crucial method for researching natural phenomena. Galileo's experiments at the Leaning Tower of Pisa in Italy debunked Aristotle's long-standing notion about falling bodies, which had endured for nearly 2000 years. In pioneering the age of scientific experimentation, Galileo played a pivotal role in advancing the development of natural science.

The establishment of early universities in Europe marked the third societal development that fostered the progress of modern science. The University of Oxford and Paris University, founded in the twelfth century, and the University of Florence in the fourteenth century, played a significant role in the Renaissance and beyond. These early universities served as distinct and independent institutions for research, offering a space where individuals could engage in thoughtful exploration of natural science. Despite being simpler than modern universities, their key attribute was independence, representing the vitality of a university.

Modern science owes its development to these three societal developments that first appeared around the Mediterranean. Could the occurrence of these developments be a coincidence? No! This fact firmly proves the importance of the Mediterranean, being the birthplace of these three developments, which gradually spread worldwide.

Hence, the origin of numeral symbols and the concept of zero, papermaking technology, printing, and astronomical and mathematical achievements originated from the same continent with its favorable natural conditions. Drawing on continued advancements, including Aristotelian logic, and against the backdrop of the great Renaissance, which marked the first awakening of humanity, early modern civilization finally emerged in Europe through the wisdom and diligence of numerous scientists. The significance of the geography of the Mediterranean region, particularly the two major peninsulas, cannot be understated in providing the social and natural background for the foothold of the Renaissance in Greece and Rome.

The following chapters will explore in detail the importance of the Renaissance in the birth of modern science and the shaping of the Mediterranean, examining various perspectives.

Returning to Eurasia

The development of civilization primarily hinges on the accumulation of knowledge and the exchange of diverse thoughts. The expansive terrain of Eurasia facilitates cultural exchanges between Asia and Europe. Benefiting from favorable conditions, Europe absorbed the accomplishments of ancient Asian civilizations through the Mediterranean Sea during the Renaissance. Modern civilization, therefore, emerged in Europe against this backdrop. If an ocean separated Asia and Europe, there would only be isolated continents around the world, cutting off the channels of information exchange among peoples. It follows that Eurasia is the only supercontinent on Earth, and its importance in the origin and development of human civilization cannot be replaced by any other continent.

Therefore, it seems an ideal natural arrangement that the Asian continent with high mountains and great rivers, the temperate European plain, and the immense Mediterranean Sea together form an excellent platform for the birth and development of human civilization (see The Mysterious Plate Movement, below).

Several ancient civilizations stood for thousands of years, long before the Renaissance. These civilizations excelled in architecture, mathematics, and astronomy and surpassed the rest of the world. However, modern science failed to originate from those civilizations, and some even disappeared from the world long ago.

The innate curiosity of human nature compels us to sail the seas to explore the unknown. The Mediterranean acts as a colossal magnet, drawing people from its surroundings to navigate its waters and engage in maritime trade, which flourished more prominently around the Mediterranean than in any other part of the world.

Now, let's simplify the above process into a logical diagram as follows:

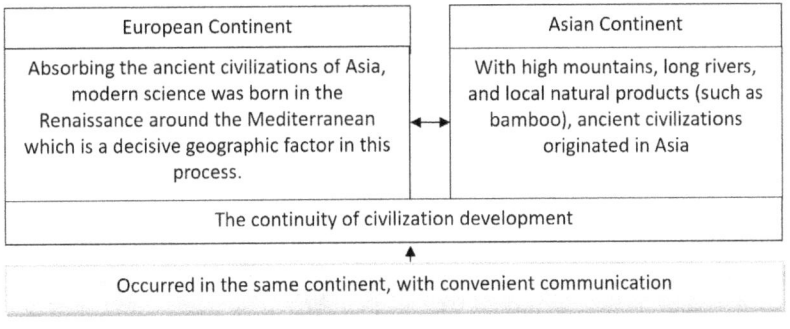

In the following chapters, we will further study the importance of the Mediterranean in human development, and you will see many significant problems in the human world fundamentally resulting from Eurasia. The Eurasian Continent is too important for humanity. The presence of the Eurasian Continent is mysterious and thought-provoking!

The Logical Step 4: The Large-scale Advancement of Modern Civilization

Europe's superior natural conditions contributed significantly to the birth of modern civilization. However, the small size of European countries historically led to conflicts, making them less peaceful lands. To some extent, these factors constrained the further large-scale advancement of modern science and technology. The question arises: how should modern science and technology progress from here?

To address this, we must first understand the broad scope of today's modern science and technology. Its central purpose is the profound exploration of nature. If we liken early modern science to the gateway, modern science represents the depth achieved after entering that gateway. Its primary objective is the comprehensive exploration of the natural world, encompassing the subatomic realm, the universe, various Earth-related fields (such as deep-sea exploration, global atmospheric movement, and deep-earth geology), molecular-level research on life, new materials, new forms of energy, as well as the investigation and application of communication and information systems (including artificial satellites).

Achieving these extensive explorations and applications demanded a solid foundation in modern science, vast territory, an enlightened populace, and geographical stability free from disturbances by neighboring countries—attributes that Europe did not fully possess during that time.

Modern science can exert a potent global influence only when it takes the lead in a large country with an expansive land area and a substantial population. The reason for this is that a multitude of enlightened individuals and vast land serve as the "fertile soil" for the advancement of science and technology, forming the foundation of a scientifically powerful nation. Consequently, it's logical to infer that humanity requires a stable, expansive, untapped new land with a large population as a stage for the further development of modern civilization.

The location that meets these requirements is North America!

Besides being located in the temperate zone, North America's most significant geographic advantage is that it is separated from the war-ridden continents of Europe and Asia by the expansive ocean on both sides. This vast continent is only half the size of Eurasia, but its natural conditions are more advantageous (especially in the mid-latitudes of North America) than those of Asia. North America was originally sparsely populated and abundant in lakes, rivers, and alluvial plains, grasslands and forests, and mineral and energy resources. If the Americas had been as densely populated as Eastern Asia, Europeans would not have been able to supplant them!

The four essential conditions for a leap toward modern science include being distant from Eurasia, having a temperate climate, being sparsely populated, and possessing a vast territory with abundant natural resources. Meeting all these criteria, North America is the ideal location for humanity to advance towards modern science. This natural arrangement aligns well with human development and is a highly logical configuration, satisfying all four conditions!

Despite their expansive territories, the northern regions of Eurasia and North America are primarily situated at high latitudes, characterized by cold climates and sparse native populations. Their existence appears to hold considerable importance for global climate patterns and the availability of natural resources.

It's worth noting that nature has bestowed the American continent for the progress of human development. However, acquiring this ideal land and the crucial responsibility of advancing modern science depend on human efforts.

Considering human nature, Europeans, equipped with modern science and technology, started to recognize the potential of this vast and fertile land as early as the fifteenth century, when its population was not actively involved in advancing science and technology. Over more than 200 years, characterized by conflicts and competition among European invaders, the present-day geopolitical landscape of eastern North America took shape by the eighteenth century, with westward expansion continuing into the nineteenth century. This once scientifically dormant and relatively sparsely populated land eventually became the domain of the United States.

Here, we can't help but ask, what would human history be like without the North American continent? This is a thought-provoking question. From the perspective of human history in the past two hundred years, could the Eurasian continent have solved the problem of human development, including scientific innovations, without North America?

Lastly, it is crucial to understand why Europeans could navigate across the Atlantic Ocean to the other side of the world in the late fifteenth century, while coastal countries in Asia did not embark on similar explorations. In addressing this question, it's essential to recognize that Columbus's discovery of the New World resulted from the prevailing conditions in European society during that era. While Columbus has often been associated with a spirit of adventure, his motivations for exploring were multi-faceted. He sought a westward route to Asia, primarily driven by a desire to establish a direct trade route. His venture was also influenced by the economic potential of finding valuable resources and spices. The accumulation of scientific knowledge would have been a byproduct of his efforts.

Compared to the Pacific, the Atlantic Ocean is significantly smaller. During the age of exploration, people lacked knowledge about the extent and duration of a transatlantic voyage, making the successful crossing a welcomed surprise. Would Columbus have succeeded if the Atlantic were as vast as the Pacific?

Awakened Europe, especially coastal Western Europe, encountering a relatively smaller Atlantic Ocean appears to be a logical "natural arrangement" for human development. The connection between the two—the Atlantic and seafaring Europeans—seems inherent.

Consider the marvel of this natural arrangement: Eurasia and the Mediterranean Sea on one side of the globe, the Americas on the other, and the comparatively narrow Atlantic Ocean in between. To put it in perspective, the maximum width of the Atlantic is 1,740 mi (2,800 km), while the Pacific spans 13,235 mi (21,300 km).

The Logical Step 5: Land in the Southern Hemisphere

The logical steps outlined above illustrate that the birth and development of humans necessitated a vast territory with diverse natural environments, conditions met by the temperate and tropical regions of the Northern Hemisphere. However, the land-sea ratio on Earth remains constant. Excessive or insufficient land would impact the available space for human activities and alter the global climate, including ocean currents. Such changes may render the environment unsuitable for the survival, evolution, and development of humans and all life. Given this, it logically follows that there would be limited land in the Southern Hemisphere based on the global land distribution. Let's now examine how nature has distributed land in the Southern Hemisphere.

The Defining of the African Continent

In the Southern Hemisphere, considering Africa as the birthplace of humanity and the site of the earliest human activities, it follows that the area of Africa should not be too small. It extends south of the equator, naturally forming the triangular continent of Africa, with a significant portion being tropical. Notably, the GRV spans almost the entire length of Africa from north to south, suggesting that the north-south length of Africa is closely tied to the extent of the GRV (Figure 2–28). Consequently, tropical Africa appears limited in its ability to support human development.

The African continent features one of the world's greatest rivers, the Nile, which was crucial in constructing the pyramids. The pyramids represent the highest levels of human knowledge achieved over 4,000 years ago. Without metal tools, ancient people made incredible pyramids that continue to puzzle and astound us today. Could we characterize this as forward-thinking on the part of ancient people? In this context, forward-thinking refers to the development of theories or technologies before their prevalence. Recognizing the presence of forward-thinking may hold significant value in studying the nature of human intelligence and even the structure of the human brain.

We may never fully solve the mysteries of the pyramids, so what does that imply for science? Will it inspire a new understanding of the wisdom of early humans in Africa? If so, the significance of the pyramids to humanity cannot be undervalued!

If the Nile were not long enough, there would be no Nile Delta, no fertile soil on its banks, and no pyramids. Hence, the length of the Nile plays a pivotal role in human development, underscoring the Mediterranean's significance!

All these factors appear to underscore the importance of the expansive extent of the African continent, emphasizing the need for it to extend sufficiently southward. However, most of the African continent is tropical (Figure 2–28), and the consistently high temperatures throughout the year are not conducive to human development. Therefore, the inevitable choice for a natural arrangement is Africa's connection with the temperate continent of Europe (see The Logical Step 3).

Finally, let's consider an assumption that warrants exploration. The GRV and the Nile are remarkably close to each other east to west and are roughly at the same latitudes (Figure 2–28). While we are uncertain about their specific connection to human evolution and development, it appears plausible that the indigenous people in Egypt thousands of years ago were likely early humans originating in Africa. This potential association could be a significant factor in the construction of the pyramids. During that era, Egypt's civilization stood out as the most advanced in Africa, Europe, and Western Asia.

Considering the great revelation of the pyramids to humanity and their potential value to modern science, is it a unique natural arrangement for their close link to the GRV? Would some of those who once designed or constructed the pyramids over 4,000 years ago have entered Europe via the Mediterranean or the Suez land bridge? These are thought-provoking questions. The migration of early humans from Africa into Europe should be said to be a regular occurrence. In the future, perhaps, scientists will find that the natural arrangement of the Mediterranean

Sea, the pyramids, and the GRV are of profound significance for humans to develop (see The Mysterious Formation of the Great Rift Valley and Red Sea, below).

In summary, tropical Africa, the Africa encompassing the GRV, the Africa with the Nile and pyramids, and the Africa connected to Europe fulfill the requirements for human birth and early development, significantly shaping the destiny of Africa. This arrangement appears to be a natural outcome, devoid of arbitrary choices.

Now, let's turn to the remaining continents of the Southern Hemisphere, South America, Australia, and Antarctica.

The Defining of the Antarctic and South American Continents

From a global climate perspective, the South Pole "must" be associated with a sufficiently large continent, Antarctica, which plays a crucial role in forming the planetary wind system. The Antarctic continent and the Earth's rotation with the obliquity of the ecliptic are the two fundamental factors determining the formation of the planetary wind system (as we learned in previous sections). One can envision that as the Antarctic continent shifted from lower latitudes of the Southern Hemisphere toward the South Pole (during the Cenozoic Era) and the ice sheet formation began (without the vast Antarctic continent, extensive ice sheets would not exist), the planetary wind system gradually intensified to its current scope.

Here, we encounter the West Wind Drift induced by the Mid-Latitude West Wind of the planetary wind system. The West Wind Drift is the strongest in the global ocean current system. It serves the critical purpose of isolating warm water from Antarctica, contributing significantly to maintaining the continent's expansive ice sheet. The East Wind Drift (Figure 2–30) further supports the formation of the ice sheet.

The Drake Passage between South America and Antarctica adds to the surprises. The Drake Passage is a major ocean passage connecting the Pacific and Atlantic Oceans. Notably, it is the widest and deepest ocean passage, spanning 559 to 590 mi (900 to 950 km). Through this passage, the flow from the Pacific to the Atlantic reaches a substantial 5.3 billion ft^3/s (15 million m^3/s). The surface water temperature varies from 43°F (6°C) in the north to 30°F (-1°C) in the south.

According to modern science, the presence of the Drake Passage (Figure 2-30, approximately at latitude 60°S) is crucial in regulating the global climate and forming the vast Antarctic ice sheet over millions of years. Therefore, it is considered a vital component of the global climate system. Researchers have extensively studied its role in the global climate, particularly since the application of satellite technology.

With the location of the Antarctic Continent defined, we can logically and correspondingly determine the definition of the South American continent. Interestingly, the continents of South and North America belong to one plate, or the American plate. (Figure 2-19). So, the location of the American continents is necessary from both a geological and a climatic viewpoint (we'll explore this further in the next section).

After delving into the significance of North America in human development, it logically follows, from a human development perspective, that South America would be situated on the opposite side of the world. As the Northern Hemisphere has fulfilled its natural mission of human birth and development, the Southern Hemisphere continents naturally become an extension of the Northern Hemisphere's human civilization.

Historically, much like the North American continent, South America's vast and fertile lands intrigued Europeans for centuries. However, the conquest of South America was, to some extent, a haphazard process. After inflicting centuries of tumultuous and often brutal historical events, Europeans, notably the Spanish and Portuguese, eventually seized control of the vast continent. In the nineteenth century, they established numerous countries, ushering in a new multi-ethnic era with predominantly European, European-Indians, Native Americans, Indians, Asians, and other populations contributing to the diverse fabric of the region.

The Significance of the American Continent Layout

Imagine yourself as the designer responsible for the Earth's distribution of land and sea. What essential factors would guide your decisions in laying out the North and South American continents? The primary consideration should be the arrangement of ocean currents, which serve as the principal conveyors of global heat. Warm currents contribute to increased temperature and humidity, while cold currents have the opposite effect. Therefore, your hypothetical design aims to create a global climate through land-sea distribution that is most conducive to the survival and evolution of all life, including humans.

Given Africa and Eurasia (and their natural missions), which must exist and remain in their positions, the American continent is logically placed on the opposite side of the world, extending continuously from the Southern Hemisphere to the Northern Hemisphere. This configuration is crucial for the efficient formation of two oceanic current systems (Pacific and Atlantic) that play a significant role in regulating the temperature and humidity of the continents. For instance, the Gulf of Mexico Warm Current is a vital ocean current influencing the climate of the east coast of North America and Western Europe.

The current distribution of land and sea appears to be the optimal solution, surpassing any alternative scenarios we can conceive. As mentioned earlier, the Drake Passage is necessary for this design.

For North America to fulfill its role in forming the global climate, it "must" possess enough land at high latitudes, akin to Eurasia. This requirement dictates the position of the land in mid-latitudes accordingly. Only through the proper division of the ocean by landmasses can the creation of oceanic currents occur, contributing to the shaping of global climate patterns.

The present layout of North and South America, driven by considerations of global climate (specifically ocean currents) and human development, is deemed necessary. Contemplating an alternate reality where the American continent did not exist or was in a different location prompts uncertainties about the fate of humans and the global climate.

The Mysterious Formation of the Panama Land Bridge (Isthmus of Panama)

As the positions of North and South America formed, two possibilities arose: one was to create a land bridge between the continents, and the other was to maintain a broad strait between them (including the narrow strip of land in southern Mexico). Nature ultimately opted for the former. While the natural formation of this land bridge remains unknown, certain facts hold true:

- North and South America existed before the Isthmus of Panama formed millions of years ago.

- The formation of the Panamanian Land Bridge altered global ocean currents (Figure 2–30), subsequently influencing the global climate to some extent. On this shape of the Earth's surface, the Quaternary Great Ice Age started 2.5 million years ago, covering our planet. Throughout the Ice Age, between the glacial and interglacial periods, the global climate underwent alternating changes that fostered the evolution of life, contributing to significant developments such as the widespread emergence of polyploidy and the emergence of humans.

However, the relationship between the formation of the Panamanian Land Bridge, the Quaternary Ice Age, and the transformation from tropical rainforests to savannas in East Africa millions of years ago remains a topic for further scientific exploration. We believe there is likely a connection between these factors. Hence, the formation of the Panama Land Bridge was essential and occurred around three million years ago, earning it the label the mysterious Panama Land Bridge.

The Panama Land Bridge and the Drake Passage appear to be fundamental. Together with the American Continent, they constitute the land and sea distribution opposite the Eurasian and African continents, forming a logical chain connecting all continents and oceans, crucial for human development. The absence of the Panama Land Bridge and the Drake Passage would dramatically alter the global climatic pattern.

Regarding the question of why South America and Africa have triangular shapes, it remains unanswered. Perhaps this shape is most advantageous for global ocean currents and climate under a constant land-to-sea ratio.

The Defining of the Australian Continent

Let's investigate the last continent, Australia. Approximately 70 percent of its land is arid or semi-arid, with around 20percent being desert. The coastal regions, particularly the east, offer a more favorable natural environment for human habitation and development. Similar to South America, Australia acted as an extension of the modern civilization of the Northern Hemisphere and was eventually colonized by the British before gaining independence in the twentieth century.

Contrary to the seemingly logical inevitability of the continents discussed earlier, the existence of Australia is not without its own unique considerations. Far from it! The proportion of land to sea plays a crucial role in its existence. As part of the Indian Ocean plate (Figure 2–19), the positioning of the Australian continent was defined as the plate drifted northward, situating it between Antarctica and Eurasia. While this location may lead to meteorological challenges like droughts and deserts, it minimally impacts ocean currents and the global climate, which is conducive to a favorable environment for the survival and development of humans and other life (Figure 2–30).

In essence, the land distribution in the Southern Hemisphere appears to balance the requirements of both human development and global oceanic currents.

The Logic of the Land-Sea Distribution

Let us assume that the natural mission of land-sea distribution is to create the conditions needed for the birth and development of humans and human civilization. So, nature made the following rational arrangement.

According to the view stated above, it was essential for the location of Africa to be near the equator to form equatorial rainforests, which then transformed into savannas due to global climate change. These environmental conditions were needed for primates to evolve into humans (see The Logical Step 1). Next, the location of Africa logically determined the location of the Eurasian Continent and the Mediterranean. Only in this way could early humans conveniently migrate to the Eurasian Continent from Africa and facilitate the origin of early civilization followed by modern civilization. On the other side of the globe, the presence of the American Continent created the conditions necessary for the emergence of today's civilization and, finally, the determination of land-sea distribution in the Southern Hemisphere, including the formation of the vast Antarctic ice sheet.

The characteristics of this distribution are perfectly consistent with the conditions required for the birth and development of humanity and the global climate pattern. Again, it should emphasized that everything happens conditionally. Thus, we see the logic of the global distribution of land and sea. Do you have any better ideas for facilitating the birth and development of humans?

THE RATIONALITY OF THE EARTH SYSTEM

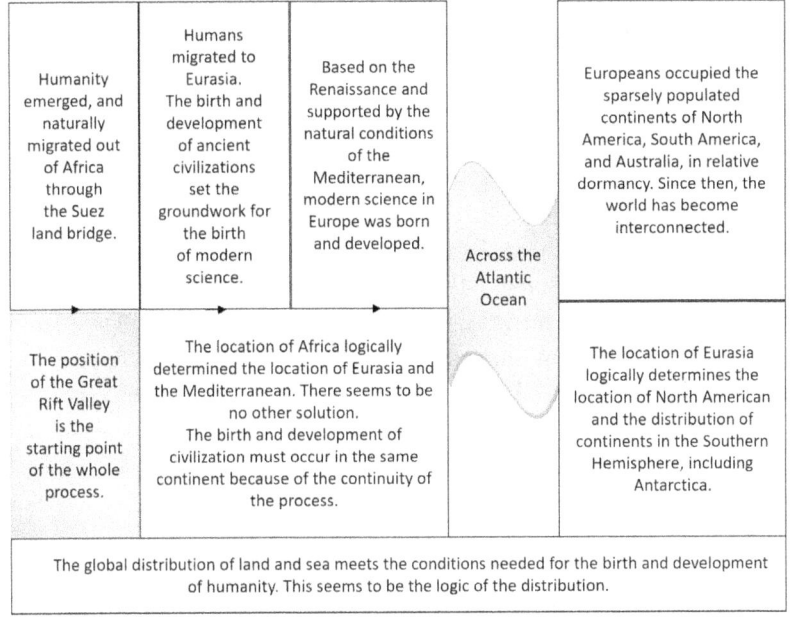

→ Logical direction

Hence, the arrangement of land and sea is not arbitrary; it follows a deliberate order. "God does not play dice."

Now, consolidating all the relevant factors to understand the importance of the distribution of land and sea, it appears to be a comprehensive collaboration that undeniably reflects the rationality of nature, assuring the emergence of civilization in a limited time (see the last chapter).

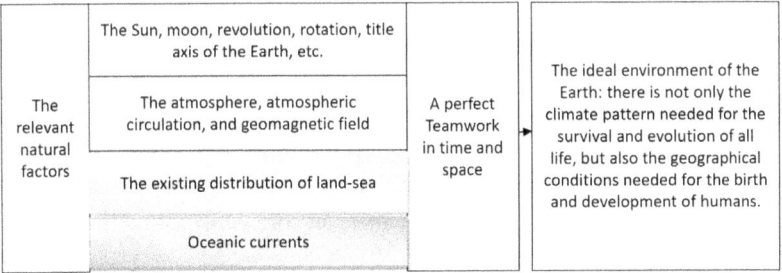

The above explanation illustrates how nature guided human intelligence toward human civilization using the "tool" of land-sea distribution. We can liken this process to a person ascending a tall building. A baby is born on the first floor, the baby grows into childhood, and the child climbs the stairs. As the child grows to adulthood, continuing to climb, the adult reaches the top. There, the person witnesses and appreciates the vastness of space and the vibrant world. Throughout this journey, fueled by human curiosity, each step holds a powerful allure. However, without stairs, even a wise person would struggle to comprehend the greatness and wonder of the world.

The current land-sea distribution is rational for the birth and development of civilization and the formation of global climate. These two results are its natural mission. If the land-sea distribution differed from its established distribution, would it meet all the conditions required to fulfill this natural mission? All in all, the development of civilization requires certain conditions and continuity. These two points are essential.

But there's something even more amazing about the distribution of land and sea. We'll go on to describe it in the following chapters.

THE MYSTERIOUS DISTRIBUTION OF ISLANDS IN THE WORLD

We'll now shift our focus from the arrangement of continents on the Earth to exploring the significance of islands. With their distinct natural mission, islands form an integral part of the Earth's land. While continents represent the primary landmass, as an alternative landform, islands hold equal importance. Although their global land area is minimal, their role in human development is undeniable. From ancient times, islands have played a unique and crucial role in shaping human development.

This chapter investigates the significance of two essential islands on opposite sides of the "maternal" Eurasian continent: Britain in Europe and Japan in Asia. The presence and development of these islands are contingent upon the temperate zone, a prerequisite we will explore further. The discussion will highlight the profound importance of these islands and the intricacies of their formation. Notably, no other significant islands are near a continent in the temperate zone, prompting thoughtful consideration.

Britain—The Only Continental Island in Europe

The Significance of the Existence of the English Channel

The ancient European continent experienced more conflicts and less peace, which hindered the advancement of science. In contrast, Britain enjoyed a relatively peaceful period, particularly during the Elizabethan Era (1558–1603), leading to unprecedented prosperity known as the English Renaissance and the Golden Age of England.

Furthermore, in the early thirteenth century, the British established the Great Charter, which limited the power of the King. Subsequently, during the seventeenth century, ideas concerning freedom, equality, and human rights emerged in Europe, playing a significant role in social progress and contributing significantly to the enlightenment of humanity in Europe. These factors fundamentally explain why Britain became the world's first industrialized country.

It's worth noting that the presence of the English Channel played a crucial role in preserving order on the island (Figure 2-31). Without it, the island might have descended into the same warlike state as the continent.

Figure 2-31: Britain and the English Channel (Source: NASA, public domain). Its maximum width is 149 mi (240 km), and its narrowest is 21 mi (33.8 km).

The Mysterious Formation of the English Channel

Now, let's examine a more profound question: why did the Islands of Great Britain emerge during the development of the European continent? From a geological perspective, the islands of Great Britain are categorized as continental islands, serving as extensions of the mainland. The answer lies in understanding how the English Channel formed.

It's essential to recognize that the strait's formation is precisely positioned in width between the island and the continent. The subtlety here is that it needed to be detached from the continent but not too far. If the strait were too narrow, Britain would have been caught up in the chaos of Europe. On the other hand, if it were too wide, cultural and economic communication with the continent would have been challenging, hindering the absorption of the spiritual nourishment produced by the continent since the Renaissance. Britain's current distance strikes the right balance.

Moreover, the island of Great Britain is situated entirely between 50° and 60° N latitude. In the Asian continent, this latitude corresponds to a frigid region unsuitable for human survival and development. However, Britain enjoys a temperate maritime climate, characterized by mild and humid conditions throughout the year, without drastic seasonal temperature changes (see The Logical Step 3).

The word "magic" seems apt to describe the formation of the Island of Great Britain. While various theories, such as the strata subsidence of the English Channel, can explain the process, it's more fitting to consider it as serving the positional arrangement of Great Britain. This strategic detachment proves crucial, significantly influencing human development history and forming the social backdrop for the birth of the United States.

One might ask: without the British islands, would Newtonian mechanics have come into existence? Numerous questions arise, emphasizing the conditional nature of the world's development. The social progress and scientific advancements in Britain during the seventeenth century, coupled with Newton's genius and industrious qualities, contributed to the emergence of this great scientist. The absence of the English Channel might have delayed scientific progress, leading to unpredictable consequences.

Japanese Islands—Asia's Only Continental Island in the Temperate Zone

The Japanese Islands and Nation

Scientists propose that during the recent ice age, early humans on the mainland might have reached the area of the Japanese islands through the northeastern tip of Asia. The interglacial period arrived as the recent ice age concluded, causing rising sea levels and forming the Japanese islands.

Japan stands out as one of the most disaster-prone regions in the world. Situated in the Earth's seismic belt, it harbors many large and small volcanoes, constituting a tenth of the world's total volcanoes, many of them active. The country also experiences many earthquakes, with the devastating Kanto earthquake of 1923. In 2011, a catastrophic earthquake and massive tsunami in Fukushima caused significant damage.

Beyond seismic activities, annual typhoons and floods challenge Japan. Given its compact size, these natural disasters can easily lead to nationwide crises. Moreover, 71 percent of Japan's land area is occupied by mountains and hills, resulting in a scarcity of arable land.

The natural conditions of the Japanese islands are notably harsh compared to Great Britain. However, these difficult conditions have shaped a nation characterized by continuous self-improvement. Faced with frequent natural disasters and limited territory, the Japanese have skillfully faced these challenges, actively learning from the outside world to enhance their strength and survival abilities. This adaptive approach becomes crucial given the vast and frequent natural disasters that could otherwise lead to destruction.

Historically, Japan started as a poor and backward country. However, as Europe embraced modern science and technology, Japan's aptitude for learning allowed it to catch up in the nineteenth century and eventually become a world power. Japan has left a lasting impact on world history and significantly contributed to modern science and technology.

Turning our attention to land distribution, the formation of the Japanese archipelago becomes a point of interest.

The Mysterious Formation of the Japanese Islands

The formation of the Japanese islands is a complex geological process. Without expertise in geology, the following simplified explanation is offered. According to the theory of plate tectonics, the collision boundary between the Pacific and Eurasian plates (Figure 2-33) is located in the western Pacific. The Pacific plate, being lower than the continental plate, subducted beneath the Eurasian plate when it moved westward, resulting in the compression of the continental plate. The elevated portion gave rise to the Japanese Islands, while the submerged section formed the Sea of Japan on the west side of Japan. The deep trench is situated on the east side of the Japanese islands (Figure 2-32, left). This geological process commenced in the middle Tertiary Period.

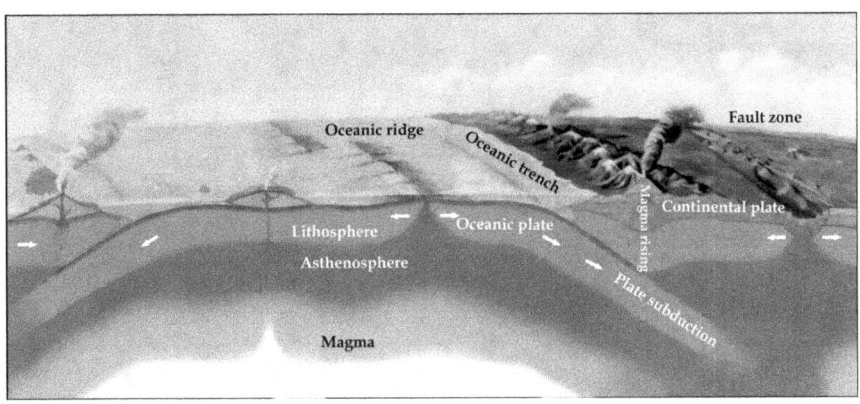

Figure 2-32: Plate collision (see Figure 2-20) (Source: USGS, public domain).

Remarkably, the global plate boundary is thought to extend for many thousands of miles (Figure 2-33). However, only Eurasia features islands on both sides, while other continents generally lack significant islands on both sides.

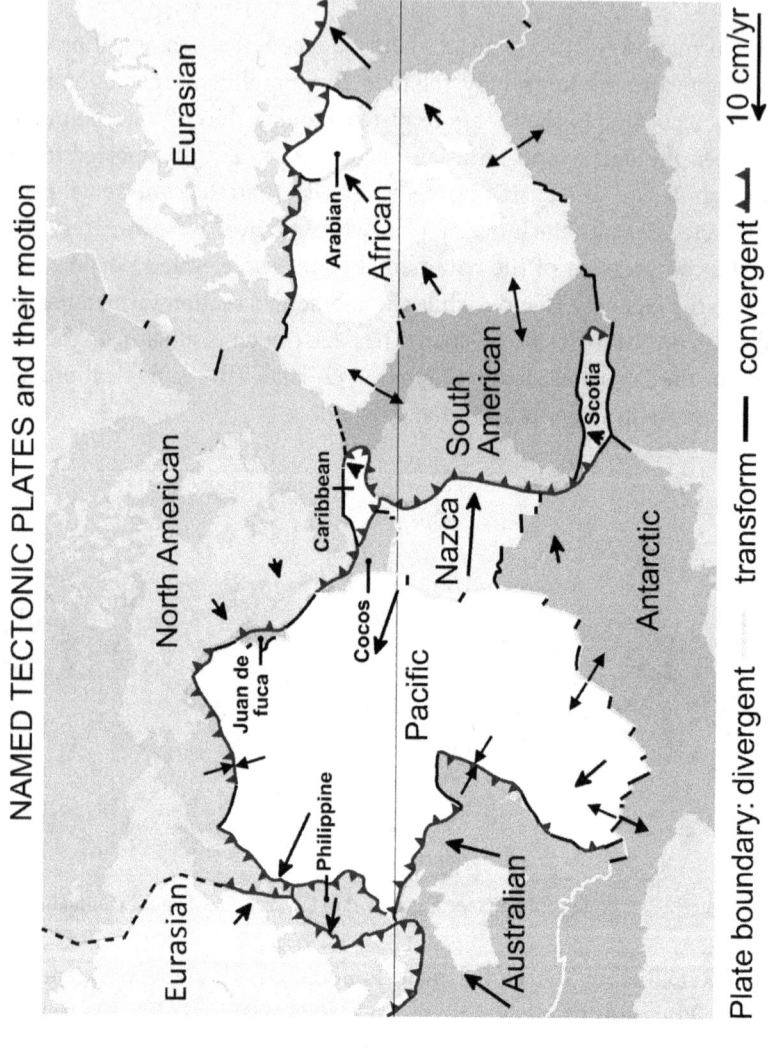

THE RATIONALITY OF THE EARTH SYSTEM

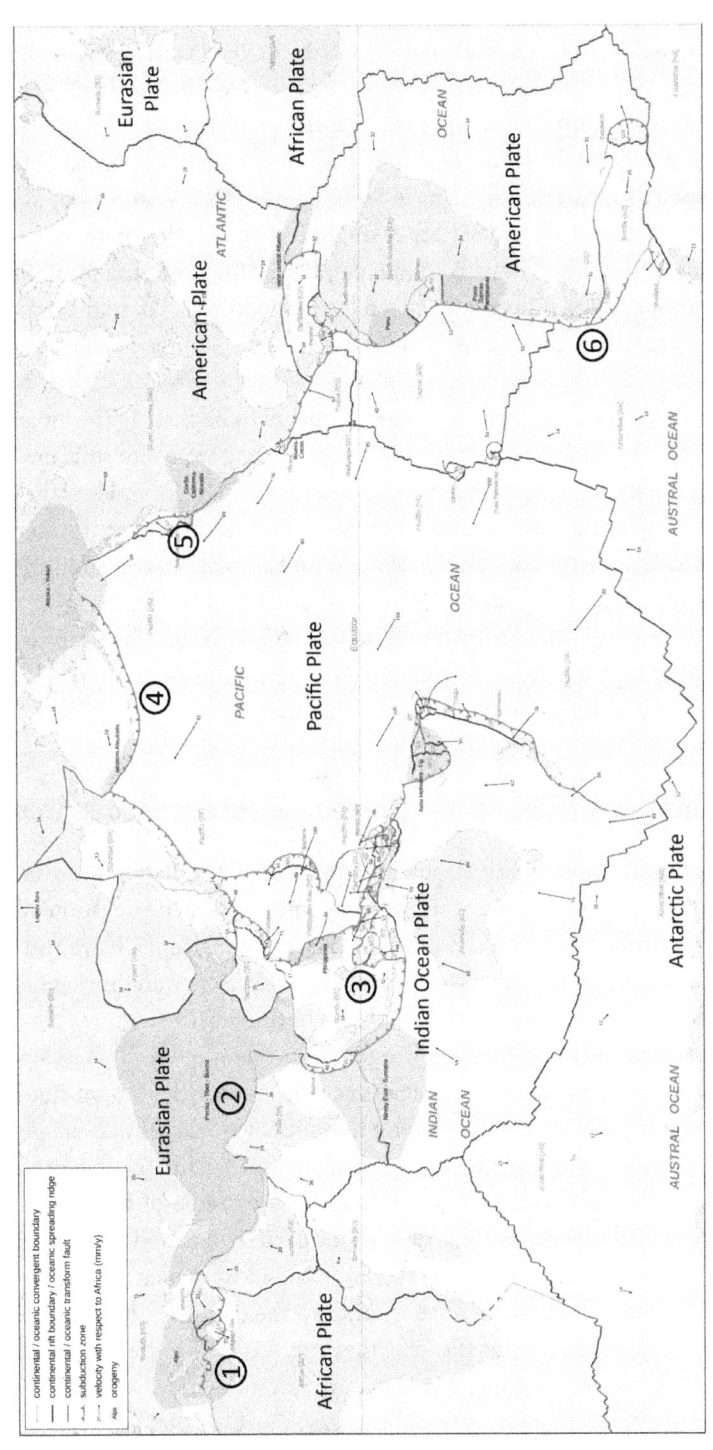

Figure 2-33: Plates (see Figure 2-19) (Source: [Top] Hughrance, CC BY 4.0, plate boundary arrows added, [Bottom] Eric Gaba, CC BY-SA 2.5, labels added).

The location of the Japanese islands is surprisingly clever. Despite being continental islands, had they been much closer to the mainland, the ancient Japanese might have assimilated into the larger population on the mainland. Conversely, they might have had little connection if the islands were farther away from the mainland. However, Japanese cultural development was indeed influenced by the mainland. Therefore, the current position of the islands strikes a balance—neither too far nor too close.

The size of the Japanese islands appears to be fitting. They may not have developed an influential civilization if they were any smaller. They could not be any larger since they are surrounded by four plates: the Eurasian plate to the west, the Pacific plate to the east, the North American plate to the northeast, and the Philippine Sea plate to the southwest (a smaller plate) leaving no room for substantial growth (Figure 2-33).

In essence, the formation of the Japanese islands is far from simple. Considering their role and influence in human history, one might perceive the geological process as somewhat magical.

An Imaginary Third Island in the Temperate Zone and Its Fate

Intriguingly, large archipelagos are situated in the temperate zone on Eurasia's east and west sides. This arrangement raises the question of why similar configurations did not occur on other continents. Have you ever wondered about the potential impact of an imaginary archipelago in the temperate region associated with other continents?

Human history provides insights into this inquiry. It suggests that if such island groups were near other continents, they might not have played a significant role in shaping the human world, unlike the pivotal roles played by the islands of Great Britain and Japan. This hypothetical situation is not difficult to grasp. Eurasia, the cradle of both early and modern civilizations, is the origin of modern civilization for other continents and islands worldwide. If a large island were near another continent, it would likely share a fate similar to the continent itself. There are only two exceptions globally, and they are located in the tropics rather than the temperate zone: Madagascar, east of Africa, and Sri Lanka, east of the Indian Peninsula. Madagascar's official language is European French, and Europeans once occupied Sri Lanka.

In essence, substantial islands in the temperate zone seem to hold significance primarily around the "maternal" Eurasian continent, which nurtures them. On the one hand, they can easily absorb the accumulated civilization of the mainland over thousands of years. On the other hand, they can develop their own distinct maritime cultures, contributing significantly to human civilization in ways different from mainland cultures.

The placement of these two archipelagos on either side of Eurasia, the birthplace of diverse human cultures, appears to bear special significance, explaining their existence. Nature seems to attach value or importance to the existence of every piece of land. The secret here is the location of the two islands; moving from their current position, they lose the significance of their existence. Don't you agree?

Eurasia constitutes 36.3 percent of the world's total land area and about 55 percent of the Northern Hemisphere's area. The predominance of land in the Northern Hemisphere is attributed to the requirement of the GRV to be near the equator (a necessity for primate evolution). This logical arrangement positions Eurasia in the north, defining the locations of the significant islands and the American Continent.

While plate tectonics may offer an explanation, observing the current distribution of land and sea leads to the acknowledgment that each continent, island, and land bridge on the Earth seems to have a natural mission, whether for establishing global oceanic currents or fostering human development. From a modern scientific standpoint, the entire land arrangement appears highly rational, leaving little room for arbitrary change.

Do you not find this perspective compelling? Establishing Earth's current land and sea distribution involved numerous factors, including tectonic plate movements and astronomical conditions, resulting in a highly complex and harmonious natural development process across space and time for humanity.

THE MYSTERIOUS PLATE MOVEMENT

Background Information

After reading the two chapters above, you may feel that things on the Earth are intriguing, not as simple as commonly understood. In this chapter, we will focus on how plate movement created many of the natural conditions necessary for the birth and development of human beings, such as the Mediterranean Sea. The process is complex and subtle.

In essence, these phenomena can be attributed to the theory of plate tectonics. The collisions between Earth's plates, as they shift, not only shaped the current land-sea distribution but also formed many unique geographical environments. These environments are crucial for human survival, operating under specific conditions.

If the previous chapters hinted at the underlying "intent" behind the Earth's development, this chapter will illustrate how nature carried out this "intent" through intricate plate movements. This process facilitated the creation of geographical conditions necessary for the birth and development of humanity. It appears to be a carefully planned sequence, demonstrating the rationality inherent in nature's development—unveiling the mystery of plate movement.

The Theory of Plate Tectonics

The theory of plate tectonics is a remarkable scientific achievement of the twentieth century, and it is wise to acknowledge the contributions of those who contributed to its research. The German scientist Alfred Lothar Wegener (1880–930) proposed the initial concept of continental drift in the early twentieth century. Tragically, while on a solo expedition seeking further evidence for his theory on the Greenland ice sheet in November 1930, his 50th birthday, Wegener met an untimely end in an accident. Subsequently, the theory of plate tectonics took shape in the late 1960s through the persistent efforts of American and Canadian scientists, building upon Wegener's continental drift theory and the sea-floor spreading hypothesis. The pivotal role played by the Glomar Challenger, a large American scientific research ship, was noteworthy in this process.

Modern science and technology not only confirm the existence of tectonic plates but also enable the accurate measurement of the velocity of plate movement. Earth's surface layer consists of six major tectonic plates, which subsequently consist of smaller plates. These plates "float" on the asthenosphere and are constantly moving. Two types of boundaries exist between plates: divergent boundaries (←→ growth boundary, or ocean ridges), where hot magma rises, spreads, and forms new ocean plates at the ocean ridges, and collision boundaries (→← convergent boundary, subduction boundary), where ocean plates subduct beneath continental plates, creating trenches and islands. The ocean plate spreads from divergent boundaries on the ocean floor or mid-ocean ridges and converges at the other side, known as the continental margin (Figure 2-33, Figure 2-32). However, the scientific understanding of the tremendous forces propelling ocean plates sideways (←→) remains unknown.

Through the movement of tectonic plates and other natural factors, the Earth forms many diverse natural environments on its surface. These environments include various landforms, climates, material compositions (deserts, rocks, soil, water, etc.), and freshwater distribution. Each of these natural environments holds distinct values for evolution and the development of civilization.

To explore these unique environments with special significance, as a long journey, let's start from the western end of the Mediterranean Sea and travel east along the plate boundaries to the southern tip of South America (Figure 2-33, →←).

The Mysterious Formation of the Mediterranean and Its Significance

The Mediterranean Sea holds immense significance in human history and development. Its formation through geological processes is crucial to understanding its role in shaping human civilization.

The Mediterranean lies just near the collision boundary between the two tectonic plates. When the African plate moved north and collided with the western part of the Eurasian plate during the Tertiary Period (Figure 2-38, left), there was a massive compression between the two plates so great that it formed the Alps just north of the boundary (Figure 2-29). Miraculously, the Mediterranean (including the two peninsulas) remained. It would not be surprising if the Mediterranean had completely disappeared.

The exact reasons behind the northward drift of the African plate remain a mystery, but without it, the Mediterranean as we know it would not have formed. Initially sealed off by surrounding land, the ancient Mediterranean eventually transformed into a desert basin through massive evaporation around six million years ago. Then, approximately 5.5 million years ago, the Strait of Gibraltar opened, allowing vast amounts of seawater from the Atlantic Ocean to flood in, giving rise to the modern Mediterranean Sea. This entire process seems almost magical in its complexity.

Before becoming the Mediterranean Sea, the basin was approximately 9,800 ft (3,000 m) lower than the sea level of the Atlantic Ocean. Even today, the Mediterranean is notably deep, with an average depth of 4,800 ft (1,450 m) and reaching depths exceeding 16,400 ft (5,000 m). This depth, combined with its vast expanse of water, significantly influences the climate of southern Europe, giving rise to the "Mediterranean climate," characterized by mild winters and dry, sunny summers. This climate, along with the region's abundant natural resources and the Mediterranean's importance as a trade route, played a significant role in the development of ancient civilizations. In contrast, Asia at 40°N has cold winters and hot summers (Figure 2-27).

In this and the following chapters, we will dive deeper into the mysterious formation of the Mediterranean, recognizing its profound importance to humanity throughout history.

The Mysterious Formation of the Strait of Gibraltar

The Mediterranean owes its existence to the Strait of Gibraltar, a crucial waterway. Originally, this strait was a narrow land bridge where the European and African plates converged during the Tertiary Period (Figure 2-29). Today, the strait measures approximately 17 mi (28 km) wide north-south, 40 mi (65 km) long east-west, and averages 1230 ft (375 m) deep.

Picture the original Gibraltar land bridge as a substantial rock wall, separating the Atlantic Ocean from the Mediterranean (Figure 2-34).

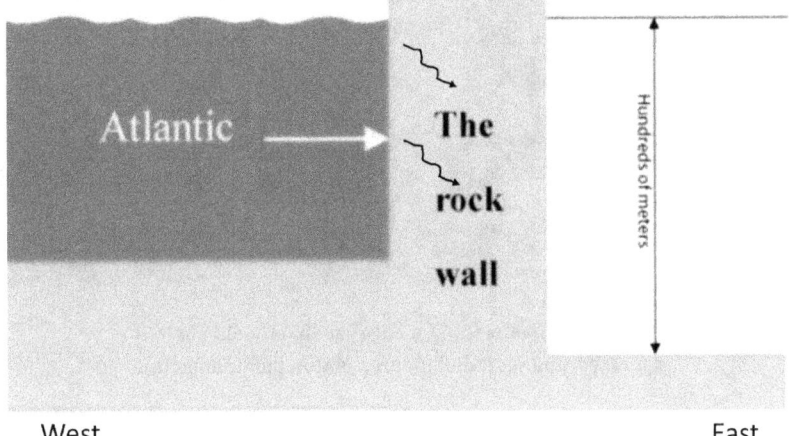

Figure 2-34: Profile of potential configuration, with jagged arrows showing fracture flow in the rock wall, and the white arrow showing the water pressure from the Atlantic (Source: Created by the author).

But what natural processes dismantled this imposing barrier around 5.5 million years ago? The answer remains elusive. Some scientists speculate that over five million years ago, a meteorite impact may have shattered this immense rocky barrier (Figure 2-35, right), yet this theory awaits confirmation.

However, delving deeper into this mystery reveals three potential configurations of the pre-strait land bridge (Figure 2-35):

Form 1: The Eurasian and African plates pressed together, completely joining the north and south sides of the strait, stretching over approximately 621 mi (1,000 km) from east to west (Figure 2-35, left).

Form 2: Only a segment of the strait was joined, spanning approximately 40 mi (65 km) from east to west (Figure 2-35, middle).

Form 3: The north and south sides connected at a single point, extending about 12–19 mi (20–30 km) from east to west (Figure 2-35, right).

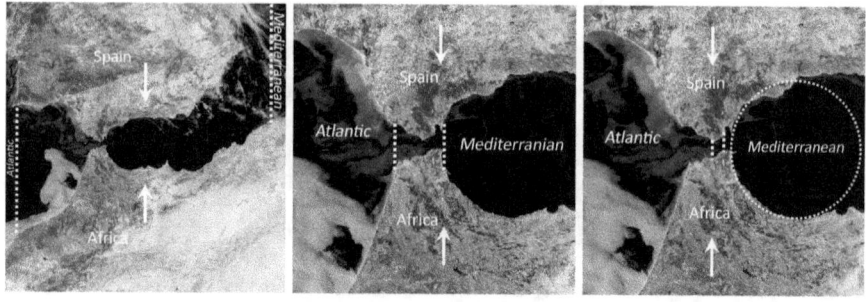

Figure 2-35: Geologically, a huge gap that should not have appeared, and yet it did! (Source: NASA, public domain).

All three potential configurations share a common feature: each would form a natural rock barrier, with the Atlantic Ocean to the west and a dry Mediterranean lowland several hundred meters deep to the east. Essentially, the result is a rock dam under significant water pressure.

What natural forces caused the breach in this dam, allowing Atlantic waters to spill into the Mediterranean lowland? Likely causes include weathering and seawater infiltration weakening the rock structure. The squeezing action between the two plates may have induced numerous cracks and fractures in the dam, facilitating penetration. However, for Forms 1 and 2 to have occurred, these factors were insufficient due to the dam's thickness.

Form 3 was more susceptible to destructive forces, with dimensions of about 9.3 mi (15 km) from south to north and 14.3 mi (23 km) from west to east at its narrowest point. Over hundreds of thousands of years, weathering and other effects gradually widened cracks and fractures in the dam, allowing Atlantic and rainwater to penetrate and erode it under high water pressure. As the cracks increased in number and size, the flow of water intensified, ultimately leading to the gradual opening of the Gibraltar land bridge and the formation of the strait.

Thus, a combination of factors, including high water pressure, weathering, penetration, and erosion, worked together over time to dismantle the massive rock barrier. The resulting pressure from the Atlantic waters likely triggered a monumental flood event known as the Zanclean Flood.

The estimated width of the dam, approximately 14.3 mi (23 km), is an assumption; it could have been narrower. This width might be observable underwater, along with its geological features such as rock types and weathering conditions.

Even more intriguing is the timing of the rock wall's collapse, roughly 5.5 million years ago. This timing appears to coincide with the period it took for the Mediterranean to fill with water over hundreds of thousands of years and the subsequent development of the Mediterranean climate.

The formation of the Mediterranean climate is a complex process influenced by geographical factors, terrain, and the characteristics of Mediterranean water. Initially sourced from the warmer Atlantic, Mediterranean waters likely cooled over hundreds of thousands of years, shaping the unique ecosystem and local products surrounding the sea. However, transitioning from cooling seawater to establishing the Mediterranean climate may have been a lengthy process, possibly spanning a million or even two million years due to the water's high specific heat capacity.

Ultimately, the Mediterranean offered humans its distinctive geographic and ecological environment at precisely the right time. The complementary relationship between Mediterranean and Atlantic waters is crucial in preventing the Mediterranean from drying up.

As you can see, the formation of the Strait of Gibraltar appears ingeniously timed and executed, illustrating the intricate interplay of geological processes and climatic developments (Figure 2-36).

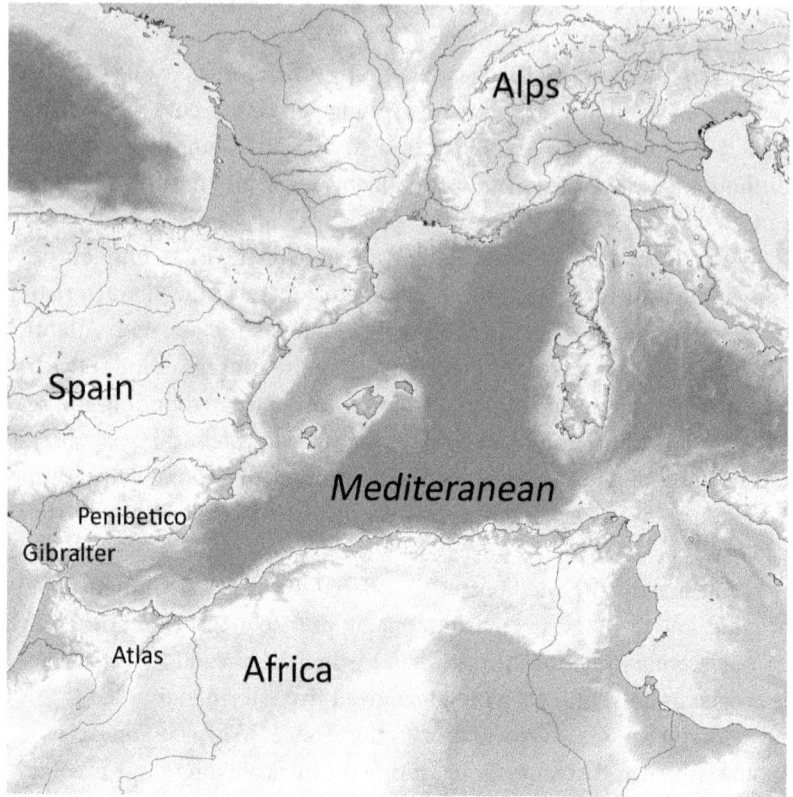

Figure 2-36: Relief map of the Straight of Gibralter and the Mediterranean Sea (Source: Nzeemin, CC BY-SA 4.0, labels added).

But the story of the Mediterranean Sea's creation doesn't end here; it gets even more exciting. Let's take a deeper look into the formation of the Strait of Gibraltar from another perspective.

The Penibetico Mountains System (*Sistema Penibetico*, Figure 2-36) in southern Spain, situated on the north bank of the strait, primarily consists of marble. Conversely, the Atlas Mountains (Figure 2-36) in northern Africa, on the south bank of the strait, are primarily composed of limestone, marlstone, and claystone, indicating a diverse composition of relatively soft rocks. These rock types likely contributed to the formation of the Gibraltar land bridge. Marble in the Penibetico Mountains System may have formed through metamorphosis induced by the collision process of the two plates, as marble is metamorphized from limestone. On the south bank, the main chemical composition is calcium carbonate.

THE RATIONALITY OF THE EARTH SYSTEM

When exposed to water and carbon dioxide, calcium carbonate converts into calcium bicarbonate, dissolving in water and being carried away by currents. This process contributes to the formation of limestone caves (Karst caves) seen worldwide, with extensive lengths. Greater water pressure accelerates rock dissolution, hastening Karst cave formation.

From a hydraulics perspective, the substantial difference in water levels between the Atlantic and Mediterranean lowlands suggests high water pressure within the rock dam, potentially accelerating Karst cave formation.

Based on the rock types on either side of the Strait, we can deduce that the north side (the Asia-Europe plate, composed of marble) was subjected to compression, while the south side (the African plate, comprising calcium carbonate) was the compressing side. This implies that pressure resulted from the northward drift of the African plate, aligning with contemporary tectonic understanding.

The presence of significant calcium carbonate (a sedimentary rock) in the Gibraltar area raises questions about its initial abundance. Would the Strait of Gibraltar have formed if the rock were not water-soluble? It's noteworthy that limestone caves are more prevalent globally compared to marble caves, such as those found in Coyhaique, Chile.

Is the above argument reliable? Yes, it is indeed reliable! The rock types on both sides of the strait are the best evidence to explain how the African continent collided with the European continent.

If we accept that the Strait of Gibraltar opened widely around 5.5 million years ago, there would have been ample time before then for numerous large caverns to form and weaken the land bridge (the collision between plates occurred 40 to 50 million years ago). Water from the Atlantic Ocean and rainfall would have continuously infiltrated the land bridge (the rock dam). At the same time, the Mediterranean lowland to the east remained dry and deep. In addition, atmospheric air would have entered fissures and fracture zones in the land bridge from the rocky surface on the eastern side (see Figure 2–34). With water on one side and air on the other, high-pressure water and air would have filled the rock wall, inevitably leading to collapse.

This mechanism appears ideal for breaking the rock wall, contingent upon the initial formation of a "small" rock dam when the north and south plates pressed against each other. This crucial aspect is delicate and precise, and any deviation from this process would result in the disappearance of the small rock dam.

Furthermore, we can also envision that the bell-mouth-shaped shoreline on the east and west sides of the strait likely formed through the scouring action of rapid water flow and erosion over many years as vast quantities of water passed through the strait. Consequently, cracks and openings in the rock wall would have accelerated the formation of this distinctive coastline. From a hydraulic perspective, compared to Forms 1 and 2, Form 3 is most likely to produce the bell-mouth shape on the east and west sides of the strait.

However, the original shoreline would not have exhibited this flared shape. The process is highly complex, involving intricate interactions between water and rock, with the above explanation providing only a simplified description.

Moreover, due to the significant force between the two plates, Forms 1 and 2 are the most probable, while Form 3 seems less likely. Nevertheless, Form 3 might be the only way for the wall to break, highlighting the subtlety and wonder of the Mediterranean Sea's formation.

The statements above explore the various possible forms of the land bridge, the formation of the rock wall, with its appropriate chemical composition, the depth of the Mediterranean lowland, the effects of high water pressure and precipitation on the rock wall, seawater dissolution and penetration of calcium carbonate, and the scouring of banks by currents and weathering. Together, these demonstrate that the formation of the Strait of Gibraltar is a complex and lengthy natural process involving the interplay of many factors. Any change or absence of these factors would hinder the formation of the strait.

Nature's processes are far from simple! We'll explore this topic further from a broader perspective later on, promising an even more exciting discussion.

The Mysterious Formation of the Suez Land Bridge

Before the construction of the Suez Canal, which stretches 100 mi (161 km) in length, the region between Africa and the Arabian Peninsula could be considered a land bridge, or the Suez land bridge. At the divergent boundary of the Red Sea, there's a notable shift in direction towards the northeast at its end, where it narrows, effectively maintaining the Suez land bridge. It would have been plausible for the Red Sea to expand northwestward directly into the Mediterranean (Figure 2–37). Had this occurred, the Sinai Peninsula would not have formed, resulting in different oceanic currents in the Atlantic, Mediterranean, Red Sea, and Indian Ocean. The high temperature and salinity of the Red Sea would have significantly influenced this altered oceanic system, affecting the climate and ecological environment of the Mediterranean. This uncertainty would cast doubt on the stability we experience today.

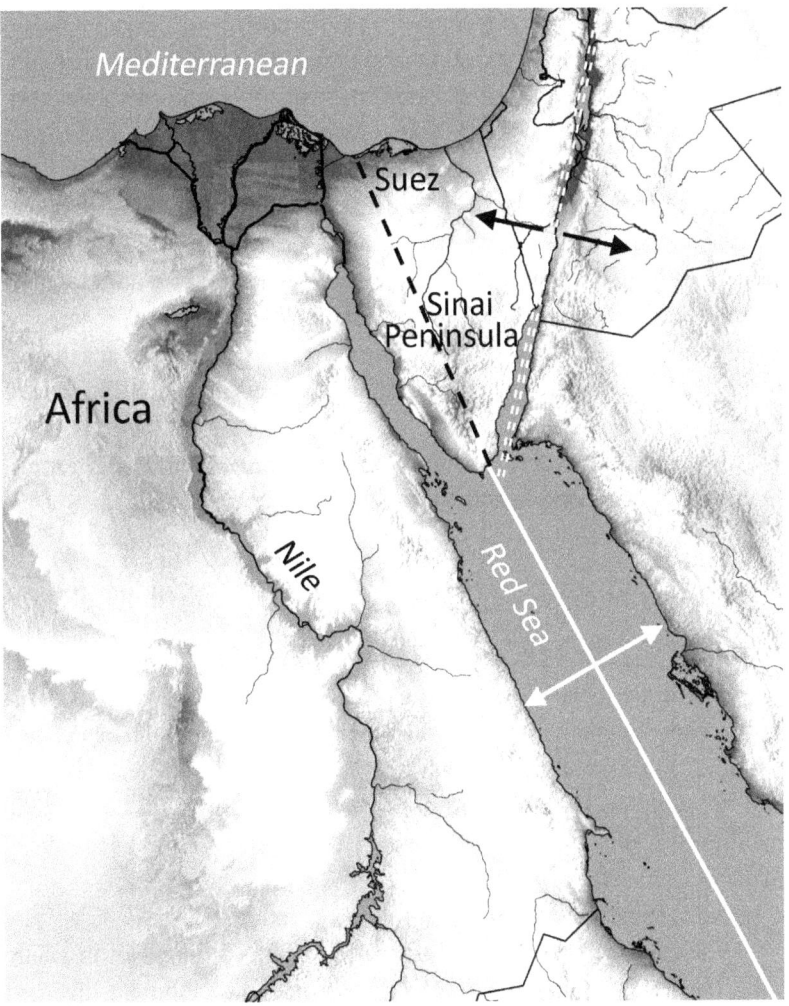

Figure 2-37: Geologically, the emergence of the Suez land bridge defies expectations, yet it inexplicably materialized! The solid line denotes the growth boundary, while the arrows indicate the direction of plate movement (seafloor spreading). The dashed line signifies the probable direction of the growth boundary (Source: Public domain).

However, a more significant consequence would have been that early humans remained confined to Africa, as the Red Sea spans about 124 mi (200 km) in width, making it challenging for migration. The tropical climate prevalent in Africa is less conducive to human development than in the temperate zones, as human history shows. The ramifications of early humans being trapped in Africa would have been profound, drastically altering humanity's destiny.

Fortunately, the Suez land bridge prevented this scenario, establishing the Mediterranean as a distinct and independent inland sea. The construction of the Suez Canal had minimal impact on the Mediterranean environment due to its relatively small size. Presently, water flows from the Atlantic through the Strait of Gibraltar into the Mediterranean, then through the Suez Canal into the Red Sea, maintaining a slightly lower surface temperature than the Atlantic and the Red Sea.

From the standpoint of human intellectual evolution and development (see The Logical Step 1), the existence of the Red Sea is imperative—it "must" exist and "cannot" merge with the Mediterranean Sea. Plate movements fulfill these conditions and preserve the Suez land bridge.

We cannot overstate the significance of the Suez land bridge for humanity's development. It suggests a sense of purpose in nature.

The Mediterranean—An Exciting Geographic Assembling

The Mediterranean region appears as a masterpiece of nature's design. At North Latitude 40° on the European plain, the region sits within the temperate zone, encompassing pivotal geographic features such as the two peninsulas, the Strait of Gibraltar, the Suez land bridge, the Nile Delta, and the Red Sea. It's a convergence of elements rarely found elsewhere, making it a vital natural arrangement, especially for human development. Sounds unbelievable!

Now, let's explore the mystery of its formation.

The roles played by the Gibraltar and Suez land bridges in encircling the Mediterranean were crucial, yet their development diverged significantly.

The Gibraltar land bridge, situated within the collision boundary of two continental plates, should have naturally closed. However, at the opportune moment, it fissured widely (!), allowing a deluge of water to inundate the area. Conversely, the Suez land bridge, located within the divergent boundary of the Red Sea, should have similarly cracked open but instead remained closed, sealing off the Mediterranean. Why does the seafloor of the Red Sea continue to spread while the Suez land bridge remains intact? This set the stage for early humans to migrate out of Africa. Isn't it intriguing when you contrast the natural progression of these two land bridges?

During the formation of the Mediterranean Sea, amidst the collision of two plates, the narrow land bridges of Gibraltar and Suez emerged. Their positioning and dimensions were so precise that the Gibraltar land bridge could open while the Suez land bridge remained closed. This fact alone defies explanation in modern science. Had the Gibraltar land bridge (23 km, west-east) been as wide as the Suez land bridge (173 km, north-south), its opening would have been implausible. The very formation of the Mediterranean might have been jeopardized had either of these land bridges differed. When considering the Mediterranean's significance in human development, as detailed in subsequent sections, isn't the remarkable formation of these two land bridges astonishing?

We know that early humans crossed the Suez land bridge from Africa into Eurasia. However, crossing into Europe via the Strait of Gibraltar posed a different challenge altogether. The Suez land bridge is the sole connecting point between these significant continents. Upon careful consideration, this "natural arrangement" is remarkably nuanced, as it ultimately determined the course of human development and fate.

The above explanation shows that the "ring-fence" of the Mediterranean is not merely a complex geological process but also appears to be a deliberate arrangement. When the two continents collided, a narrow land bridge emerged at the western end and opened precisely at the right time. Similarly, the Suez land bridge appears to be an intentional arrangement, stemming from the "design" of the boundary trend between the Indian and African plates. This encompasses the formation of the GRV and the Red Sea, which we will explore further in the next section.

Let's briefly review the factors contributing to the formation of the Mediterranean: the collision of two plates at the appropriate position (Figure 2-38, below, and Figure 2-33), close proximity to the Atlantic, the Gibraltar land bridge, its rock composition, the directional trend of the growth boundary along the Red Sea, and the presence of the Suez land bridge.

The Mysterious Formation of the Great Rift Valley and the Red Sea

The formation of the GRV has been a topic of extensive discussion, given its significance. To gain a comprehensive understanding, let's explore its mystery from the perspective of plate movement. An intriguing anomaly arises in the directional trend of the oceanic ridges between the Indian Ocean Plate and the African Plate.

Typically, divergent boundaries (oceanic ridges, growth boundaries) are situated at the ocean floor, where sea-floor spreading occurs. However, an exception occurs with the divergent boundary between the African plate and the Indian Ocean plate. This boundary alters its direction, transitioning sharply from north to northwest (Figure 2-38, left), before traversing onto the African continent in an "L" shape, effectively "tearing off" a corner of the continent to form the narrow sea known as the Red Sea and the Arabian peninsula (Figure 2-38, right).

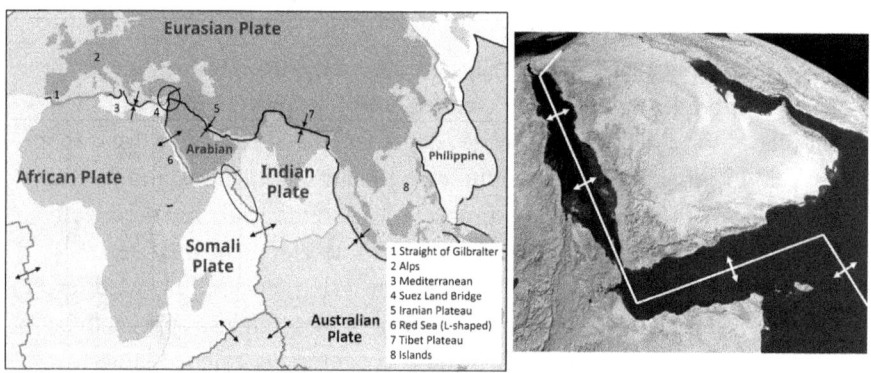

Figure 2-38: The left image shows changes in fault direction circled (Source: M.Bitton, CC BY-SA 3.0, arrows and circles added). On the right is the simplified fault boundary along the Red Sea, with a unique L shape, which seems to be on purpose because the Red Sea is a necessary part of the GRV (Source: Przemek Pietrak, CC BY 3.0, fault line added).

The smooth directional trend of ocean ridges is generally consistent and lacks abrupt changes. So, why does the ocean ridge between these two plates repeatedly change direction, seemingly intent on penetrating the African continent to form the Red Sea? Scientists have yet to provide a convincing explanation, but they speculate that the Red Sea will likely continue to expand into a vast new ocean over millions of years.

In nature, cracking is a common phenomenon (glass crack, ice crack, earthquake crack, etc.), and each crack typically follows a fixed direction of development without sharp changes. The spreading of the ocean floor between the Indian Ocean plate and the African plate can be likened to the cracking of a rock layer. However, its significant change in direction seems abnormal and defies the general law of crack development. Nevertheless, it appears to have occurred deliberately to create the GRV and the Red Sea—a region crucial for the emergence of humans.

It's important to note that we're solely exploring the paths taken by the boundaries, not the causes behind their formation.

The formation of the GRV and the Red Sea can be viewed as two interrelated geological processes. The GRV originates from underground magma upwelling, followed by the elevation of the Earth's surface in the region, resulting in the formation of the East African Plateau, faults, and ultimately the graben (depression formed between fault lines; see Chapter I.1, The Timely Appearance of the Great Rift Valley and the Red Sea) that defines the GRV. This process commenced approximately 30 million years ago, likely predating the formation of the Red Sea.

The geological timeline likely unfolded as follows: around 50 to 40 million years ago, the collision between the African plate and the Eurasian plate induced magma upwelling in the northeastern part of the African plate, leading to the formation of the GRV and, subsequently, the Red Sea (Figure 2-28).

Overview

Finishing this section, let's ponder an intriguing question that modern science has yet to fully unravel: why did the African plate drift northward and eventually collide with the Eurasian plate? While the precise causal link remains elusive, we can speculate on the complex relationship between the African plate's northward movement, the seafloor spreading along its divergent boundaries to the east, west, and south, and the consequent formation of the GRV and the Red Sea (Figure 2-33, Figure 2-38). Despite lacking a definitive answer, we recognize that this phenomenon produced four significant results essential for the emergence and development of humanity:

- The birth of the Mediterranean Sea, the cradle of modern civilization.
- The birth of the GRV, the "womb" of humans.

- The formation of the Red Sea, an essential component of the "womb."
- The formation of the Suez land bridge passage of humans out of Africa.

Without these geological formations, the existence of humanity as we know it today would be improbable. Remember, every event unfolds within specific conditions (Figure 2–39 below).

You could also consider that the divergent boundary between the two plates repeatedly changed its direction as if it aimed to form the Red Sea and the GRV. With this as the starting point according to the needs of human development, the positions of the Eurasian and African continents are determined in turn (see the sections above). Don't you think so?

If we regard the GRV, the Red Sea, and the Mediterranean as pivotal to humanity's birth and development, then it's reasonable to view the above phenomena not as random chance but as elements of a grand design.

Below is a chart depicting this process:

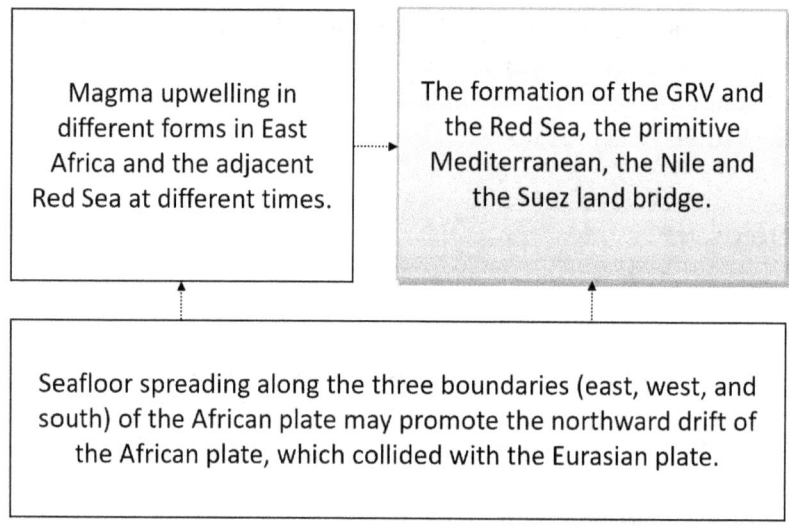

Another intriguing question arises after delving into the Mediterranean and the GRV: Why did nature put these two distinct geographical regions together? Let's explore the advantages of their convergence.

One can envision how early *Homo sapiens* conveniently migrated into Europe around 150,000 years ago by crossing the Suez land bridge. With its fertile land and favorable climate, Europe became a natural habitat for settlement and development. It's plausible to suggest that human civilization likely took its initial steps in Europe. The Mediterranean region, as an expansive "calm" sea, may have lured many early human settlers, fostering their development and igniting their adventurous spirit through navigation. This spirit, integral to curiosity and deeply rooted in human spirituality, became a driving force for progress. It propelled generations of early humans across the globe and may have contributed to the Renaissance, which dawned in the fourteenth century. So, the Mediterranean Sea is like a vast training ground for humanity to venture boldly out to sea.

If these two crucial continents were not connected, if Europe were a dry, cold continent (a likely alternative), and if there were no Mediterranean, humanity would certainly not be where it is today. Earth's complexities unfold in intricate ways!

The Atlantic, the Mediterranean, the GRV, the Red Sea, the Suez land bridge, the Strait of Gibraltar, and the Nile Delta all join together in the temperate zone, forming the most critical and mysterious region on Earth (Figure 2-27, Figure 2-38 left, and Chapter I). This region served as the primary stage for humanity's birth and development, extending its influence to other corners of the world.

Reflecting on the above discussion, one can appreciate the meticulous workings of nature in crafting such a unique and profoundly significant map for humanity's journey.

THE ANCIENT CIVILIZATION SITES ON THE COLLISION BOUNDARY

The Relevant Natural Background

Traveling eastward from the Mediterranean, you'll find three ancient cradles of civilization: Mesopotamia (Sumer), the ancient Indus Valley civilization, and the birthplace of Arabic numerals in the Punjab region (which still retains its name in present-day India and Pakistan). These regions share a common natural backdrop as follows (Figures 2–33 and 2–26):

In light of plate tectonics theory, we can imagine the collision between the Indian Ocean plate and the Eurasian plate suggests varying compressive strengths and areas along the extensive geo-suture, the boundary line of the two plates, spanning thousands of miles. This dynamic has generated different landforms north of the geo-suture (Figure 2–33), including the Alps, the Iranian plateau, and the Himalayas (Figure 2–26).

Subsequently, precipitation through these landforms gradually formed many rivers and alluvial plains, among which the ancient civilizations formed. These civilizations, situated near the 30°N latitude, benefited from favorable climates, fertile soil, and abundant water resources—scarce along the geo-suture—making them ideal for early human habitation. Moreover, their strategic location in the heart of the Eurasian continent facilitated cultural exchanges between East and West. It's conceivable that without these high mountain ranges, the rivers and alluvial plains at their southern foothills would not have existed, rendering the Eurasian continent less diverse and less conducive to developing varied cultures.

Below, we outline the natural processes that led to the formation of these three cradles of ancient civilization.

The Mesopotamian Alluvial Plain

The Tigris and Euphrates, originating from the mountains in Turkey and Iran, were the lifeblood of ancient civilizations. While the precise origins of Mesopotamian civilization remain a mystery, the emergence of the Mesopotamian plain is linked to the complex geological processes associated with the formation of the Red Sea—an outcome of the northward drift of the African plate.

To grasp the significance of the Mesopotamian plain, we must understand the geology of shields. Shields are characterized by extensive outcrops of igneous and metamorphic rocks, often lacking significant sedimentary cover. They are generally stable and tectonically inactive regions, lacking the intense geological activity associated with younger orogenic (mountain-building) belts. The western Arabian Peninsula is a prime example of a shield, boasting a flat, gently undulating surface shaped over 570 to 800 million years.

The Arabian Peninsula's formation, intertwined with the development of the GRV and the subsequent emergence of the Red Sea (Figure 2–38 and preceding sections), includes the expansive shield on both sides of the Red Sea, with the remainder comprising vast desert expanses. Consequently, the shield formation, the collision of plates, the presence of the Iranian Plateau, desertification, and sedimentation collectively contributed to the formation of the narrow yet significant Mesopotamian alluvial plain (Figure 2–39). This plain, defined by the Tigris and Euphrates, is situated amidst harsh natural surroundings and is an outcome of the complex interplay of geological forces.

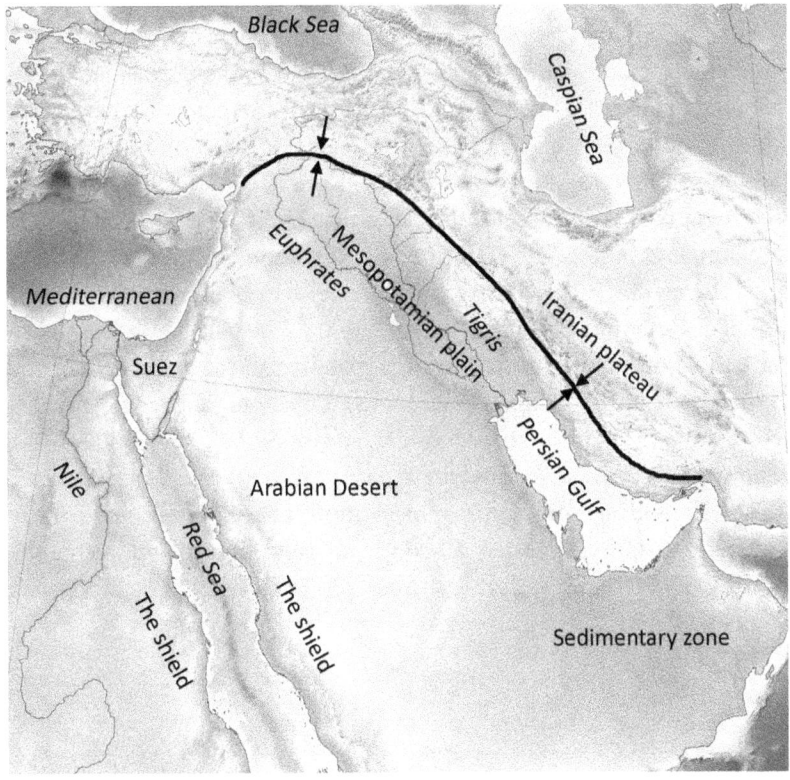

Figure 2-39: The formation of the Mesopotamian alluvial plain. The solid line refers to the collision boundary between the two plates (Source: Sémhur, CC BY-SA 4.0, labels added).

The existence of the Arabian Desert undoubtedly constrained the development of ancient Mesopotamian civilization. While modern-day Iraq has a subtropical desert climate, Mesopotamia likely enjoyed a more favorable environment 6,000 years ago, characterized by milder temperatures and higher precipitation. This era coincided with the emergence of early civilizations along the Nile.

Climate remains a pivotal factor influencing human activities throughout history. Variations in global temperatures and sea levels profoundly shape the global environment, significantly impacting human civilizations worldwide.

The Indus, the Ganges, Punjab, and the Qinghai-Tibet Plateau

Geographically, the civilizations of the Indus and Ganges basins emerged from environments sustained by towering mountains, subsequently developing in locations like Harappa, Mohenjo-Daro (center of the ancient Indus River civilization), and Punjab (the birthplace of numerical symbols, including zero). Beyond these two river systems, the Qinghai-Tibet Plateau also gave rise to several other crucial rivers that have shaped life and human progress in the South Asian subcontinent.

The significance of the Tibetan Plateau to Southeast Asia's environment cannot be overstated (Figure 2-41). The collision between the Indian Ocean plate and the Eurasian plate led to the formation of the Himalayas and the Gangdis mountains, fold mountain ranges parallel to each other (Figure 2-40). Among the peaks of the Gangdis stands the renowned Kangrinboqe, towering at 21,837 ft (6,656 m) above sea level. This peak holds not only geographical importance but is also revered as a sacred site in South Asian subcontinent culture, boasting one of the world's most famous Buddhist shrines and drawing countless pilgrims annually.

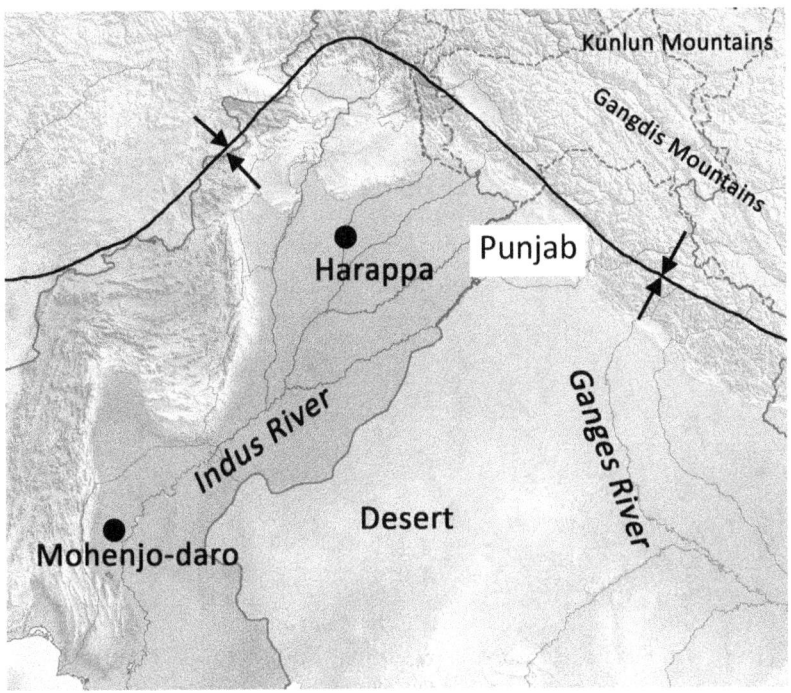

Figure 2-40: The location of the Indus River. Punjab represents one possible area where ancient civilizations once arose. (Source: Carport, CC BY-SA 3.0, labels added).

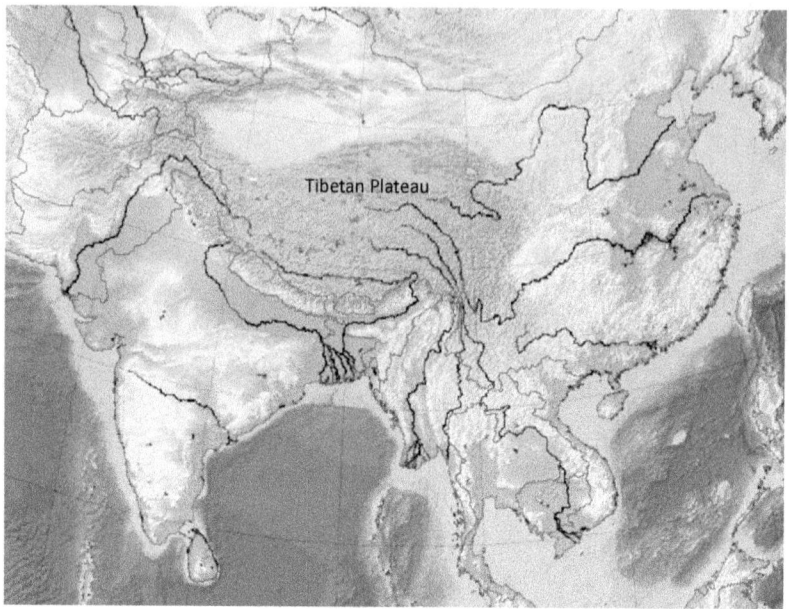

Figure 2–41: The main rivers originating from the Tibetan Plateau nourish the land in eastern Asia and the South Asian Subcontinent (Source: Uwe Dedering, CC BY-SA 3.0, cropped, labels added).

Without the Qinghai-Tibet Plateau, one might speculate about human survival and development, but this hypothetical scenario could not be an isolated event. Without the collision between the African/Indian plates and the Eurasian plate, essential for forming the Mediterranean and Himalayas, the plateau would not have occurred. Consequently, the necessary geographical conditions for civilization's birth and progress would have been absent.

The significant role of continental drift in shaping geographical conditions underscores its non-random nature. For instance, the Tibetan Plateau exhibits a "water tower effect," sustaining life in the Eastern Asian continent and the South Asian subcontinent. Without this abundant water resource, the viability of life in these regions would be significantly compromised.

It's evident that the collision between the Indian Ocean plate and the Eurasian plate profoundly influenced the origin and development of civilization, making it one of the most impactful collisions in Earth's history.

The Environment of East Asia and its Influence on Humanity

Continuing eastward along this collision boundary, we encounter the Himalayas and the Qinghai-Xizang Plateau (Tibetan Plateau), often referred to as the "roof of the world" (Figure 2-33, Figure 2-41). These geographical features result from the collision between the Eurasian plate and the northward-moving Indian Ocean plate, known as the Himalaya Orogeny (Young Alpedic). Viewed globally, the Himalayas mark the eastern terminus of a vast mountain range in southern Eurasia that begins with the Alps (Figure 2-26, Figure 2-38, left). Spanning thousands of miles like a wall, these mountains obstruct the northward flow of warm, moist air from the Indian Ocean, giving rise to the barren environment of inner Asia characterized by extensive arid and semi-arid regions, including vast deserts and plateaus, some towering at elevations exceeding 13,000 ft (4000 m)—conditions inhospitable to life. Roughly one-third of China's land, comprising high plateaus, deserts, and the Gobi, falls under this category.

Thankfully, despite the harsh environment of eastern Asia, the mainland boasts an expansive habitable landmass (Figure 2-25, Figure 2-41). Positioned at mid-latitudes and flanked by the sea to the east and south, this region benefits from abundant annual rainfall from monsoons and numerous rivers originating from the Tibetan Plateau. These rivers, including the Chang-jiang River (Yangtze River) and the Yellow River (Figure 2-41), have formed countless alluvial plains, known as the Central Plains, characterized by fertile soil, temperate climates, ample water sources, and a variety of natural resources. The ancient Chinese civilization flourished within this fertile environment, the most populous in the world.

However, despite the natural advantages of the Central Plains, they are encircled by unforgiving environments (Figure 2-25, Figure 2-41). The Central Plains' closed natural setting provided the foundation for an insular society, a closed society, or a non-dissipative society in ancient China, fostering a self-sufficient economy that profoundly shaped the character and culture of the Chinese nation. The self-reliant economic model and the centralized imperial power structure (social system) were mutually reinforcing, sustaining a rigid hierarchical system for over 2000 years. At its peak, the emperor wielded absolute authority, with officials deeply integrated into the corpulent hierarchical structure, controlling all facets of society, including public discourse—an arrangement that perpetuated generational ignorance and apathy. Dissent against the emperor's ideology and power structure was considered the gravest of offenses.

To underscore the societal impacts of the environment, we may contrast Eastern Asia's closed geographic confines with the Mediterranean's open expanses, each nurturing distinct cultures and destinies. Thus, we may ask: Who determined the course of human development?

THE UNIQUENESS OF THE ISLANDS OF EASTERN ASIA

Continuing eastward along the collision boundary of the Indian Ocean and Eurasian plates, the Asian islands span from high latitudes to the equatorial region (Figure 2-33, Figure 2-38 left, →←). Asia's eastern and southern coasts are surrounded by over 10,000 islands—a unique phenomenon. The islands stretch from the Japanese archipelago to the Indonesian islands, with more than a dozen major islands in this region. The formation of these islands is linked to the complex processes of collision of the Indian Ocean and Pacific plates with the Eurasian plate (the creation of the Japanese islands has been previously discussed). Numerous volcanic islands dot the seascape as well.

Nature seems to have paired two contrasting environments in Eurasia. On the one hand, there is the inland expanse encompassing much of Eurasia, while on the other, there are the encircling islands. The marine environment surrounding these islands stands in stark contrast to the mainland's closed environment, fostering a distinct island culture, particularly evident in temperate zones. We do not know why nature has arranged these two environments together. Still, the existence of the islands around the mainland (including Western Europe facing the sea) resulted in varying national characters and cultures formed by the marine environment. This maritime culture undoubtedly enriches human history and holds potential for the future. A hallmark of maritime culture is the spirit of adventure inspired by the sea (see Chapter II.3, The Mysterious Plate Movement: Overview), which profoundly influences human progress and innovation.

As indicated in the previous chapter, had these islands emerged on the periphery of other continents, we might have missed their significance. This point, gleaned from the annals of human development, is readily apparent.

NORTH AND SOUTH AMERICA

The "Empty" Continent

Continuing our eastward journey (Figure 2-33, →←), we pass the Kuril Islands, produced by the collision of the Pacific and Eurasian plates, and then we reach the Pacific Ocean. Crossing this vast expanse to the north of the American plate, we trace along its western coast to its southernmost tip, the Drake Passage. During this long journey, scant evidence of very early human habitation has been found on the vast American continent.

For example, according to maps in *The World to 1500: A Global History*:[1]

- Global Distribution of Hominids and Homo Sapiens (Map I): Australopithecines (4 million to 1 million years ago) and Homo erectus (1 million to 200,000 years ago) were nonexistent on the American continent.

- Early Human Migrations (Map II): Early humans entered the Americas through the Bering Strait about 50,000 years ago.

- Expansion of Agriculturists (Map IV): The expansion of agriculturists (3000 BC) did not occur on the American continent.

Does this revelation surprise you? Not really! Instead, it underscores the pivotal role of Eurasia's environmental conditions in shaping human development. One theory suggests that the early inhabitants of America migrated from the Asian continent, having reached a certain level of development before venturing into the Americas.

While the history of early human development in the Americas remains a subject of archaeological inquiry, what's pertinent for our discussion is the sparsely populated nature of prehistoric America, a condition crucial for the eventual development of modern civilization (see Chapter II.3, The Logical Step 4).

This reality underscores, once more, that the origin of humanity and the development of civilization is a complex process contingent on numerous factors. It's erroneous to assume that the mere presence of flora and fauna in the Americas would have naturally led to the evolution of primates into humans and the subsequent emergence of modern

---

1. Stavrianos, *The World to 1500*

civilization. The truth is far more complex—everything unfolds under certain conditions!

The Complicated and Subtle Movement of the Pacific Plate

More intriguing than the sparse population of early America is the correlation between the growth boundaries (ridges) in the Pacific and the geographical features of the two coasts: the east coast of Asia and the west coast of the Americas.

Undoubtedly, the current coastal geography of each side of the Pacific Ocean is tied to the location of the Pacific growth boundary (Figure 2–33, ←→). These features include the chain of islands along the East Asian coast, the series of coastal mountains running north to south along the western coast of the Americas, and, notably, the absence of islands off the American coast.

But why is this so? Drawing upon tectonic theory, we can logically infer that islands such as those found in Japan and Taiwan result from the collision between the Pacific and Eurasian plates—a seemingly straightforward process. In contrast, the formation of mountain ranges along the American continent's western coast appears more complex, with no corresponding island chains.

For instance, the formation of the Rocky Mountains in western North America involved many natural factors over vast stretches of time and space, unlike the relatively simpler process of island formation in Japan. Similarly, the formation of the Andes involved interactions with the Antarctic plate. It's conceivable that if the growth boundary of the Pacific Ocean were positioned differently, the coastal geography of the western/eastern Pacific Ocean would undergo drastic alterations—location matters.

The topography of the United States is characterized by elevated terrain on its east and west sides, with plains in the middle. The western plateau, formed by the coastal mountains, extends across the country's western region, nearing its central region. These mountains give rise to numerous rivers that nourish the central plains, contributing to the country's favorable geographical environment. The coastal mountains of the United States hold significant value for its environmental landscape.

We suppose that if the west coast were dotted with islands like those in Asia, thereby replacing the existing coastal mountains, the country

THE RATIONALITY OF THE EARTH SYSTEM

would lose its advantageous geographical setting, and the imagined islands would serve little purpose. We term this "the natural arrangement," though its realization demands precision, subtlety, and complexity. This natural arrangement is entirely necessary from the point of view of human history.

To summarize our journey, we begin from the Mediterranean Sea, moving eastward along subduction boundaries formed by continental plate collisions, passing through the South Asian Islands, arriving at the collision zone of the Pacific plate with continental plates, and ending in South America (Figure 2–33). These geological processes started roughly in the middle of the Cenozoic Era and continue to this day, giving rise to mountains in Turkey, the Iranian Plateau, the Tibetan Plateau, the South Asian islands, the Japanese islands, and the western coastal mountains of North and South America—all formed by plate collisions during this time.

OVERVIEW

If we view landmasses as raw materials, the collision of plates resulting from plate movement can be likened to the processing of each landmass. The collision between the African and Indian Ocean plates with the Eurasian plate has fundamentally shaped the Eurasian continent into its present form, giving rise to features such as the Mediterranean Sea, mountain ranges, and plateaus. Similarly, the Japanese islands, the coastal mountains of North and South America, and the Drake Passage were formed through the spreading of the seafloor of the Pacific plate.

This process can be traced back to the continental drift that occurred following the breakup of Pangea approximately 200 million years ago, with individual landmasses gradually drifting toward their current positions. Information regarding these geological processes is readily available through various resources for those interested in learning more (see, for example, Figure 1–3). Throughout this extensive process, it becomes apparent that a global system of seafloor spreading and continental drift must have existed. It appears that the current distribution of land and sea was shaped by this system, particularly for the emergence and development of humanity.

Let's take a deeper look. It appears that for such significant results, nature has prepared as follows:

Firstly, the Indian Ocean plate and the Asian Continent amassed sufficient "raw material" and energy beforehand to give rise to the towering Himalayas and the Tibetan Plateau (Qingzang Plateau), effectively sealing off the geographical environment in eastern Asia. The formation of the Mediterranean Sea, as previously examined, entails an intricate, subtle, and precisely orchestrated process. Each step in this process, from the chemical composition of the rock forming the land bridge of Gibraltar to the changes in the direction of the ridge at the end of the Red Sea, appears planned in both time and space, culminating in the forming of the vital Suez Land Bridge. The repeated changes in the direction (trend) of the growth boundary between the Indian Ocean plate and the African plate also seem purposeful, aimed at forming the GRV and the peculiar L-shaped Red Sea, both necessary for the birth and development of humanity (Figure 2-33, Figure 2-38).

Similarly, the direction (trend) and location of the growth boundary of the Pacific plate (the ridge) appear to be a subtle and precise design, determining the geographic features on both sides. While one side features significant island groups, the other lacks islands altogether, indicating a purposeful allocation of geographical characteristics to meet the needs of human development without "wasting" valuable land.

Given the significance and suitability of these outcomes in global climate and human development, they do not seem to be products of mere chance; rather, every process involved appears to exhibit "the Reason" or "rationality." As Einstein said, "God doesn't play dice"!

And that concludes our narrative on plate movement. Perhaps you've realized what an incredible story this is. This tale illustrates how these plate collisions are indispensable for shaping the global climate, fostering the evolution of life, catalyzing the emergence of humanity, fueling the development of civilization, and shaping the fabric of today's society. Some may label it the great plate collision; others may see it as an intricate "design," while others may call it a series of natural arrangements. Such descriptions are not exaggerated. But what lies beneath this sequence of geological processes? Why is it orchestrated in this manner? That we will discuss in the last chapter.

Indeed, one could hypothesize an alternative arrangement for distributing the Earth's land and sea. Yet, after careful study (aided by modern computer simulations), one might find that the existing distribution of land and sea—comprising islands and peninsulas—is optimally suited for human development and the global climate system. It's plausible that no better solution could be devised.

Let's reiterate: everything unfolds conditionally, and without the conditions (arrangements) formed by plate collisions, the progress of global climate and human history would have been stymied.

In the expanse of the universe, the emergence of the human realm is far from straightforward; it requires the coordination of numerous natural conditions. We've only explored the conditions required from the land and sea distribution perspective. It's as if God cannot play dice to achieve these conditions.

While various theories may seek to explain these phenomena, one could argue that these theories merely "serve" these arrangements—the occurrence of these conditions. Thus, the distribution of the Earth's plates does not appear to be arbitrary.

NATURE SET THE WAY FOR THE DEVELOPMENT OF HUMAN CIVILIZATION

The chapters above argue that the Earth's distribution appears planned. The following is a further exploration of this issue. We have examined the origin and development of ancient and modern civilizations, but one may ask, what is the force that drives the ongoing development of human civilization? How does nature solve this problem?

Next, you will see how nature created the conditions that led to rapid human development, including the forming of nations and the competition between them. The whole process is inseparable from the principles of natural philosophy and rationality.

The Emergence of Nations: An Essential Element for Human Development

Understanding the unprecedented pace of modern scientific advancement is a complex endeavor, especially when considering human nature, the historical backdrop of perpetual competition among nations, disparities in land distribution and natural resources, and the diversity of cultures.

A nation's culture and religion are closely tied to its natural environment, with nature serving as the primary determinant shaping the essence of a nation. The varied natural environments across the globe give rise to distinct nations and cultures, each with its unique characteristics.

Inevitably, competition between nations, whether in trade or military endeavors, is a driving force fueled by human curiosity and ambition, driving the pursuit of scientific and technological development. Indeed, the rapid development of modern civilization owes much to the competitive dynamics among multiple nations—an indisputable historical reality.

Several key factors influence terrestrial environments, including latitude (frigid, temperate, tropical), terrain (plateaus, plains, mountains, hills), ecology (grasslands, forests, deserts), and geography (continents, coastal areas, islands, peninsulas). These factors intertwine to shape diverse local conditions conducive to fostering distinct national cultures and characters.

While climate plays a significant role in the formation of nations, it is not the sole determinant. For instance, the Indian peninsula, despite sharing a general climate type, hosts several distinct ethnic groups.

The factors contributing to these conditions, excluding latitude, are closely intertwined with plate movements. As humans migrated out of Africa and settled in new territories during the current interglacial period roughly 11,000 years ago, they encountered diverse natural environments, leading to the emergence of unique societies and cultures.

It's evident in history that populations adapted to challenging conditions, such as those in colder regions, often developed physical resilience that was advantageous for survival and defense. For instance, the nomadic tribes inhabiting the northern Chinese grasslands developed physical resilience and bred superior horses over millennia, prompting their southern neighbors to construct the Great Wall of China over 2,000 years ago as a defensive measure.

Conversely, those inhabiting more favorable environments, such as the temperate alluvial plains along the Yangtze and Yellow Rivers in China, exhibited intellectual prowess conducive to forming ancient civilizations. While they may have been physically smaller on average, their ancient civilization consistently displayed technological advancements beyond their neighboring regions.

So far, we examined the formation and characteristics of nations. Now, let's shift our focus to another aspect: the distribution of natural resources on the Earth.

Natural resources encompass both surface and subsurface reserves. Surface resources include freshwater, forests, grasslands, and fertile land suitable for cultivation and habitation. Meanwhile, subsurface resources primarily consist of energy sources like coal, oil, and natural gas, as well as various metallic and non-metallic mineral deposits.

These resources are not evenly distributed worldwide. Regions with limited surface resources, such as deserts, may possess abundant underground reserves, while areas rich in surface resources, like alluvial plains, may lack significant subsurface deposits. Furthermore, the diversity of underground resources means that only a few types, such as oil or metallic minerals, may be plentiful in a particular area.

Considering the diverse cultures and characteristics of nations and the environmental differences, the uneven distribution of resources globally, conflicts, competition, and even wars for survival and development are inevitable. These tensions serve as driving forces behind the advancement of science and technology among nations.

While wars throughout history have led to the destruction of wealth, they have also spurred technological innovation and cultural exchange, such as the Crusades era. However, it's important to recognize that war is not always the optimal solution, as noted by Chinese military strategist Sun Tzu over 2,000 years ago, and may lead to more losses than gains.

It's essential to acknowledge that continued warfare would ultimately lead to self-destruction (Figure 2-42). Instead, conflicts and disputes must be resolved through rationality and wisdom, focusing on promoting social progress rather than conquest.

Figure 2-42: Mosaic found in 1831 AD from the ruins of Pompeii, painted in 310 BC, depicting the history of Alexander the Great and Darius III, the last king of the Persian Empire, in the autumn of 333 BC, at the battle of Issus (Source: Lucas, CC BY-SA 2.0).

In summary, nature appears to play a crucial role as the guiding force in human history. On one hand, it has shaped human nature and the diversity of nationalities through evolution. On the other hand, nature has created various environments and unevenly distributed natural resources, leading to competition and the global advancement of science and technology. This intertwined relationship has defined human history and seems to have been the primary method through which humanity has developed over the millennia.

Human nature is the inner root of human conflict. Our exploration of human nature will continue in the next chapter. Do you have any alternative methods for rapidly enhancing human abilities?

It appears that suffering is a fundamental catalyst for rapidly enhancing human abilities, a phenomenon deeply rooted in human nature. However, broad religious differences among nations often serve as a significant factor leading to conflicts. Nevertheless, as people become more enlightened and civilization progresses, these religious disparities may gradually diminish, potentially fostering a world where conflicts occur less frequently.

Geographic Conditions Assist Human Development

In the preceding chapters, we examined the emergence of the Mediterranean Sea, detailing the collision of continental plates, the formation of the Strait of Gibraltar, the presence of the Suez Land Bridge, and the unique features of the GRV and the Red Sea. This process appears meticulously orchestrated by nature, laying the groundwork for exploring the Mediterranean's profound significance to humanity.

Indeed, it's arguable that the human world as we know it today would not exist without the Mediterranean. By examining the formation of the Mediterranean Sea and its impact on humanity, we can unravel the enigmatic relationship between humans and nature, gaining deeper insights into the essence of the natural world.

The Mediterranean Sea serves as a starting point for social progress and the cultivation of human capabilities. Next, we will explore how the Mediterranean Sea is the natural foundation for human progress, and you will learn why nature invested considerable effort in creating the Mediterranean Sea.

Mediterranean Sea—The Cradle of the Renaissance

With its blend of reason and greed, human nature was created by nature itself. The two are inextricable and woven into the fabric of human existence, profoundly shaping our history. Throughout our exploration, we'll revisit this concept of human nature and how nature has dramatically assisted human progress and development, beginning with the Mediterranean Sea.

Human society, in essence, comprises both the physical and spiritual aspects of humanity, with thought serving as the conduit for expressing human wisdom and spirit.

Throughout history, ancient civilizations, primarily situated in the Eurasian continent (including ancient Egypt), independently adopted autocratic systems despite their geographical and cultural distances. This autocratic governance reflects an inherent stage of societal development rooted in human nature and prevailing ignorance, although people were instinctively intelligent and productive.

In these early societies, characterized by limited knowledge and survival skills, the power structure was pivotal in mobilizing resources for monumental undertakings and navigating major crises. However, this concentration of power inevitably led rulers to control thought to maintain their authority indefinitely, resulting in societal stagnation and intellectual numbness over time.

This power-dominated society, often summarized by the dictum "The king is the law" (as articulated by Karl Marx), fostered a culture where rulers remained unchallenged and unaccountable, regardless of the consequences of their actions. Rulers became a powerful interest group with absolute authority, regardless of how they chose to enhance their image. Such autocratic systems, driven by the perpetuation of power through hereditary succession, endured for extended periods throughout history, fueled by fundamental aspects of human nature.

However, societies governed by such absolute authority lacked the diversity of voices necessary for progress and evolution. Consequently, the simple linear system of autocratic rule hindered societal advancement, as evidenced by humanity's long history.

The transition from societal numbness to awakening marked a pivotal stage in human development, albeit characterized by uneven progress for centuries. In this context, the Mediterranean region, with its distinctive geographic features, served as a catalyst for breaking free from the stagnation of the Middle Ages, heralding the dawn of a new era: the Renaissance!

The Renaissance, with its emphasis on intellectual and cultural revival, underscores the pivotal role of the Mediterranean in driving societal progress and fostering the principles of human progress.

Next, we will look further into these points.

The Mediterranean Started an Awakening and Human Progress

If we consider nature as the foundation of human existence, then science becomes the ladder toward a brighter future, with wisdom as the material used to construct this ladder. The pinnacle of wisdom lies in innovation, and the catalyst for innovation is freedom of expression and the legal system that comes with civilization. Freedom, inherent in all forms of life, including human spirituality, fuels progress.

The emergence of the Renaissance, aided by the Mediterranean, exemplifies this interplay. Geographically, the Mediterranean possesses two distinct advantages. Firstly, it offers convenient transportation, with relatively calm seas due to the protection the Italian and Greek Peninsulas provide, facilitating safe and efficient maritime travel. Secondly, the Mediterranean climate boasts pleasant conditions year-round, conducive to human activities and agriculture, attracting people to its shores.

Like a giant magnet or hub, the Mediterranean region bustled with trade and cultural interactions beginning centuries ago. Although no cameras captured these scenes, one can envision diverse people traversing land and sea, exchanging goods and ideas year-round across Europe, Asia, and Africa. Despite language barriers, a system of trusted exchange developed, fostering prosperous commerce and trade. Without centralized power over commerce, these interactions spurred ideological vitality, laying the groundwork for the modern ideals of freedom and equality, enriching humanity's inner dignity.

Renowned scientist Dr. Joseph Needham (1900–95) asserted that modern science owes its existence to thriving trade. Intellectual exchanges among scholars gave rise to humanism, serving as the theoretical foundation of the Renaissance. This era marked a profound awakening in the human spirit after millennia of philosophical dormancy.

The Crusades, spanning nearly two centuries (1096–1291), paved the way for cultural interchange across Eurasia. The capture of

Constantinople by the Turks in 1453 led scholars to flee to Italy, carrying with them ancient Greek texts that became pivotal for the Renaissance. Meanwhile, other parts of the world remained entrenched in power-controlled societies. From the fourteenth to the sixteenth centuries, the European Renaissance emerged as a beacon of enlightenment, with the Mediterranean serving as its cradle.

Could the Renaissance have flourished if only one peninsula, such as the Greek peninsula, existed in the Mediterranean? History suggests otherwise. Both peninsulas played crucial roles. Dante, Petrarch, and Boccaccio, pioneers of humanism, hailed from the Italian peninsula.

Thus, we may attribute the modern concepts of freedom and equality to the free trade and markets flourishing around the Mediterranean. These ideals evolved over centuries of cultural exchange among people from three continents.

Although few paintings directly depict Mediterranean activities during the Renaissance, some renowned artworks offer glimpses into the bustling Mediterranean life of the eighteenth century (Figure 2–43).

Figure 2–43: The Molo with the Library and the Entrance to the Grand Canal (Italy, 1740) by Canaletto (1697–1768). This painting reflects the prosperity of the Mediterranean in the eighteenth century (Source: Public domain).

The humanism championed by the European Renaissance ignited the human intellect, paving the way for the emergence of human rights as a challenge to prevailing power structures. Viewing each individual's mind as a variable, a society founded on human rights represents a highly intricate *nonlinear system* (strictly speaking, the dissipative structure of nonlinear systems; see Chapter IV, No Other Identical Natural Process in the Universe), progressing from disorder to order over time, perhaps spanning centuries. Throughout this societal evolution, the freedom of expression and the open exchange of ideas play pivotal roles, fostering the release and application of wisdom and continual understanding of the world.

Centralized power structures and human rights ideology both stem from human nature. When a power structure (social system) suppresses rights, it solidifies into absolute control, resulting in a simplistic linear societal framework. The humanism of the Renaissance, by championing free thought, laid the groundwork for the emergence and advancement of modern science and served as a defining divide between democracy and autocracy.

As the vitality of complex nonlinear systems began to increase across Europe and past the Renaissance, people naturally asserted themselves as collective sovereigns of their nations, fostering the democratic ideal. Over time, democracy, despite its inherent fluctuations, steers politics towards rationality by curbing power. Conversely, autocracy hinges entirely on the whims and moral fiber of the dictator, with the state merely serving as a tool for their unchecked ambitions.

If we set the total energy for promoting social progress at 100, modern societies distribute this energy evenly among all people. However, in a society controlled by absolute power, this distribution is compromised, resulting in a loss of energy for social progress, including laws and credit.

Just as the natural world operates as a complex nonlinear system, human society, as a product of nature, follows suit. A profound point.

The Renaissance Demonstrates the Principles of Human Development

It's widely recognized that true creativity flourishes in an environment of unfettered freedom, a condition determined by the intricate workings of the human brain. Dignity, pride, and self-respect form integral facets of human nature, universally appreciated and revered. Preserving one's dignity is paramount; it stems from our ability to reason.

THE RATIONALITY OF THE EARTH SYSTEM

In essence, dignity lies in using human rights to restrict power structures. Without rights, dignity fades. In ancient times, only rulers had dignity, and the oppressed people had none. The ideal of human rights spread globally after the Enlightenment (seventeenth to eighteenth century in Europe), aligning with innate human needs and embodying dignity.

Historical evidence underscores the role of freedom and rights in fostering creativity within societies. Most modern science and technology innovations, particularly original scientific theories, emerged within societies fostering human rights, subsequently spreading worldwide. This example underscores the stifling effect of absolute power on the expression of thoughts, stifling wisdom.

The Renaissance heralded a transformative era of social progress, propelling humanity towards theoretical scientific innovation, a pivotal departure from mere accumulation of knowledge. This seismic shift marked the elevation of the human intellect, defining a nation's strength not by population or territory but by scientific and technological advancements.

Through the market, widespread scientific and technical breakthroughs drive economic prosperity and human progress, paving the path to sustainable global development. History demonstrates continual scientific innovation, underscoring its indispensable role in human development.

So, the exercise of rights is the most basic symbol of modern society and the source of its vigor and vitality. Only through the exercise of rights is it possible to realize sustainable development and establish a legal system.

Now, we can simplify the principles of human development shown in human history as follows:

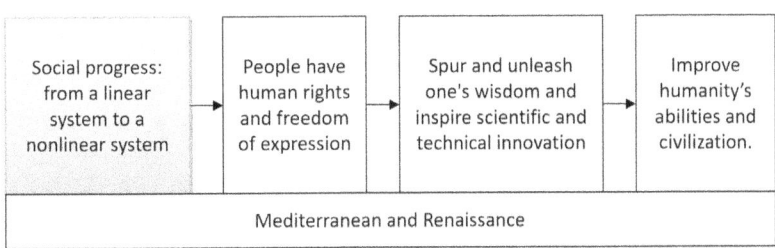

As human rights flourished in Europe, their influence sparked an awakening that naturally began to expand throughout the globe. However, history reveals the challenges inherent in social progress. Social progress hinges on advancing the societal framework, a journey fraught with obstacles even within democratic societies.

Scientific and technological innovation stands as humanity's worth, with an innovative society representing a profound outcome of social progress rooted in enhanced public literacy and a legal system.

Democracy and science are the most significant enlightenment of history to humanity since the Renaissance. Democracy is essential for building an innovative society, yet deeply ingrained cultural norms, traditions, and beliefs—shaped over centuries—exert a formidable influence on individuals' minds. Overcoming this historical inertia, particularly in populous societies, poses a formidable challenge, impacting the nurturing of innate positive traits like curiosity, exploration, and enterprise.

Fostering a renewed awakening and elevating literacy levels are crucial but time-intensive endeavors. Moreover, varying economic and educational levels of development further complicate social progress, rendering advancement towards an innovative society a complex and lengthy process. Promoting cultural progress with awakening is crucial.

In contrast, authoritarian regimes, by controlling thought and fostering complacency, hinder the evolution towards a civilized society characterized by innovation. Hence, global innovation remains uneven, reflecting the enduring struggle inherent in human progress.

Establishing an innovative society is anything but straightforward!

A Global Marathon Race

We've examined the complexity of societal progress, which results in significant differences across the globe. These disparities are determined primarily by national cultures, themselves deriving from complex processes influenced considerably by the natural environment.

Every nation has its distinct culture, natural environment, land, and population. Like participants in a marathon, nations struggle for survival and development. Some lead the pack, others trailing, and many somewhere in between, all navigating the "race track" of human civilization.

But what determines the winner? Innovation!

As the only species capable of altering the natural landscape with impunity, humanity's progress impacts the environment, depleting resources and ravaging ecosystems. Yet, we can surmount these challenges through scientific and technological innovation, expanding our capabilities and ensuring sustainable development, aligning with the natural order of progression.

Hence, a nation's stature on the global stage hinges not on the size of its territory or the magnitude of its economy but on its contributions to human civilization through groundbreaking innovation. Indeed, scientific and technological advancement is the primary catalyst for economic growth, a principle exemplified in the Mediterranean and Europe post-Renaissance. In this global marathon, the nations fostering substantial value through innovation lead the pack, their power effectively tempered by legal and ethical frameworks. Such transformative innovations pave the way for humanity's ascent towards a brighter future (see Chapters V and VI).

Therefore, the global marathon is the road, set by nature, of rapid development of science and technology. This path combines human nature (internal causes) with natural conditions (external causes) to form an inevitable and enduring process toward human survival. You may not find a better way to spur humanity on the path of rapid development.

The Scientists Whose Names Should Be Known

Scientists are a valuable asset to humanity, serving as the driving force behind the advancement of civilization. Newtonian mechanics alone facilitated the industrialization of the world. However, the growth of a modern scientist is a complex journey that requires many preconditions.

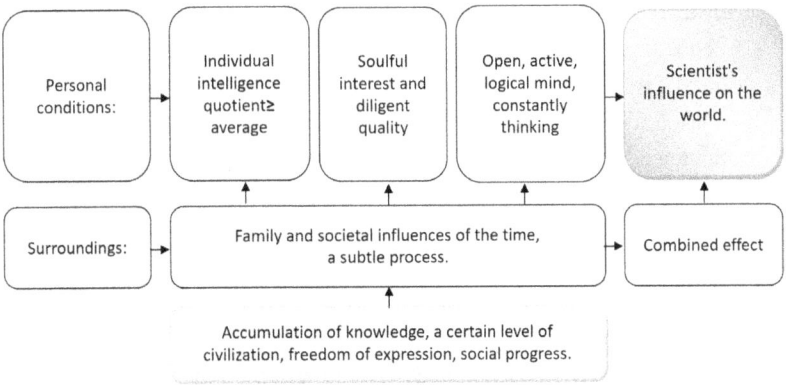

A soulful interest is distinct from a typical interest, such as music or sports. It arises from an innate, lifelong, and selfless passion. Passion is rooted in human nature. Without passion, innovation fades. Discovering one's soulful interest early in life is advantageous. For example, Newton and Einstein both held their unique passions since childhood.

The humanism promoted during the Renaissance stimulated intellectual curiosity and passion, unlocking the depths of people's souls and raising scientists whose outstanding contributions laid the foundation for modern science. Let us honor their illustrious names:

- Leonardo di ser Pieroda Vinci (1452–1519), the perfect representative of the Renaissance. An Italian, he made many contributions to science and engineering, though he may be most famous for his art.
- Nicolaus Copernicus (1473–1543), a great Polish scientist who advanced the history-making theory of the heliocentric system (and also an excellent doctor).
- Francis Bacon (1561–1626), an English philosopher. In his book *Novum Organum*, he introduced inductive reasoning, making outstanding contributions to science and logic in later generations.
- Tycho Brahe (1546–1601), a Danish astronomer, considered the founder of astronomy. For 20 years, he meticulously observed the night skies using only his naked eyes.
- Johannes Kepler (1571–1630), a German astronomer who used Brahe's observations to establish the three laws of planetary motion.
- Galileo Galilei (1564–1642), a great Italian scientist and a vigorous advocate of heliocentric theory. He was the first to use experimentation to verify a theory when he experimented with gravity at the Leaning Tower of Pisa and disproved Aristotle's faulty theory, which had been accepted for nearly 2,000 years. He also invented the first telescope to observe and study the Moon and other celestial bodies. Because he openly supported Copernicus's heliocentric theory, he was confined to his home during the Inquisition, showing that the Renaissance was a period of friction between new and old thoughts.
- Rene Descartes (1596–1650), a French scientist and philosopher. He created Descartes' Methodology and the Cartesian Coordinates system.
- Blaise Pascal (1623–62), a French physicist, mathematician, and philosopher. The Pa (Pascal) unit of pressure is in his name.

THE RATIONALITY OF THE EARTH SYSTEM

- Robert Boyle (1627–91), a British chemist, regarded as the founder of chemistry. He discovered Boyle's law, although the notion of chemical elements was not established until the nineteenth century.
- Robert Hooke (1635–1703), a British scientist who discovered Hooke's Law. He also invented the microscope, which revealed the existence of cells, marking a significant milestone in biology.
- Isaac Newton (1643–1727), probably the most outstanding scientist ever who profoundly impacted humanity's development with his groundbreaking contributions.
- Gottfried Wilhelm Leibniz (1646–1716), a renowned German mathematician who independently founded calculus while Newton was developing similar theorems.
- Immanuel Kant (1724–1804), a famous German philosopher and originator of the nebular hypothesis.
- Charles-Augustin de Coulomb (1736–1806), a French engineer and physicist. The unit of quantity of electric charge in physics, C (Coulomb), is named for him.
- Pierre-Simon Marquis de Laplace (1749–1827), a well-known French scientist and developer of the Kant-Laplace nebula hypothesis.
- Michael Faraday (1781–1867), a famous British physicist and chemist. Born into poverty, he only went to primary school. In 1831, he constructed the first electric generator based on his previous discoveries regarding the relationship between electric currents and magnetic fields. This groundbreaking achievement paved the way for the modern electric power era.

These scientists emerged in Europe around 200–300 years ago, vividly illustrating the relative dormancy of the rest of the world and highlighting the pivotal role of the Mediterranean. Physically, all human brains share the same structure, yet the concentration of scientists in Europe underscores the importance of social progress.

These are the pioneers of modern science. Their spirit and dedication to science serve as an inspiration for later generations. Their names will forever shine in the history of human progress.

History also attests to their scientific and technical innovations significantly advancing material civilization, underscoring that civilization's evolution hinged on such innovation. Scientific and technical advancement is the engine of civilization's development, essential for humanity's sustainable progress.

These luminaries represent the precursors to the birth of modern science (Figure 2-44):

Figure 2-44: From left to right, Nicolaus Copernicus, Galileo Galilei, and Isaac Newton (Source: Public domain).

Before proceeding, it's crucial to reiterate the following points.

Globally, human intelligence is inherently similar, yet the societal context in which intellect flourishes varies greatly. During the sixteenth and seventeenth centuries, scientists predominantly emerged in Europe. Their upbringing would likely have differed if they had been born elsewhere, turning them away from scientific pursuits. At this time, Europe was fertile with social soil favorable to nurturing human intelligence, while the rest of the world resembled social "deserts," lacking the environment necessary for scientific genius to grow. This distinction underscores the distinction between possessing intelligence and maximizing its potential.

NATURE, THE CHIEF DIRECTOR OF THE HUMAN WORLD

Geographical determinism, also known as environmental or climatic determinism, traces its roots back to ancient Greek thinkers like Plato and Aristotle, who recognized the significant influence of the natural environment. This theory continues to develop today, highlighting the interconnectedness between an individual or a nation's growth and their surrounding environment. This principle is widely accepted.

Let's consider the birth of modern science. If human intelligence and the emergence of modern science were natural inevitabilities occurring without the need for specific environmental conditions, then modern science would have spontaneously arisen worldwide around 300 years ago instead of only in the Mediterranean region. However, given the global disparities in natural resources, environmental conditions, national character, and culture, this scenario would likely have led to perpetual conflict and war. Human greed, coupled with these differences, would have hindered the possibility of global peace.

Is this the optimal path for human progress? Can progress be achieved under such circumstances? Certainly not!

We must recognize a fundamental fact once more: while everyone possesses intelligence, not everyone can fully utilize it to become a scientist or an inventor. Such achievements depend on various conditions. Utilizing wisdom to its fullest extent is a complex process that demands the coordination of numerous factors. Without these conditions, individuals or nations cannot fully realize their potential.

Therefore, conditions are crucial. Nature seems to understand this, as it has created diverse environmental conditions not only for the birth and development of humans but also for the full expression of human intellect through competition. This dynamic is evident throughout human history.

We have examined how nature has created this diversity of environments previously. Now, let's take a look at the existing distribution of land and sea.

Two primary geographical conditions stand out: the natural environment of the GRV and the Mediterranean Sea. These environments respectively serve the birth of humanity and the birth of modern science and technology. It seems that other geographical conditions are determined based on these two. Does this notion strike you as odd?

According to the Earth's structure, the shape of its surface could have numerous possibilities. Still, according to what we have described above, for the development of land and sea distribution, no other option could support the evolutionary process from the earliest life to primates to humans and modern civilization, as determined above.

Complex processes determined our current land-sea distribution a few million years ago. This land and sea distribution has significantly influenced human development. Could this distribution be altered? And if so, what ramifications might ensue? Consider the following scenarios:

- If the African and Indian Ocean plates had not drifted to the north and collided with Eurasia.
- If the northern section of the oceanic boundary between the Indian and African plates hadn't changed direction toward Africa.
- If the subduction (collision) boundary along the Mediterranean had completely fused.
- If the rock wall at Gibraltar had not widely cracked.
- If the Drake Passage had not formed.
- If the Pacific and Atlantic were switched, or if the Atlantic were as large as the Pacific.
- If the Antarctic continent had not formed, or if it emerged at other latitudes.
- If an ocean divided the Eurasian continent.
- If the Red Sea directly penetrated the Mediterranean (which would have been likely).
- If the American continent hadn't drifted to the other side of the world.
- If there were only one continent on Earth.
- And numerous other possible scenarios.

According to the chapters above, if any of these statements were true, humanity would never have reached this point, even if humans had emerged.

Can you deny the significance of the GRV during this process? Can you deny the role of the Suez land bridge? Can you deny the role of the Strait of Gibraltar? Can you deny the significance of the Mediterranean?

Can you deny the role of Eurasia? Can you deny the role of Antarctica? ... They are all necessary conditions for human development that no one can deny. We believe that everything happens with conditions. The Earth's development is a process of establishing various conditions in both time and space, ultimately paving the way for the emergence of humanity and human civilization.

If you disagree with the view above, you could devise another solution and then see whether it can meet all the conditions required for the birth of humanity, the birth and development of modern civilization, and the global climate suitable for the survival of humans and all living organisms.

The development of nature is complex, subtle, and precise, and it seems not arbitrary.

WHY THE NATURAL WORLD SUPPORTS HUMANITY

The Mysterious Relationship between Nature and Human Nature

In essence, the natural world supports humanity due to inherent human nature. The environment serves as the primary catalyst for human behavior. This chapter summarizes the preceding chapters, unveiling the mysterious bond between humans and nature, thereby revealing the essence of nature itself.

The Mysterious Relationship between the Natural World and Human Nature

The emergence of human intelligence may be one of nature's most profound developments. Yet, it seems improbable that the universe, described by Einstein as possessing "the highest wisdom," would bestow human intelligence without a purpose. Humans may have their own ultimate natural mission, thus prompting the natural next step, the optimal utilization of human intelligence. However, this step poses a fundamental problem: navigating the inevitable complexities of human nature.

Why?

Human nature constitutes an essential survival and developmental trait, reflecting the commonalities across all life forms to a certain extent. Deeply ingrained and unchangeable within the human mind, human nature is the essence of humanity. Without it, the human spirit would regress to mere instinct, stifling the full development of human intelligence and robbing the human world of vitality, rendering human existence meaningless.

Human nature appears to be a powerful tool created by the natural world, driving competition and the struggle for survival while also fostering wisdom. The distribution of land and sea, alongside various geographical conditions, seems to have been arranged according to this logic, propelling humanity towards enlightenment and innovation.

However, given the complexities of human nature, waking from millennia of slumber is anything but straightforward! This complexity perhaps shows why nature took hundreds of millions of years to create the existing distribution of land and sea on the Earth's surface—the two natural philosophy principles.

Now, let's review the process of human progress induced by nature, showing the logic of natural development, or the relationship between human nature and the natural world.

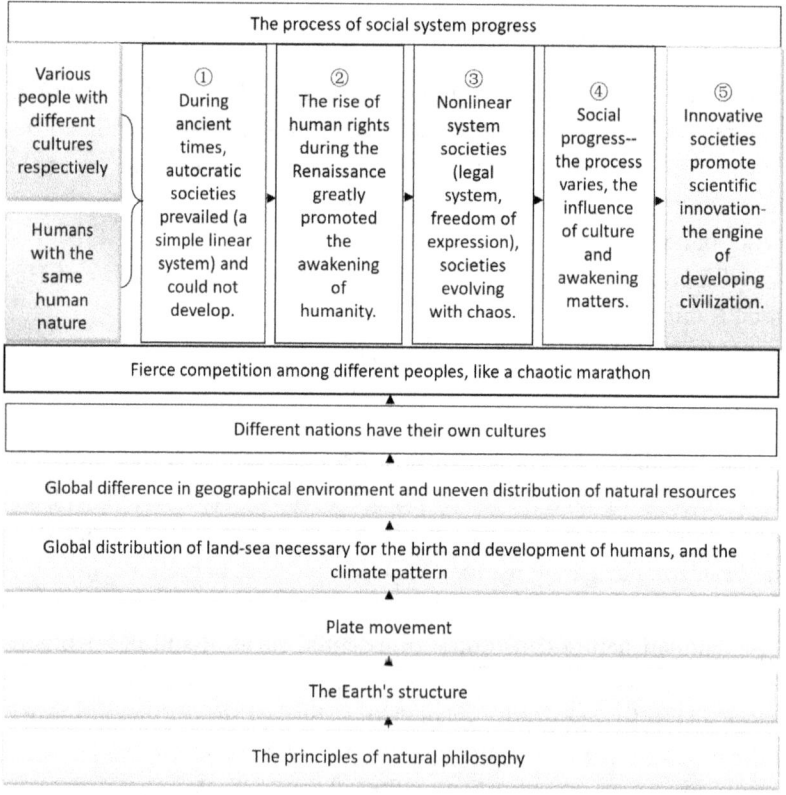

THE RATIONALITY OF THE EARTH SYSTEM

Please don't take the diagram above for granted. Humans could not have come this far without the set of natural developments at the bottom of the diagram.

It took the Earth 4.6 billion years of complex development to achieve its current environment, which are the natural conditions required for human emergence and development. History has proved that only in this way could the natural world use human nature to develop human intelligence fully.

The value of human existence lies in scientific innovation. However, on the road to an innovative society, different cultures lead to various processes of social progress, so nations are in different historical stages, from stage (1) to stage (5), much like a marathon.

Please note: what stage of social progress are you currently in? Without social progress, there can be no innovative society, and innovation is the value of human survival.

The following natural philosophy principles govern the diagram above:

- Everything happens conditionally.
- The consistency of function and structure.
- The complexity of human nature.
- Human intelligence and its full use are two different things.

This diagram shows that the Earth's environment meets the following conditions:

- The natural conditions required for establishing the global climate patterns to support human beings and all life, including the area of land needed by the temperate zone, tropical and cold zones, etc.
- The natural conditions required for the birth and development of humans, including the GRV, the Mediterranean, the global distribution of land and sea, various landforms, various environments with different climatic characteristics, the uneven distribution of natural resources, etc.

Even with intelligence, humanity could not have come this far without these environmental conditions. Carefully review the diagram above. Don't you think it shows the mysterious relationship between the natural world and human nature?

The Natural World and Human Nature: Another Significance

The diagram above illustrates the relationship between human rights and chaos, characteristic of nonlinear systems. Chaos, inherent in human nature's blend of reason and greed, tends to precede the formation of a new social order, which may be social progress. The law checks the innate greed in human nature, yet effectively implementing it is challenging, although democracy aids in its enforcement. However, as a complex nonlinear system, the legal system is an important regulatory factor for a democratic society, making it move continuously to a higher stage, although it is a slow process.

In this context, chaos refers to a transitional phase under certain conditions, representing the development of a nonlinear system from disorder to order, rather than a constant state. In contrast, societies governed by a simple linear social system, with the ruler as the lawmaker, exhibit less chaos but lack the capacity for progression and innovative social functions, a pattern evident throughout history.

Considering the characteristics of human nature, it becomes apparent that there's no superior solution to harness human intelligence fully. The logical distribution of land and sea, particularly with the Mediterranean, emerges as the optimal arrangement by nature. If you review the chapters above, you will have a deeper understanding of the relationship between the nature of humans and the environment.

For example, the Earth's surface comprises both ocean and land, leading to the development of distinct marine and continental cultures in human history. Each culture formed against its own natural background. Interestingly, the Mediterranean Sea roughly marks the dividing line between these two cultures. Historically, in the vast Eurasian continent, the farther east you go from the Mediterranean, the less influence the Renaissance had. In the eastern part of Eurasia, due to geographical barriers and significant historical inertia, the influence of the Renaissance was almost nonexistent.

Everything happens under specific conditions. The conditions necessary for the origin of early civilizations differ significantly from those required for modern civilizations. However, both types of civilizations inevitably emerged on the same continent (see Chapter II.3), Eurasia, highlighting its significance. The coincidence of these different conditions on a single continent inevitably led to imbalances in the progression of civilizations. Addressing these imbalances and the resulting challenges requires the lengthy and often painful process of human awakening and advancements in science and technology.

Reflecting on the development of civilization, it becomes evident that every part of Eurasia, from west to east, holds significance; none can be eliminated. The natural arrangement is inherently rational, guided by the principles of natural philosophy. Eurasia's role cannot be replaced by any other solution. Don't you agree?

Overview

Like a mother understands her child, nature appears to understand humanity and human nature. Nature has not only given rise to the human species but also created a series of natural conditions tailored to the characteristics of human nature to significantly promote the progress and development of humans, as shown in the diagram above. This seems to be the logical progression of our planet's development.

We believe that the universe, with its "highest wisdom," has imbued humanity with a purpose, hinting at an ultimate natural mission for humans in the distant future—a topic we shall explore later in this book. In this sense, an innovative society progresses and aligns with the developmental logic of nature. Thus, the more progressive a society is, the more it innovates scientifically and technologically, becoming impactful globally.

Historically, scientific innovation has been pivotal in propelling human development. However, nurturing a social environment conducive to innovation has been complex. For millennia, populations subdued by absolute power struggled to cultivate fertile ground for innovation. Only in recent centuries has modern humanism fostered the fertile social soil necessary for significant scientific advancements that shape the world and humanity's future. Nonetheless, this historical process has been silently supported by nature, as shown in the diagram above. It represents a merging of human nature and natural conditions, compelling human beings by their very nature to step onto the road of innovation.

Finally, let's take one more look at this mysterious relationship. Under certain astronomical conditions, nature first built a planet with a unique structure. Through millions of years of plate movement, the current distribution of land and sea formed, and humans emerged and developed. The natural world supported human wisdom and the rapid development of science and technology throughout this marathon. Note that hundreds of millions of years of plate movement also facilitated the evolution of life, based on the two principles of natural philosophy, underscoring that natural development follows a directed rather than arbitrary path.

The mysterious and complex story of the natural world and human nature continues to unfold, embodying the truth that "God does not play dice with the universe" and revealing "the Reason that manifests itself in nature." Expanding upon the principles of natural philosophy, the transition from developing human intelligence to harnessing its full potential is inevitably complex!

Chapter III

The Mysterious Periodic Table

The periodic table is one of nature's remarkable creations. The chemical elements of the table, with their rational abundance, are undoubtedly the material basis of the existence of the Earth System, human beings, and civilization. Understanding it scientifically and philosophically is of major significance. In this chapter, we reveal the characteristics of the periodic table that scientists and philosophers haven't considered but are necessary for the formation of a civilized planet.

The periodic table embodies the principle of natural philosophy, especially "the highest wisdom and the most beauty" or "the profoundest reason and the most radiant beauty" that humans may never fully understand.

THE MAGICAL ORDER OF ELEMENTS IN THE PERIODIC TABLE

Each element in the periodic table has a unique electron arrangement (Figure 3-1). This arrangement is precisely regulated, and modern science has begun to understand these laws. Due to this precise electron arrangement, each element exhibits unique functional characteristics. It is striking that the properties and functions of the elements, resulting from their specific electron arrangements, create the vibrant and diverse nature that sustains life, humanity, and civilization, and they seem to hold infinite potential. For example, on Earth, carbon exists primarily as a solid,

THE RATIONAL UNIVERSE EVOLVING FOR HUMANS

while oxygen and nitrogen are gases. The states of these elements—and the compounds they form—are essential to the development of life and civilization. Gaseous carbon dioxide and liquid water, for instance, are indispensable. These phenomena reflect a remarkable alignment with the periodic table's structure, as though tailored to Earth's unique conditions.

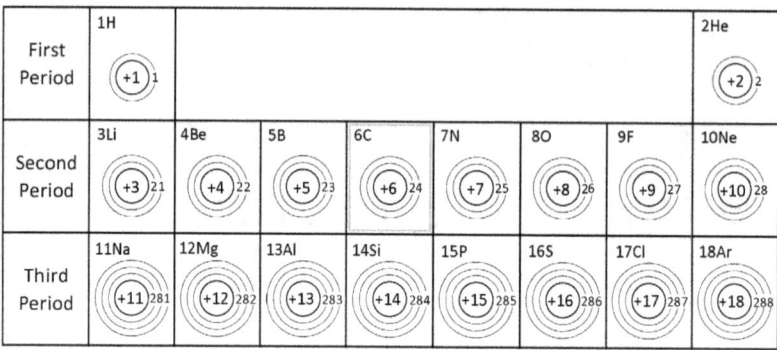

Figure 3-1: Electron arrangement of the first eighteen elements
(Source: The author's book in Chinese, published 2016).

Hydrogen, the simplest element with atomic number 1, is followed by carbon, nitrogen, and oxygen, listed as numbers 6, 7, and 8, respectively. Pay close attention: carbon is fundamental to forming life (macromolecules) and combustion. Combined with hydrogen, carbon forms methane (CH_4), the most important fuel for modern civilization. Interestingly, carbon dioxide, the oxide of carbon, is a gas (!) rather than a solid (see Understanding the Element Carbon, below). Nitrogen, an inactive gas essential for life, is the ideal major component of air, making it safe for combustion.

Oxygen, necessary for breathing, is an excellent partner to almost every element, enhancing their functionality. For example, oxygen combines with hydrogen to form water, which has numerous significant properties supporting the Earth System; with carbon and hydrogen to enable photosynthesis; with carbon to form carbon dioxide in the air through combustion; to form the ozone layer that absorbs ultraviolet rays in the atmosphere; and with silicon, aluminum, and magnesium to form rocks and soil. Also, oxygen is slightly soluble in water, just enough to support aquatic life. If oxygen were not soluble, life in water would be impossible. Conversely, if oxygen were more soluble in water, the atmospheric oxygen content could not maintain the necessary 21 percent due to the extensive surface of the oceans.

It seems Nature has meticulously arranged every detail to form the Earth System. Don't you think so? Why do elements 6, 7, and 8 possess such critical properties? Notably, oxygen's subtle and beneficial properties precisely maintain an environment suitable for life on land and in water. Isn't it amazing?

Why are the outermost electrons in an atom's shell eight in saturation rather than another number? This question, along with many others about atomic structure and the periodic table, begs for answers. Why? Why? Why? While these questions may not have immediate answers, reflecting on them and studying Figure 3-1 again might make you realize that the elements with "so many important properties" and the periodic table as a whole are not coincidental but indicative of a deliberate superdesign. Don't you think so?

The periodic table also illustrates the logic behind the existence of matter in atomic form, highlighting its unlimited potential. It seems unlikely that any other combination of elements could have formed the natural world and humans with civilization. Each chemical element's unique functions and properties suggest a purposeful design defined by its specific electron arrangement. The physical constants and atomic structures, including quarks, appear to exist to serve this extraordinary table.

UNDERSTANDING OF THE PERIODIC TABLE FROM NATURAL PHILOSOPHY

The above provides a physical and chemical perspective on the periodic table and its electron arrangements. Now, let us explore a philosophical understanding of the periodic table and the insights it offers.

Just as everything in the natural world has its own natural mission, what is the underlying theme of the periodic table? What is humanity's ultimate natural mission? The answers may be hidden within the periodic table itself.

The history of science tells us every chemical element contributes to the development of civilization, a shared mission between all elements in the table. On the other hand, each element has its own major role in the natural world, which can be divided into three primary roles (see the table below).

The first role is to facilitate the formation of life. The elements primarily involved in this process are carbon, hydrogen, oxygen, nitrogen, sulfur, and phosphorus.

The second role primarily serves the Earth's formation, fulfilled by silicon, aluminum, magnesium, nickel, iron, etc., even including radioactive elements (for energy inside the Earth).

The third role is to serve the development of civilization only. All the elements in this group have nothing to do with the formation of life or the formation of the Earth, although they are stored in mineral deposits. They seem to serve the development of civilization throughout history with different missions at different times. Gold, silver, lead, and hydrargyrum, for example, are not necessary for the first or second roles, but they are useful in the development of early civilization.

Similarly, many chemical elements are not necessary for the first two roles but are useful for the development of modern civilization, such as titanium (for space technology, etc.), niobium (high melting-point, superconductor), rhenium (high melting-point, high resistance), tantalum (noncorrosive), zirconium (high melting-point, noncorrosive), hafnium (high melting-point, noncorrosive), germanium (semi-conductor), tellurium (cadmium telluride in solar energy), etc. Combined with other elements, they can make various alloys with excellent properties beneficial in space, nuclear, electronics, and other industries. We could say that without these rare-earth metals, there would be no modern civilization. However, they are rare in the universe and difficult to refine, so their utilization in ancient times was impossible.

Saying that the periodic table, in a sense, represents the development of civilization may not be nonsense but an indisputable fact. Earth is extremely fortunate; it not only contains every natural chemical element in appropriate abundance but also harnesses their full potential in both the inanimate and animate worlds. Consequently, our planet, humans, and civilization represent the essence of the periodic table, which could be considered "cosmic information." In other words, the table signifies the meaning, implication, intention, or suggestion of the awe-inspiring universe.

With its unique arrangement of electrons, each element combines with others to form many molecules serving those three roles. From atoms to molecules to the three roles seems a magical but ideal process.

As a tool to clearly understand the periodic table, the following chart outlines the natural roles of each element with abbreviations:

L = Life
E = Earth
C = Civilization
Cm = Modern civilization
Ci = Inceptive civilization

In the beginning, the earliest life forms (prokaryotic cells) are believed to have been composed of only a few elements necessary for forming proteins and nucleic acids. As life evolved from simple to complex organisms with more intricate physiological structures, additional elements, including trace elements, were absorbed into their bodies. Civilization includes modern and inceptive civilization, but Cm indicates elements unknown before modern times. The natural mission refers to the role of each element in the natural world and the development of civilization.

Atomic number	Chemical symbol	Major natural missions	Other natural missions
1	H-Hydrogen	L, E, C	The simplest element
2	He-Helium	C	
3	Li-Lithium	Cm	L; useful in modern civilization
4	Be-Beryllium	Cm	Useful in modern civilization
5	B-Boron	C	L (maybe)
6	C-Carbon	L, C	Ci, Cm; carbon is the basis of the table
7	N-Nitrogen	L	C
8	O-Oxygen	L, E, C	Oxygen is an excellent helper to almost every element, assisting them in their ability to play more roles
9	F-Fluorine	C	
10	Ne-Neon	C	
11	Na-Sodium	C	L, E
12	Mg-Magnesium	E	L, C
13	Al-Aluminum	E	C
14	Si-Silicon	E	L, Ci; silicon is the main element in rock and soil used to make stoneware and pottery in the old and new Stone Ages; very useful in modern civilization
15	P-Phosphorus	L	C
16	S-Sulfur	L	C

Atomic number	Chemical symbol	Major natural missions	Other natural missions
17	Cl-Chlorine	C	L
18	Ar-Argon	C	
19	K-Potassium	C	L, E
20	Ca-Calcium	C	L, E
21	Sc-Scandium	Cm	Useful in modern civilization
22	Ti-Titanium	Cm	Useful in modern civilization; an ideal material
23	V-Vanadium	Cm	L; useful in modern civilization
24	Cr-Chromium	C	L; useful in modern civilization
25	Mn-Manganese	C	L
26	Fe-Iron	E	L, Ci; iron wares were made over two thousand years ago
27	Co-Cobalt	C	L; useful in modern civilization.
28	Ni-Nickel	E	L, C
29	Cu-Copper	Ci	L; bronze wares were the first metal wares made in ancient times
30	Zn-Zinc	Ci	L; zinc is one of the metals used in ancient times
31	Ga-Gallium	Cm	Useful in modern civilization
32	Ge-Germanium	Cm	Useful in modern civilization
33	As-Arsenic	C	L (for some species of animals)
34	Se-Selenium	C	L
35	Br-Bromine	C	
36	Kr-Krypton	Cm	Useful in modern civilization
37	Rb-Rubidium	Cm	Useful in modern civilization
38	Sr-Strontium	Cm	L; useful in modern civilization
39	Y-Yttrium	Cm	Useful in modern civilization
40	Zr-Zirconium	Cm	Useful in modern civilization
41	Nb-Niobium	Cm	Useful in modern civilization
42	Mo-Molybdenum	Cm	L; useful in modern civilization
43	Tc-Technetium	Cm	Useful in modern civilization
44	Ru-Ruthenium	Cm	Noble metal, extremely rare
45	Rh-Rhodium	Cm	Noble metal, extremely rare
46	Pd-Palladium	Cm	Noble metal, rare
47	Ag-Silver	C	Noble metal

THE MYSTERIOUS PERIODIC TABLE

Atomic number	Chemical symbol	Major natural missions	Other natural missions
48	Cd-Cadmium	Cm	Useful in modern civilization
49	In-Indium	Cm	Useful in modern civilization
50	Sn-Tin	Ci	L; tin is a chemical component of bronze wares
51	Sb-Antimony	C	Used in ancient times
52	Te-Tellurium	Cm	Useful in modern civilization
53	I-Iodine	C	L
54	Xe-Xenon	C	
55	Cs-Cesium	Cm	Useful in modern civilization
56	Ba-Barium	C	
57	La-Lanthanum	Cm	Useful in modern civilization
58	Ce-Cerium	Cm	Useful in modern civilization
59	Pr-Praseodymium	Cm	Useful in modern civilization
60	Nd-Neodymium	Cm	Useful in modern civilization
61	Pm-Promethium	Cm	Useful in modern civilization
62	Sm-Samarium	Cm	Useful in modern civilization
63	Eu-Europium	Cm	Useful in modern civilization
64	Gd-Gadolinium	Cm	Useful in modern civilization
65	Tb-Terbium	Cm	Useful in modern civilization
66	Dy-Dysprosium	Cm	Useful in modern civilization
67	Ho-Holmium	Cm	Useful in modern civilization
68	Er-Erbium	Cm	Useful in modern civilization
69	Tm-Thulium	Cm	Useful in modern civilization
70	Yb-Ytterbium	Cm	Useful in modern civilization
71	Lu-Lutetium	Cm	Useful in modern civilization
72	Hf-Hafnium	Cm	Useful in modern civilization
73	Ta-Tantalum	Cm	Useful in modern civilization
74	Wu-Tungsten	C	
75	Re-Rhenium	Cm	Useful in modern civilization
76	Os-Osmium	Cm	Noble metal, extremely rare
77	Ir-Iridium	Cm	Noble metal, extremely rare
78	Pt-Platinum	C	Noble metal, rare
79	Au-Gold	C	Noble metal, rare
80	Hg-Mercury	C	Mercury is one of the metals used in ancient times

Atomic number	Chemical symbol	Major natural missions	Other natural missions
81	Tl-Thallium	Cm	Useful in modern civilization
82	Pb-Lead	Ci	Lead is a chemical component of bronze wares
83	Bi-Bismuth	Cm	Useful in modern civilization
84	Po-Polonium	Cm	Radioactivity
85	At-Astatine	Cm	Radioactivity
86	Rn-Radon	Cm	Radioactivity
87	Fr-Francium	Cm	Radioactivity
88	Ra-Radium	Cm	Radioactivity
89–103	Ac-Lr: Ar series	Cm	Radioactivity

With an understanding of the table above, let's examine the relationship between the periodic table and the natural world, including humans and civilization.

Approximately 28 elements are essential for forming the human body (though this number can be controversial). About eight elements—oxygen, silicon, aluminum, iron, calcium, sodium, potassium, and magnesium—constitute 99 percent of the Earth's crust by weight and are crucial for forming our solid planet. While we temporarily set aside radioactive elements (which provide the Earth's internal power), all of these elements, except aluminum, are also necessary for life.

Ninety-two natural chemical elements (up to uranium, U) are stored as minerals in the shallow layer of the Earth's crust. Approximately 29 of these elements are essential for forming life and the Earth. The remaining 63 (about 68 percent) are not necessary for forming life or the Earth but are crucial for supporting civilization.

Every element, from hydrogen (atomic number 1) to uranium (atomic number 92), has untapped potential values to explore for civilization. For instance, graphene, derived from graphite, has proven highly useful in modern times. Future materials could be developed from various elements in the periodic table. Elements from hydrogen (H) to iron (Fe, atomic number 26) are believed to have been formed through nuclear fusion in stars, a continuous and stochastic natural process that is relatively easy to understand. However, the formation of elements beyond atomic number 26 is much more complex, and these elements are rarer in the universe. Earth is fortunate to possess all the elements of the periodic table in suitable abundance.

In any case, it's crucial to recognize that every element plays a role in the advancement of civilization. Approximately 60 elements, including aluminum, trace elements, and extremely rare elements, have been found in the biosphere. This leaves about 32 elements that are unique to civilization. If we exclude nine radioactive elements, there are still 23 elements exclusive to civilization. The fact that all elements in the table share a joint mission related to civilization may suggest that "the subject of the table" is civilization itself, with human beings as its overseers.

Regarding molecules apart from oxygen and nitrogen in the air, the natural world we see daily is composed of various complex molecules, such as a drop of water, a mass of soil, a grain of sand, a rock, plants and animals, etc. But there are only thirty or so different elements in those natural objects. Thus, you will find that elements are merely bricks. Every element tends to combine with others to form a molecule. The molecule is the key to making up the natural world. That seems to be a clever "design," how nature uses limited elements to create the infinitely colorful natural world! Thirty or so kinds of elements have combined to make a wide variety of materials. Consequently, the significance of atomic structure and the periodic table is to provide vast possibilities for the world, including the endless development of science. Atomic structure and the table will bring forth humans' infinite development of science and an incomparably glorious future. This is what Einstein called "the highest wisdom" of nature.

Of course, we have other instances to show the relationship between the periodic table and the natural world. Referring back to the function of the Earth System, you will find how important, necessary, and appropriate the relevant elements are in constructing a civilization-brewing machine!

So, we indeed cannot help being amazed at the various natural arrangements from the elementary property to the Earth System for fulfilling their respective natural missions!

"The subject of the table," as well as those natural arrangements, as stated above, show us that "the answer might be hidden in the periodic table." This may represent a kind of suggestion from the universe, or cosmic suggestion, that seems to be an intention of the periodic table leading to the eventual appearance of humans and civilization on Earth. Don't you think so?

Finally, let's pretend you have the power to create a natural world based on the elements in the table. What do you think about that? After a lengthy study, perhaps you might think the current world is the best solution, and no other could substitute.

The universe first created chemical elements, then molecules, followed by the formation of the Earth, the emergence of life, and eventually the rise of humans and civilization. In this sequence, Earth, humanity, and civilization have effectively embodied the "intention" of the periodic table, utilizing each element to its fullest potential and in its appropriate context. This accomplishment is truly remarkable!

Each element within the Earth System contributes to evolution and civilization, marking its profound significance in the natural world.

THE MYSTERIOUS ABUNDANCE OF ELEMENTS ON THE TABLE

As a civilization planet, Earth's abundance of elements must meet specific requirements. From four perspectives, we will investigate how nature ensures the sufficient presence of elements, again illustrating the existence of the Reason in nature.

The Difference in Element Abundance between the Sun and the Earth

Our planet is a part of the solar system, so the abundance of elements comprising the Earth and the Sun should have been similar. The Sun's mass accounts for 99.86 percent of the solar system's total mass, and the Sun's element abundance could be regarded as that of the primordial nebula 5 billion years ago.

The distribution of elements within our planet is subject to gravitational differentiation (see Chapter I.1, The Timely Differentiation of the Earth and The Timely Appearance of the Oceans). As expected, the heavy elements (Fe, Ni) sank to the globe's center, and the light elements (Si, Al, Mg) rose into its shallow layer. However, some variation in distribution is not consistent with gravitational differentiation.

Let's now compare the element abundance of the solar photosphere (regarded as the primordial nebula) and that of the Earth (discounting hydrogen and helium). The table below shows the sequence of the element abundance of the solar photosphere on the left, from greatest to least, compared with that of the Earth on the right:

THE MYSTERIOUS PERIODIC TABLE

Solar Photosphere			Earth		
Abundance	Element	Atomic Number	Abundance	Element	Atomic Number
1	Oxygen	8	1	Oxygen	8
2	Carbon	6	2	Silicon	14
3	Nitrogen	7	3	Aluminum	13
4	Silicon	14	4	Iron	26
5	Magnesium	12	5	Calcium	20
6	Neon	10	6	Sodium	11
7	Iron	26	7	Potassium	19
8	Sulfur	16	8	Magnesium	12
9	Argon	18	9	Titanium	22
10	Aluminum	13			

The comparative abundance sequences suggest that while the elements in the photosphere are sufficient for the Earth's formation, those in the Earth's shallow crust reflect a distinct natural arrangement compared to the primordial nebula. The distribution on the Earth, particularly of silicon, aluminum, and iron, appears to represent a crucial "re-arrangement." This new natural order is essential for sustaining life and fostering modern civilization on our planet. We owe gratitude to this re-arrangement; without it, our world would not have thrived as it does.

The elements found in the Earth's outer layer primarily compose the vast quantities of soil worldwide. Silicon ranks second in abundance in the crust at 27.88 percent, whereas in the photosphere, it ranks fourth, excluding hydrogen and helium. Aluminum ranks third in the crust but tenth in the photosphere. Iron is seventh in the photosphere and fourth in the crust. Oxygen dominates the crust at 46.95 percent, ranking first. When combined with silicon (27.88 percent) and aluminum (8.13 percent), the total reaches 82.96 percent. These variations in silicon, aluminum, and iron abundances extend throughout the crust, playing pivotal roles in the mantle, lithosphere, and core.

But why didn't all heavy elements (much heavier than iron!) sink into the core during the gravitational differentiation? Why did they remain in the shallow layer of the crust in an appropriate abundance?

Another significant fact will also help you further understand the abundance of silicon. As determined, heavy elements like iron and nickel sank towards the Earth's center during gravitational differentiation, while lighter elements such as silicon, aluminum, and magnesium rose to the shallow layers. However, what's intriguing is the significant disparity in abundance among these elements: silicon (density 2.3, abundance 27.88), aluminum (density 2.7, abundance 8.13), magnesium (density 1.74, abundance 2.06), calcium (density 1.55, abundance 3.65), sodium (density 0.97, abundance 2.78), and potassium (density 0.86, abundance 2.58). Despite aluminum having a density slightly greater than silicon, silicon's abundance in the crust is several times, even ten times (!) greater than other elements. This variation in the abundance of silicon and aluminum within the crust seems deliberate.

What Does the Abundance of Elements Suggest?

Regarding the periodic table, chemical element properties and element abundance are two deep concepts representing the mystery of Nature.

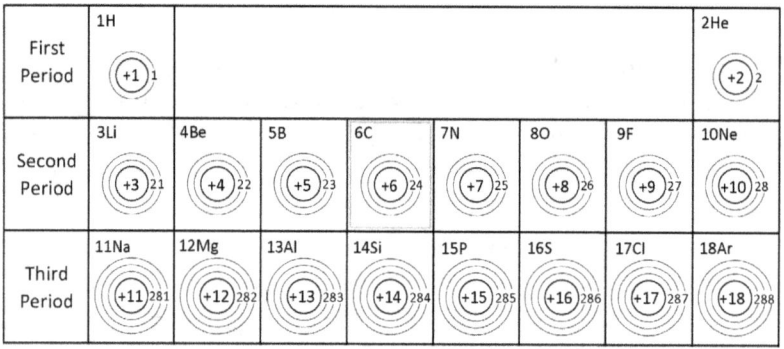

Figure 3-1: Electron arrangement of the first eighteen elements
(Source: The author's book in Chinese, published 2016).

The periodic table is repeated here for your convenience. Pay attention to the feature of the electronic arrangement of each element. Look at the first period, the second period, and the third period in the table.

You will find that the elements (H, C, O, N) in the first and the second period are the major components of life-forming substances, and the elements (Mg, Al, Si) in the third period are the major components forming the mantle and the lithosphere. The properties of these elements are irreplaceable, and the function of life substances (life-forming substances) and the mantle and the lithosphere are also rigid.

The abundance of elements in the photosphere seems to be roughly proportional to each quantity demanded for the formation of life substances and the Earth's mantle and lithosphere. In other words, every element, property, quantity required, and abundance in the Earth seems to be roughly proportionally defined in the photosphere. Except for those seven elements, all other elements in the table seem to be costars, although they are essential for forming life substances and the development of civilization. Therefore, their abundance is much less than those seven elements.

The properties and abundance of elements are critical for natural processes, from the appearance of chemical elements in the universe to the Earth's formation. The properties and abundance of elements are profound in understanding the cosmic logic.

Understandably, the quantity of silicon and aluminum needed for the formation of the crust far exceeds that of the element of carbon, although that element is the most critical one in the table. After all, the quantity of elements needed for life is much less than those needed for the massive Earth.

Don't you think that the abundance and properties of elements on Earth are designed to meet Earth's and life's needs?

The Exception of the Abundance

In general, the abundance of elements in the solar system decreases as the atomic number increases (but the abundance of even numbers is greater than that of odd numbers, as shown below). While this trend can be theoretically explained, elements like silicon and especially iron are evident exceptions to this rule.

The element abundance of silicon and iron is much greater than those adjacent to them in the periodic table. The following is the abundance of the elements in the solar photosphere:

12 Mg	13 Al	14 Si	15 P	16 S
890,000	74,000	1,000,000	7100	350,000

24 Cr	25 Mn	26 Fe	27 Co	28 Ni
11,000	5900	710,000	1800	43,000

The abundance of silicon, reaching one million (10^6 as a cardinal number), leads us to logically imagine that when elements were produced in a particular star, nature made an exception by producing unexpectedly more iron and silicon for the construction of the civilization planet Earth. Iron is a magnetic element and the major component of forming the Earth's core. Iron has also been useful in the development of civilization from the early days to now, its abundance ranking fourth in the planet. Silicon is the major element in the lithosphere and the mantle, ranking second. So, iron and silicon are needed in large quantities in the structure of the massive Earth.

It appears that nature (through stars) produces elements in quantities sufficient for constructing the Earth. However, this is a complex, subtle, and stochastic natural process without any deliberate "plan"!

Although the abundance of every element could be theoretically explained in part, don't you think that this phenomenon, the substantial existence of the abundance of every element, is thought-provoking? If the abundance of iron or silicon were less, what would happen in the natural world?

Don't you think the above is "the Reason that manifests itself in nature"?

Rational Distribution of Earth's Elemental Abundance

We can use the following chart to describe the relationship between element abundance and our planet logically:

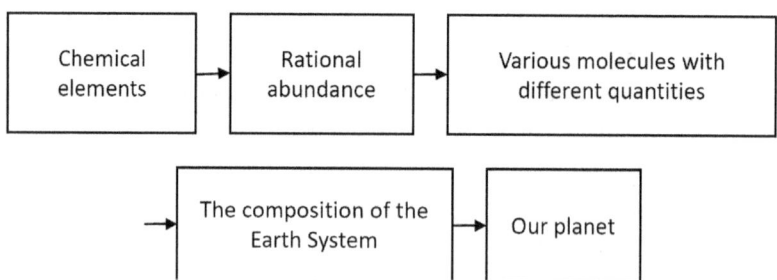

Below, we will demonstrate the rationality of the Earth's elemental abundance aligns with the needs of the various components in the Earth System.

The Distribution of Element Abundance in Earth's Layers and Their Properties

We do not know what exactly causes the distribution of element abundance in the Earth, but obviously, that distribution is favorable to the evolution of life.

To begin with, an interesting fact worth attention is that iron (Fe) is not the heaviest element in the table, yet the Earth's core is mainly composed of it, the next being nickel (Ni). Although we don't know precisely how the geomagnetic field is formed, it seems certain that its formation is closely related to iron because iron is the only magnetic element with great abundance in the table. Cobalt (Co) and nickel (Ni) are also magnetic, but their abundance is far less than iron. Further, scientists believe that nickel's magnetism is stronger than iron's, so we can logically imagine that the core as a mixture of iron and nickel is much higher in magnetism than a theoretical core made up entirely of iron. Isn't that interesting? And if the core was made entirely of iron, or if the core were smaller, could life exist on this planet?

The other layers of the Earth are the lithosphere and the mantle. Almost all elements that make up these layers are light elements. They are oxygen (O), silicon (Si), magnesium (Mg), aluminum (Al), calcium (Ca), sodium (Na), potassium (K), and iron (Fe), accounting for about 99 percent of the crust (in weight fraction). The most abundant elements in the Earth's crust include oxygen, silicon, aluminum, and magnesium. These elements collectively form what is known as the "sialsphere" (from "sial," referring to silicon and aluminum) for the upper layer, and the "simasphere" (from "sima," referring to silicon and magnesium) for the lower layer. They constitute the foundation of various types of rocks found in nature.

Oxygen, with an abundance of 46.95 percent, acts as a crucial binding agent in the Earth's crust. It combines extensively with silicon (27.88 percent), aluminum (8.13 percent), and magnesium (2.06 percent) to form the structural framework of the lithosphere, which constitutes rocks. Any other chemical elements could not possibly replace the roles of these four elements, but the elements of calcium (Ca), sodium (Na), and potassium (K) could naturally be replaced by each other. Iron, strictly speaking, is not an essential element for composing rock. So, we could imagine that only oxygen, hydrogen, silicon, aluminum, and magnesium (total abundance: 85 percent) make up a light but robust framework supporting the vast lithosphere. The other elements are merely fill materials, although they are useful to humans. Without the elements of Si, Al, and Mg, could the light lithosphere float on the mantle? Impossible!

The theory of plate tectonics explains that the lithosphere, which includes the crust, floats on the hot mantle and moves unpredictably. This movement generates diverse geological processes, forming various geographic features such as oceans, mountains, continents, islands, peninsulas, and hills on the Earth's surface. These geographic conditions play a crucial role in the evolution of life, as demonstrated by the wide variety of species inhabiting mountains, bodies of water, plains, hills, and deserts.

However, a more intriguing aspect to consider is the nature of soft rock strata. The four elements mentioned above can chemically combine into various molecules (for example, SiO_2 constitutes 57.76 percent and Al_2O_3 15.55 percent of the crust), collectively accounting for over 70 percent of the Earth's crust by weight. These combinations significantly influence the physical characteristics of the lithosphere. Despite our perception of rock as typically hard and solid, have you ever encountered stone so pliable it can be folded (Figure 2–18)? Have you witnessed geological folding? Such phenomena suggest that rock strata, in a sense, exhibit "soft" qualities.

This inherent softness of rock strata is crucial as it allows for the formation of various topographical features (landforms), such as mountains and hills, through plate movements. If rock strata were entirely inflexible, they would fracture under immense geological pressure, incapable of forming any landforms. Hence, we appreciate the light and pliable qualities of rock strata.

One might wonder if rock strata composed of elements other than oxygen (O), silicon (Si), magnesium (Mg), and aluminum (Al) would retain these same qualities. Logically, this would be impossible, as chemical properties vary between elements due to different atomic structures. Each element's distinct properties determine the qualities of rock strata. Therefore, if composed differently, rock strata would possess different physical properties and lack the lightness and softness observed in natural rock strata. It's worth noting that these folds occur under high temperature and pressure conditions.

It's essential to differentiate between minerals and rocks. Minerals are naturally occurring, homogeneous solids with specific chemical compositions, whereas rocks are naturally occurring combinations (mixtures) of various minerals with a solid appearance. The lithosphere encompasses rocks. Specifically, the simasphere represents the basaltic layer, while the sialsphere denotes the granitic layer.

In conclusion, due to their unique properties, only oxygen, hydrogen, silicon, magnesium, and aluminum can chemically combine to form light and pliable rock strata.

All the points mentioned above demonstrate that the periodic table provides the necessary elements to support the geographic conditions required for the evolution of life. Next, we will examine three essential components for the survival of life: soil and sand, the atmosphere, and water, each of which depends on a particular abundance of elements.

Soil and Sand

We have previously examined soil from the point of view of life. Now, let's understand soil in terms of chemical elements.

The core material of soil is clay minerals, and its main elements are silicon and aluminum (from the sialsphere). In other words, soil cannot be formed without silicon and aluminum. Soil is a significant component of the natural environment required for life. Clay minerals mix with other materials, including water, sands, organic matter, and microorganisms, to form soil.

Soil possesses several properties essential for the existence of plants, the most important being water retention. Soil can maintain moisture much longer than the open air, with soil moisture containing various mineral elements in the form of ions, such as potassium ions (K+) and sodium ions (Na+), which plants' roots absorb. But why does soil have good water retention?

The water-retentive property of clay is much stronger than that of sandy soil. In fact, the finer the soil particles, the more water molecules the soil absorbs. Clay particles attract water molecules most effectively, forming an absorption water layer known as the bound water layer. Bound water is difficult to separate from the surface of clay grains due to its greater density compared to free water.

The next question is, how are the fine grains of clay minerals formed? This is attributed to the elements silicon (Si) and aluminum (Al). When chemically combined with other elements, such as oxygen and hydrogen, they form fine clay grains. The size and shape of these grains can only be determined with an electron microscope. Most clay minerals are scale-like. Without Si and Al, these microscopic grains could not be formed. They are small but immensely significant. Clay minerals could not exist without these fine particles, and the natural world would lose meaning.

Next, we ask how the water absorption layer is formed. Figure 3-2 illustrates this principle:

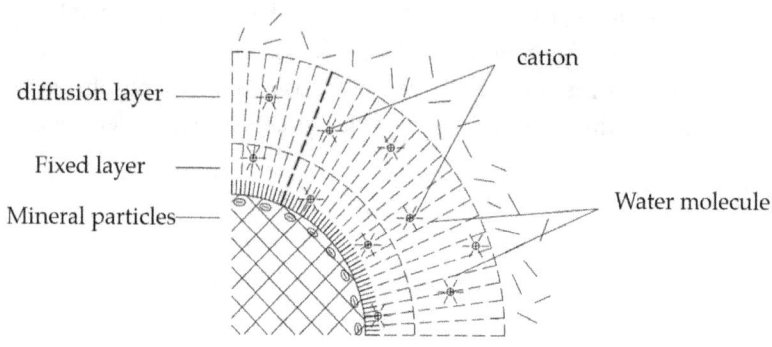

Figure 3-2: Water molecules around the clay grain
(Source: Mei, et al. *Buildings*, No. 10).

In summary, the formation of the water absorption layer is due to the electrical properties of water molecules and clay mineral particles. Typically, clay mineral particles have a negatively charged surface, causing polarized water molecules to align uniformly around them, forming the bound water layer. The finer the particle, the more bound water it can absorb. The bound water content in clay is significantly higher than in sandy soil (primarily composed of SiO_2) because the main chemical elements differ between the two soil types. Therefore, sandy soil cannot replace clay minerals. The role of silicon (Si) and aluminum (Al) in soil is irreplaceable.

The chart below shows the natural process of forming soil:

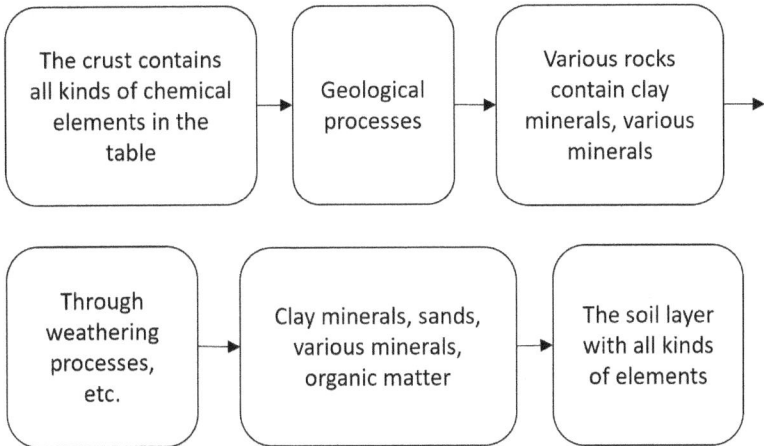

Most of the sand on Earth is composed of quartz sand, whose main chemical component is silicon dioxide (SiO_2). Quartz sand is crucial for the development of civilization; without it, concrete could not be formed. The absence of fine sand with various grain sizes, an essential component of soil, would impede plant growth, and our beautiful and valuable beaches would not exist. Silicon is as fundamental to rock as carbon is to our bodies. Silicon provides us with rock, sand, soil, glass, obsidian, and many compounds used in modern materials. Today, the global demand for sand is second only to fresh water, surpassing that of coal and oil.

So, could the human world exist without the vast amounts of soil and sand covering the Earth's surface?

The crust, particularly in the shallow layer, contains all the chemical elements in the periodic table with appropriate abundance in the form of minerals, which we can mine for many uses (see Chapter II.1, Factors Related to Civilization). Why do various heavy elements remain in the shallow layer? Why didn't they all sink into the Earth's center as the element of iron did in the process of gravitational differentiation?

Logically, we can imagine that all elements with their appropriate abundance in the shallow layer and the variation of silicon abundance in the crust are necessary for the development of civilization.

The Atmosphere

Our atmosphere is composed primarily of oxygen (O_2) and nitrogen (N_2), with smaller amounts of other gases.

Oxygen (O_2) makes up about 21 percent of the atmosphere and is a highly reactive chemical element. All other chemical elements can be combined directly or indirectly with it to form various oxides, playing a crucial role in the environment. In its gaseous state, oxygen is colorless, odorless, and non-toxic. With an atomic number of eight, oxygen possesses many significant properties essential for life and various environmental processes.

Oxygen in the atmosphere is vital for the respiration of living organisms. When combined with hydrogen, it forms water molecules, one of the most resourceful substances in the natural world (see Water, below). Oxygen also combines with carbon to form carbon dioxide (CO_2), which is critical for photosynthesis and the greenhouse effect. However, the concentration of carbon dioxide in the air must remain low (normally less than 0.03 percent). If it far exceeds this level, such as reaching 5 percent, both breathing and combustion would be severely impacted.

Oxygen is also fundamental in the formation of ozone. Under strong solar radiation in the upper atmosphere, oxygen molecules split into individual atoms, which then combine in groups of three to form ozone. Ozone effectively absorbs ultraviolet rays, protecting life on Earth. In this process, ozone absorbs and stores a significant amount of heat energy, a key factor in the formation of the troposphere.

Oxygen is slightly soluble in water, supporting aquatic life. If it were more easily dissolved, it might disappear from the air. Given its critical role in our environment, any significant change in the atmosphere's oxygen content would be catastrophic.

Carbon dioxide (CO_2) in the air will naturally dissolve in seawater, increasing seawater's acidity. Scientists believe today's growing seawater acidity will significantly influence the global environment and adversely affect marine organisms. As such, there is a delicate relationship between the carbon dioxide content in the air and the global environment (greenhouse effect), as well as the existence of marine organisms.

The diagram below shows the relationship between oxygen and its products, especially carbon dioxide and water (H_2O).

THE MYSTERIOUS PERIODIC TABLE

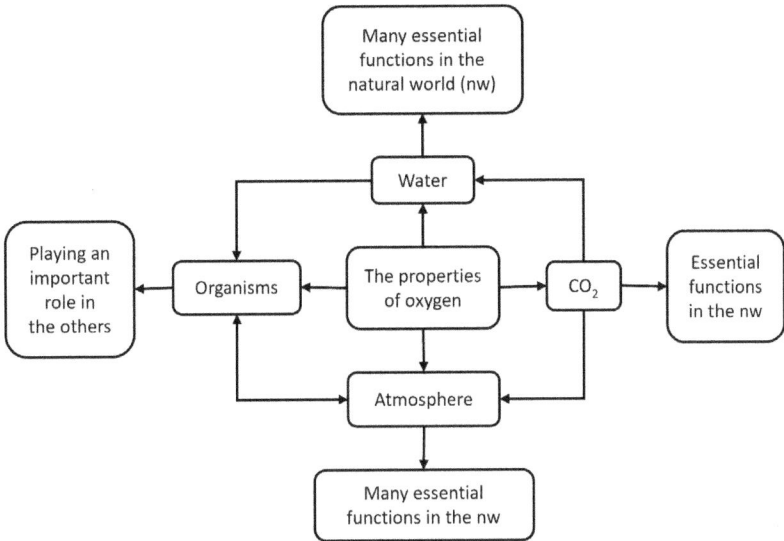

The diagram shows that oxygen is the main component that forms the environment life requires and enjoys. Oxygen is like a binding agent, forming different products of molecules, making them play their role and affect one another, in which oxygen and its products are inextricable, forming a complex and subtle environmental system. Oxygen has such an impressive, resourceful ability to make up the environment system. Isn't it amazing? Why does the atomic number eight of Oxygen have so many significant qualities?

Nitrogen (N) is a crucial element, playing pivotal roles in the formation of biological substances and environmental processes. Normally a colorless, odorless gas with stable chemical properties, nitrogen is often mistaken as inert despite its significance. It constitutes approximately 78 percent of the Earth's atmosphere, an ideal proportion that enhances the safety of combustion in the natural world. However, in the crust, its abundance is much lower than that of chlorine, a main element in seawater.

Regarding biological utilization, plants can only absorb nitrogen in the form of compounds, typically combined with hydrogen or oxygen. Processes like lightning and nitrogen-fixing bacteria convert atmospheric nitrogen into these absorbable compounds. The balance of nitrogen in the atmosphere is crucial; were it to rise to 81 percent, with oxygen correspondingly dropping to about 19 percent, human survival would be endangered due to suffocation.

Nitrogen can be likened to a guardian of the Earth's environment, ensuring the safety of natural settings where combustion occurs frequently. Its stable, seemingly inert nature paradoxically enables both safe combustion and its essential role in life processes. This dual capacity underscores nitrogen's extraordinary importance in maintaining the balance of our ecosystem.

Oxygen, nitrogen, carbon dioxide, and various trace gases are blended proportionally to form the atmosphere. They collectively serve numerous vital roles for all living organisms, highlighting the profound influence of the periodic table on life.

Water (H_2O)

Water is a fundamental component of life and the environment, crucial for regulating temperature and humidity to sustain life, particularly for humans. Life on any planet would be impossible without abundant water. Water possesses unique and invaluable qualities in the natural world. It is the only naturally occurring liquid, non-toxic, colorless, and odorless. With a high thermal capacity (specific heat capacity) and heat of vaporization, water efficiently regulates body temperature. Globally and locally, water regulates atmospheric temperatures and humidity through evaporation, especially from ocean surfaces, and precipitation.

The substantial difference between water's boiling and freezing points allows for its prevalence in liquid form, giving rise to rivers, lakes, seas, and more. At its greatest density at 39°F (4°C), water freezes at 32°F (0°C), expanding and becoming less dense, thereby forming insulating ice on water surfaces that protects aquatic life below. Water's poor thermal conductivity supports life underwater, in snow, and within ice formations.

Water is the only natural solvent capable of dissolving most inorganic salts that plants absorb. Its strong seepage force, aids weathering processes, soil formation, groundwater movement, and cellular activities. With strong surface tension, water forms droplets of various sizes, like raindrops, fog, and cloud droplets. Its ability to infiltrate and form capillary tubes is crucial for soil and organisms alike.

Water molecules have strong chemical bonds, making water the only fire-extinguishing material in the natural world, convenient and effective. Water possesses incompressibility, resisting compression or volume change under pressure. In the cold of winter, water in the cracks of rock surfaces may freeze and expand, causing the rock surfaces to fall and contributing to weathering processes. The water molecule is simple but useful, effectively supporting the workings of the natural world, both animate and inanimate, and exists almost everywhere because of its excellent properties. Nature neither requires nor possesses the ability to create an alternative natural material like water.

The presence or absence of water dictates a planet's destiny. The remarkable qualities of water, stemming from its basic yet intricate combination of oxygen and hydrogen, underscore its essential role. Any alteration to these properties would destabilize our world. This marvel is a testament to the remarkable properties afforded by the periodic table, where oxygen and hydrogen combine to form the extraordinary molecule of water with numerous incredible but absolutely necessary properties for nature, particularly for living things.

Now, let's shift our focus to another topic: seawater. Oceans constitute the predominant form of water on the Earth, comprising 96.53 percent of the global water volume. It's crucial to highlight the distinctive properties of seawater compared to freshwater found on land. As previously mentioned, seawater's composition predominantly includes chlorine and magnesium. Particularly noteworthy is how these elements exist in seawater—they naturally occur in ionic states.

Chlorine (Cl) is primarily transported by rivers to the oceans, where it ranks as the most abundant element in seawater. Despite its scarcity in the Earth's crust, chlorine's prevalence in seawater raises intriguing questions about its natural distribution. This phenomenon suggests that chemical properties play a role, yet the extent of this natural occurrence remains surprising. Interestingly, chlorine, known for its harmful effects in molecular form, exists in seawater as essential negative ions, crucial for life and civilization's development. No other element can fulfill this vital role.

Consider the implications: all forms of life, particularly humans and civilization, rely significantly on chlorine. Without this unique natural distribution, where chlorine's minimal presence in the crust transforms into abundant negative ions in seawater, our world would undoubtedly be vastly different, profoundly affecting the Earth's developmental processes (see Chapter I.2, Ancient Seawater).

So far, we have examined the main factors that make up the Earth System, and they are all based on specific amounts of element abundance.

It's easy to see how necessary and rational the abundance of the Earth's elements is!

Over 2000 years ago, Aristotle thought the natural world was only comprised of earth, air, fire, and water, which he called "elements." Although he was wrong, his idea, in a sense, indeed represents the facts. Our natural world is made up of water, soil, air, and life, interdependent in physics and chemistry, composing a dynamic system. Based on all the natural elements, particularly the elements from hydrogen to iron, this would be the only solution to the makeup of our natural world.

Finally, let us say that Earth is very lucky to have all the elements in the periodic table, including many of those listed after iron, especially the rare and valuable elements necessary for modern science development. We don't know if other planets have such a complete array of elements, but we can say that the probability is very low, even close to zero.

As a civilization planet, Earth has strict requirements for the abundance of chemical elements. It has carefully fulfilled these requirements. However, we emphasize not to take this fulfillment for granted; it was, in fact, a thought-provoking process.

During its formation, Earth selectively absorbed essential elements needed for a civilized planet and stored them in optimal quantities. For instance, silicon and aluminum are much more abundant here than in the primordial nebula. Moreover, Earth acquired elements that are exceedingly rare in the universe but crucial for the development of modern civilization. This process underscores the presence of "the Reason in nature," wouldn't you agree?

The Periodic Table Determines Earth Life is the Unique Form of Life in the Universe

Life substances, also known as life-forming substances, are exceedingly intricate and delicate. They consist of elements from the periodic table, where each element's structure and function are irreplaceable, governed by the principles of natural philosophy. Hence, there exists only one pathway from elements to life substances. Consequently, it is impossible for the same periodic table to give rise to two distinct life substances composed of different elements.

The Concept of Earth life

Building on the evolutionary processes on the Earth over the past billions of years, we could summarize the concept of Earth Life as a development process, as shown in the diagram below, not individual organisms.

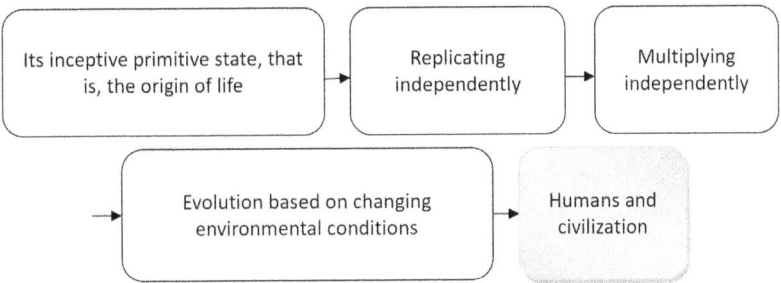

We may call the above "the whole development process of life." If life on a planet could not follow this process, it could not develop and continue to exist on that planet. The significance of the universe consists in the existence of the whole development process of life.

Replication is the basis of multiplication. Any form of life is necessarily the result of replication and multiplication, further resulting in evolution, an extremely complex, natural process, subject to variation and changing surroundings. What controls evolution is genetics, the nature of Earth Life, no matter what kind of living organisms that develop. The development of anything in the natural world must follow a course in time and space. We cannot conceive of other natural ways for life to develop.

The Energy of Earth life

Photosynthesis is the natural process of producing biological energy used to support the whole development process of life. It transforms solar energy (to be precise, photons of red light and blue-violet emission) into biological energy stored in cells as glucose, the product of photosynthesis. Powerful and continual, solar energy is Earth's only natural energy source. Photosynthesis is the one resource that always directly or indirectly supports the evolution of life.

Glucose is an organic compound with the simplified molecular formula $C_6H_{12}O_6$, fitting the general formula $C_m(H_2O)_n$. This structure resembles a compound of carbon and water, which is why glucose is called a carbohydrate. However, it's important to note that the hydrogen and oxygen in glucose are not present in the form of H_2O (water).

As an energy substance, glucose is an essential component of life-forming substances, no matter the kind of living organisms (an important natural arrangement). However, the application of glucose in an organism is a complex biochemical process in which energy can be released for use (see Understanding the Element Carbon, below). Here, it is emphasized that photosynthesis involves trapping the chemical element carbon into living organisms. The chemical structure of glucose is based on carbon and typically exists as an open-chain compound.

As glucose can be utilized by all forms of life, supporting the entire developmental process of living organisms, its formation signifies a remarkable phenomenon in the universe (see Understanding the Element Carbon, below, and Chapter I.1, The Timely Birth of Photosynthesis on Earth and The Timely Cambrian Explosion). Forming the global biosphere without sharing the same life-forming substances is impossible.

What Must Life Substance Be Like?

First, life on our planet is believed to have originated from a single remote ancestor that may have been a minute dynamic system. If this is accurate, then all life, inevitably and naturally, must consist of the same substances despite the myriad species in the natural world. It is difficult to conceive that each species is made up of its own specific and unique substance. In fact, it is impossible (as will be determined later).

As the chemical structure of glucose is based on the element carbon, the other components of life-forming substances must also be based on the same element. Otherwise, the complex organization of a body could not be formed through biochemistry. This is referred to as carbon-based life, meaning life organization based on the element carbon (see Understanding the Element Carbon, below).

Moreover, life substances should be common across all kinds of life, whether plants, animals, or microorganisms, allowing them to be transformed into one another. Without this commonality, life on Earth could not exist (as will be detailed below).

No matter how complex the structure of a building is, various buildings are constructed with only a few raw materials (bricks, concrete, etc.).

Similarly, diverse organisms are also made up of only a few substances that can meet the various physiological requirements of Earth Life for subsistence under multiple environments.

The main substances of Earth Life include inorganic and organic compounds. Inorganic compounds include water and salt. Organic compounds (organic matter) include proteins (including various enzymes), carbohydrates, lipids, and nucleic acids, the chemical structure of which is all carbon-based. Obviously, without life, there would be no organic matter.

Proteins are found in every tissue. Birds with beautiful feathers, colorful flowers, and agile beasts—all require proteins for their bodies. Each protein molecule has a complex structure tailored to its specific functions. The wide variety of physiological functions corresponds to different forms of proteins. Despite this diversity, every protein is composed of amino acids, which serve as the building blocks. Proteins are less abundant in plants compared to animals. The most remarkable protein structure in the solar system is the human brain. The evolution of life implies the enhancement of physiological functions, achieved through the roles played by proteins.

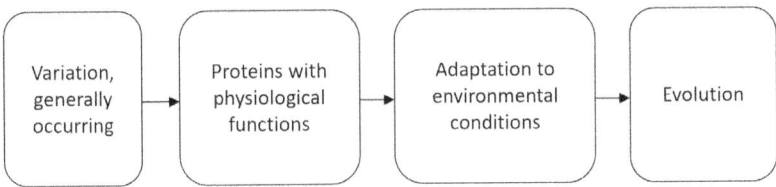

Carbohydrates are the primary source of physiological energy and are widely found in various living organisms, especially in plants, where they form the main component of cell walls. Lipids primarily function to store physiological energy. Nucleic acids are essential for heredity; without them, life on Earth could not exist.

What is particularly interesting is that all these substances can be directly or indirectly transformed into one another within a living organism. For example, grass consumed by a cow can, through complex physiological processes, be converted into various tissues and organs, such as muscle, bone, hair, and milk. Other animals can then consume these products, forming a food chain from grass to herbivores to carnivores in a grassland ecosystem. Numerous ecosystems in diverse environments make up the global biosphere, which is the foundation of evolutionary processes on Earth (see The Formation of the Global Biospheres, below).

This transformation and sharing of substances play the most significant role in the evolution of life and are credited to the elements mentioned above, specifically carbon, whose unique properties cannot be replaced by any other element (see Understanding the Element Carbon, below). The formation of the global biosphere would be impossible without the commonality of these life-forming substances.

What Must be the Form of Earth Life?

Let's start by talking about the forms of life on Earth. Based on organic matter, as stated above, life on Earth naturally developed into three kingdoms: plants, animals, and microorganisms.

Plants are the producers, using photosynthesis to create organic matter. Animals are the consumers, whose existence depends directly or indirectly on plants. Microorganisms, which include bacteria, germs, and small invertebrates, act as decomposers. Through their physiological functions, they transform all kinds of organic matter into inorganic matter. Without decomposers, the Earth's surface would be covered with waste matter and the remains of living organisms. Consequently, essential chemical elements, such as carbon, could not be circulated throughout the natural world.

Thus, the three kingdoms of life—plants, animals, and microorganisms—are rationally organized and essential for the balance of life on Earth.

Here, we may ask: is the existence of animals necessary? Is it inevitable in the evolutionary process? In fact, from pteridophytes (which use hydrophily for reproduction) to gymnosperms (which rely on anemophily), plants completed the entire evolutionary process independently, without animal involvement. However, the reproduction of angiosperms (which emerged later, during the Cenozoic era) requires animals' participation, such as bees, which exemplifies co-evolution. Human food primarily comes from angiosperms.

Therefore, in the evolution of life, the origin of animals appears inevitable, as only animals can foster intelligence and enable life to evolve in diverse environments, potentially extending into infinite space. Furthermore, following the emergence of animals in the evolutionary process, there was the Cambrian Explosion, crucial for animal evolution (see

Chapter I.1, The Timely Cambrian Explosion), laying the foundational species for the eventual emergence of human beings.

Some people speculate about the possibility of plants and animals merging into one, but this is impossible due to the fundamental characteristics of photosynthesis. Photosynthesis is a gradual, cumulative process that can only be carried out by stationary plants. Animals, on the other hand, continuously consume high levels of energy and cannot engage in photosynthesis.

Therefore, if intelligence is to emerge through the evolution of life, the existence of the three forms of Earth's life—plants, animals, and microorganisms—is inevitable. The principles of natural philosophy also determine this.

In the initial stages of the evolutionary process, more than a billion years ago, plants and animals originated from a common ancestor. These *Euglena viridis* were eukaryotic organisms (unicellular and still found in some lakes today). Subsequently, they diverged into separate paths for survival. The earliest primitive plants and animals emerged in the sea approximately 1.2 to 1.3 billion years ago.

We can easily imagine that if there were no animals in the evolutionary process, humans obviously could not possibly have emerged either. Thus, Earth Life necessarily consists of all three kingdoms. Without any one of them, the Earth-Life system could not possibly have been formed.

If life existed in another world, such life should be consistent with the concept stated above; otherwise, such life would be of little significance, let alone produce civilization.

Now, let's turn to the organization of life on Earth. As photosynthesis occurs in a cell, Earth Life, including the three kingdoms, must exist in the form of cells.

Generally, cells exhibit diverse forms (cellular morphology) in various tissues, yet they contain similar organic matter (proteins, carbohydrates, lipids, and nucleic acids) and operate under similar principles across plants, animals, and microorganisms. These shared features enable the formation of food chains and ecosystems, providing the foundation for evolutionary processes. A profound point.

In a sense, cells are like bricks, used for constructing various forms of life. Cells are also like numerous small factories in which various physiological processes occur with precision. To conceive of any other way of constructing the wide variety of living organisms we see in the natural world is difficult. The cell is resourceful and has an ideal pattern to form a wide variety of life.

A zygote, a single cell, undergoes complex cell differentiation to develop into various tissues, forming different organs within organisms. Animal tissues, such as muscular and nervous tissue, contribute to the formation of organs like lungs, heart, and liver, which are familiar to us. Plant tissues include protective, vascular, and mechanical tissue, supporting the development of organs such as roots, stems, leaves, and flowers. Each organ in animals and plants serves distinct physiological functions and continually operates to sustain the organism's life. Therefore, the absence of any organ would render a living organism unable to survive.

But what exactly does the organ system do? Here, we can summarize its function as follows: with the assistance of enzymes composed of a diverse range of proteins, Earth's life forms ingest and expel substances and energy necessary for survival. This process is known as metabolism.

In plants, metabolism encompasses water metabolism, mineral metabolism, photosynthesis, and respiration. In animals, metabolism primarily involves energy metabolism and material metabolism. Each type of metabolism in both plants and animals is an exceptionally intricate physiological process, which can be explored further in specialized sources.

The complete logic for an organism to live is as follows:

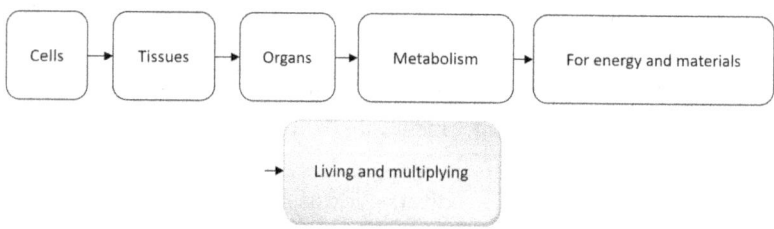

Life, from tiny insects to human beings, is generally a nonlinear dynamic system, extremely complex, precise, and subtle. We cannot conceive of any other way to maintain life. However, life span is limited, and the continued subsistence of Earth Life is based on reproduction from generation to generation.

The Formation of the Global Biospheres

Above, we have examined Earth Life, biological energy (photosynthesis), and life substances—organic matter, cells, tissues, and organs. Together, these components contribute to the formation of the Earth's biospheres. Scientists suggest that without a biosphere on a planet, the evolution of life, let alone civilization, would not be possible. Why is this the case? Let's research this question further.

As mentioned earlier, the availability of a food chain in a particular area is crucial for forming a local ecosystem (see What must life substance be like?, above). It's easy to envision that life continuously evolves under local geographic conditions. Various geographical conditions worldwide lead to the emergence of diverse species accordingly. Each species naturally tends to increase in population and geographic range, expanding into new territories and facilitating extensive species exchanges.

For instance, crops like wheat, rice, cotton, and tomatoes, as well as animals such as horses, oxen, pigs, dogs, sheep, and monkeys, originated distinctly in different geographic settings before spreading to other parts of the world and continuing to evolve. Similarly, various primate species exist globally, but those inhabiting the GRV evolved into early human ancestors only a few million years ago. This highlights how extensive species exchanges between different regions can provide greater opportunities for species development.

Therefore, the wide array of local ecosystems collectively forms the global biosphere. Without a biosphere, the evolutionary processes observed on Earth today would not exist. The diagram below summarizes the entire evolutionary process:

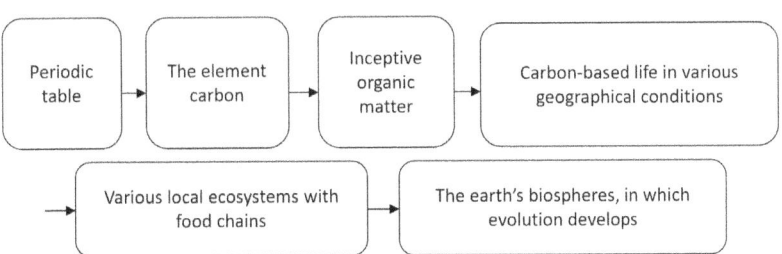

With the presence of the atmosphere, water resources, and sunlight, the biosphere integrates the evolution of life on Earth into a unified system. Within this system, millions of species are interdependent, directly or indirectly, and no single species can exist independently from this interconnected web.

One might question the relationship between organisms living in polar regions and those in equatorial regions. It's important to recognize that no species evolves in isolation; every species depends on a food chain for its existence. While it may be challenging to envision a direct relationship between elephants in grasslands of the tropical zone and polar bears in the Arctic, scientific understanding reveals an extensive biological network comprising diverse food chains spanning from tropical to temperate to frigid zones across our planet's land and oceans. These ecosystems are intricately interconnected, with no isolated components within this vast network.

It's unimaginable that civilization could arise on a distant planet lacking a global biosphere. However, Earth's immensely complex and continually evolving biosphere is fundamentally supported by the element carbon. Why is carbon so essential and abundant?

Understanding the Element Carbon

Why are proteins effective in meeting diverse physiological needs? The answer lies in the element carbon.

Carbon occupies a unique position in the periodic table, possessing properties crucial for forming life substances and fostering civilization. It plays a dual role, serving as the foundation of organic chemistry. Without carbon, the entire periodic table would seemingly lack the essential basis for its existence and significance. Carbon's ability to form exceptionally large molecules is fundamental for creating life substances.

Both carbon and hydrogen possess high heat energy, are readily combustible at normal temperatures, and are abundant in the natural world. No other elements in the periodic table exhibit these specific qualities. The substantial energy from carbon, widely available in nature, is essential for life's physiological processes and the advancement of civilization (see Chapter II.1, Utilization of the Two Primary Substance Chains).

When combined with hydrogen, oxygen, nitrogen, and trace elements like sulfur, carbon forms proteins, carbohydrates, and lipids, fulfilling every physiological requirement for the functions of organisms as described earlier. These substances, collectively known as organic matter, differ significantly from inorganic matter. Organic matter ignites easily at normal temperatures and is insoluble in water. These characteristics are crucial for physiological processes and the development of civilization. If organic matter were water-soluble, our bodies would quickly dissolve. Therefore, the inherent properties of organic matter, including its combustibility and insolubility in water, represent a clever and necessary natural arrangement.

Carbon dioxide (CO_2), abundantly present in the natural world, is the only simple molecule that participates in photosynthesis, known as CO_2 assimilation. During photosynthesis, water and carbon dioxide molecules are transformed into glucose (see The energy of Earth life, above). Thus, within the framework of the periodic table, photosynthesis represents the primary mechanism through which solar energy converts into biological energy. Note that in nature, only chloroplasts can absorb light energy (photons) with chemical reactions, and chloroplasts are composed of proteins.

Carbon exists in various forms and holds immense significance in the natural world. Amorphous carbon is commonly found as charcoal, coke (derived from coal), activated carbon, and more. These forms of carbon play a crucial role in the development of civilization (see Chapter II.1, Utilization of the Two Primary Substance Chains). Crystalline forms of carbon include graphite, diamond, and graphene at normal temperatures. Each of these forms contributes uniquely to civilization. Graphene, in particular, is a relatively new material highly prized in industry for its extensive potential applications.

As mentioned earlier, various organic substances within organisms can transform among themselves, forming diverse food chains that support the biosphere. In carbon-based life forms, both humans (along with most animals) and plants (through photosynthesis) thrive within a similar range of temperatures suitable for their existence, approximately between 62°F (17°C) to 95°F (35°C). They also endure roughly the same extreme temperatures, up to about 113°F (45°C), which they can tolerate. This temperature range is crucial for the existence of both plants and animals, and it also aligns favorably with the properties of metals.

The role of carbon in forming life and civilization is irreplaceable in the periodic table. No substitute exists. But why does the element of carbon possess these important qualities? Let's look at the periodic table below (Figure 3–3).

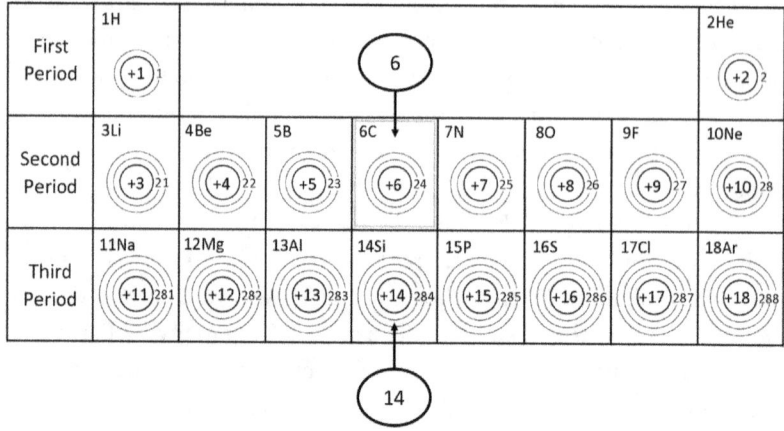

Figure 3-3: Electron arrangement of the first eighteen elements (see Figure 3-1) (Source: The author's book in Chinese, published 2016).

In general, the smaller the atomic radius, the stronger the covalent bond (atomic bond), which is advantageous for forming large molecules. The elements in the second period of the periodic table have smaller atomic radii than other periods. This period includes eight chemical elements:

1. Lithium (Li)—atomic number 3 (metal)
2. Beryllium (Be)—atomic number 4 (metal)
3. Boron (B)—atomic number 5 (solid)
4. Carbon (C)—atomic number 6 (solid)
5. Nitrogen (N)—atomic number 7 (gas)
6. Oxygen (O)—atomic number 8 (gas)
7. Fluorine (F)—atomic number 9 (gas)
8. Neon (Ne)—atomic number 10 (gas)

Beryllium is classified among the hyper-toxic elements and is exceptionally rare in the natural world. The properties of lithium and boron differ significantly from those of carbon. For instance, lithium readily reacts with water, while boron's ignition and combustion characteristics are not as beneficial for civilization. Also, both lithium and boron are relatively scarce in nature.

In contrast, carbon stands out as the only element with a small atomic radius that is crucial for forming life substances and contributing to civilization. It is abundant in the natural world. The other elements mentioned—nitrogen, oxygen, fluorine, and neon—are gases with their own distinct properties.

The carbon atom has two electron shells, as shown in the periodic table: $1s^2$ and $2s^2 2p^2$. Here, $1s^2$ and $2s^2$ represent pairs of electrons with opposite spins, while $2p^2$ indicates unpaired electrons with the same spin direction. The notation reflects the following:

- 1, 2: Number of shells
- s and p: Electron cloud shapes
- ^{2}: Number of electrons

Normally, one might expect carbon to form only two covalent bonds based on these electron configurations. However, carbon forms four covalent bonds.

The typical molecule for hydrocarbons is represented by the formula C_nH_{2n+2}, with methane (CH_4) being a prime example. To understand this structure, scientists employ the theory of hybridized orbitals, which explains carbon's ability to form four covalent bonds (further details on hybridized orbitals can be found elsewhere). Carbon undergoes hybridization in its ground state, allowing it to use its four available valence electrons to form four bonds, making it highly versatile in forming complex molecules essential for life and civilization.

Another carbon-based molecule formed through hybridized orbitals is carbon dioxide (CO_2), a gas at normal temperatures. Notably, very few solid oxides exist in a gaseous state at normal temperatures in the natural world. This unique property allows CO_2 to play a critical role in photosynthesis. In other words, carbon, through its gaseous form as CO_2, is indispensable to the process of photosynthesis.

Isn't it fascinating how carbon stands out as a special and profound element? This remarkable natural arrangement underscores carbon's pivotal role in the development of life on Earth. Its unique properties indeed shape the destiny of our planet!

From the viewpoint of forming life substances, silicon (Si) appears to be the only element in the periodic table that could challenge the role of carbon. Positioned just below carbon in the periodic table, silicon has a similar atomic structure with three shells; its outermost shell also contains four electrons ($3s^2 3p^2$). However, silicon's atomic radius is larger than carbon's (see the table above).

Through hybridized orbitals, silicon and hydrogen can form SiH_4 (silane), a hyper-toxic gas that is useful in modern civilization. However, SiH_4 is not stable and is difficult to produce. While silicon is hard to ignite, carbon ignites easily, releasing heat accordingly. Also, the oxide of silicon is SiO_2 (quartz), which is a solid rather than a gas, unlike carbon dioxide (CO_2), which plays a crucial role in photosynthesis as a gas.

The differences between silicon and carbon highlight why carbon is uniquely suited to form the basis of life, emphasizing its special and irreplaceable role in the development of life on Earth. Essentially, silicon is merely a trace element in the human body and could not possibly replace carbon in the animate world.

Some scientists imagine the possibility of silicon-based life elsewhere in the universe, but this remains speculation, given that silicon's qualities are not as practical for life as carbon's because of the extreme complexity and subtleness of the physiological structure of carbon-based life. However, silicon has many useful properties that other elements, including carbon, do not possess. Modern electronic technology, such as computer chips, relies heavily on silicon. We can even say that modern technology is silicon-based, making ours a silicon-based civilization.

Finally, it is crucial to emphasize that "the whole development process of life" is extremely complex. In this process, carbon, combined with other essential elements, forms proteins (including enzymes), carbohydrates, lipids, and nucleic acids. These processes occur effectively in the presence of water, air, and sunlight, resulting in an almost infinite number of biochemical reactions that produce the vast diversity of species on Earth. Thus, carbon is the only element capable of perfectly supporting "the whole development process of life."

To illustrate the importance of carbon, consider this: if carbon were removed from the periodic table, the natural world, including humans and civilization, would collapse immediately. However, if gold were removed, the natural world would continue to function, even though people perceive gold as more valuable than carbon.

This metaphor highlights that the roles of chemical elements in the natural world vary in significance. Some elements seem "light" (such as hydrogen, oxygen, nitrogen, etc.), and some seem relatively "heavy." Despite these perceptions, every element plays a crucial role in the development of civilization, and successful construction (both literal and metaphorical) depends on the presence of all these elements.

Before concluding this section, let's look back to the diagram, "The progression from chemical elements to life to fire and to civilization: a wonderful natural arrangement!" in the section, The Mystery of Fire, in Chapter II.1. The key to that diagram is the substantial existence of CO_2 in a gas state under normal atmospheric temperatures. Without the component of carbon dioxide, the process shown in that diagram could not be fulfilled.

Now, let's further reveal the logical relation between carbon and the natural world:

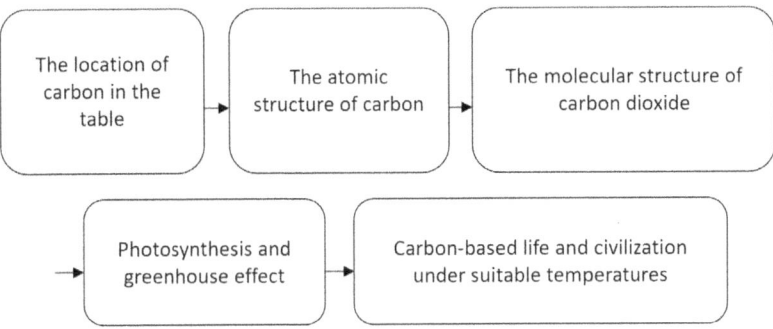

Please consider each step in the diagram above. Why does only the element located at atomic number 6 (carbon) possess the profound qualities—high heat energy, ability to form macromolecules, and its oxide existing in a gaseous state—that determine the natural world's pattern? Similarly, why are the qualities of elements at atomic numbers 7 (nitrogen), 8 (oxygen), and 1 (hydrogen) so useful that they can combine with carbon to shape the pattern of life on Earth?

Isn't it remarkable how the periodic table appears almost magical or as if it were designed with a purpose? No other element in the table can substitute for carbon dioxide in photosynthesis within this logical framework. The unique combination of these elements—carbon, nitrogen, oxygen, and hydrogen—is unparalleled in creating the conditions necessary for life.

On the other hand, nature first created the chloroplast in the earliest days and developed as follows:

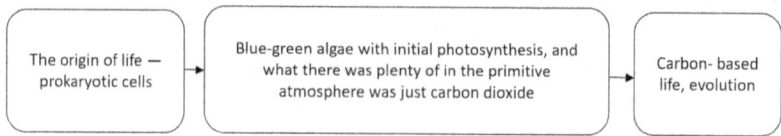

Remember, photosynthesis involves trapping carbon, incorporating it into the bodies of living organisms, thus creating carbon-based life. The chloroplast is merely a tool facilitating photosynthesis within cells. Consequently, the periodic table can be seen as the "constitution" for the behavior of Nature. Isn't it remarkable how thoughtfully nature is arranged? We are in awe of Nature!

Given the periodic table, could you select any other element besides carbon to fulfill this process perfectly? These are the reasons why Earth Life is likely the only form of life in the universe. This is determined entirely by the periodic table and the two principles of natural philosophy. Different elements have distinct electron arrangements, leading to unique properties and functions that cannot replace each other. Carbon, with its atomic number 6, is fundamental to the periodic table; its properties and functions are irreplaceable. The periodic table reveals a profound intelligence, "the highest wisdom" in nature, far beyond human imagination.

The Pattern of Life in the Universe

Building on all the arguments in this book so far, we may logically summarize as follows:

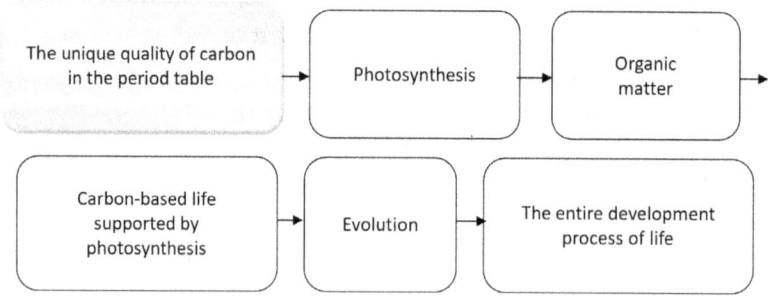

This logic is subject to the periodic table. Given the domination of the table around the universe, we can readily imagine that carbon-based life would be the only pattern of life in the universe; there seems no other possibility than carbon-based life to form the three kingdoms of life (plants, animals, and microorganism) and fulfill "the whole development process of life." If there were life on a remote planet, it must be of the same pattern as Earth Life, no matter how strange it might be.

We find it hard to imagine that there could be other kinds of creatures consisting of entirely different chemical elements from ours. That one periodic table could produce two types of creatures from completely different chemical elements seems impossible. If you disagree with this idea, you may fail to fully understand the extreme complexity of the organization of life. Of the 92 chemical elements found in the natural world, none other than carbon could be considered the basis on which both life and civilization could be perfectly formed.

The element of carbon appears to be the key to revealing the mysteries of the periodic table and is of extremely high value for us to understand the nature of the universe. Let's use a diagram to summarize the logical relationship between the table and our existing world:

THE RATIONAL UNIVERSE EVOLVING FOR HUMANS

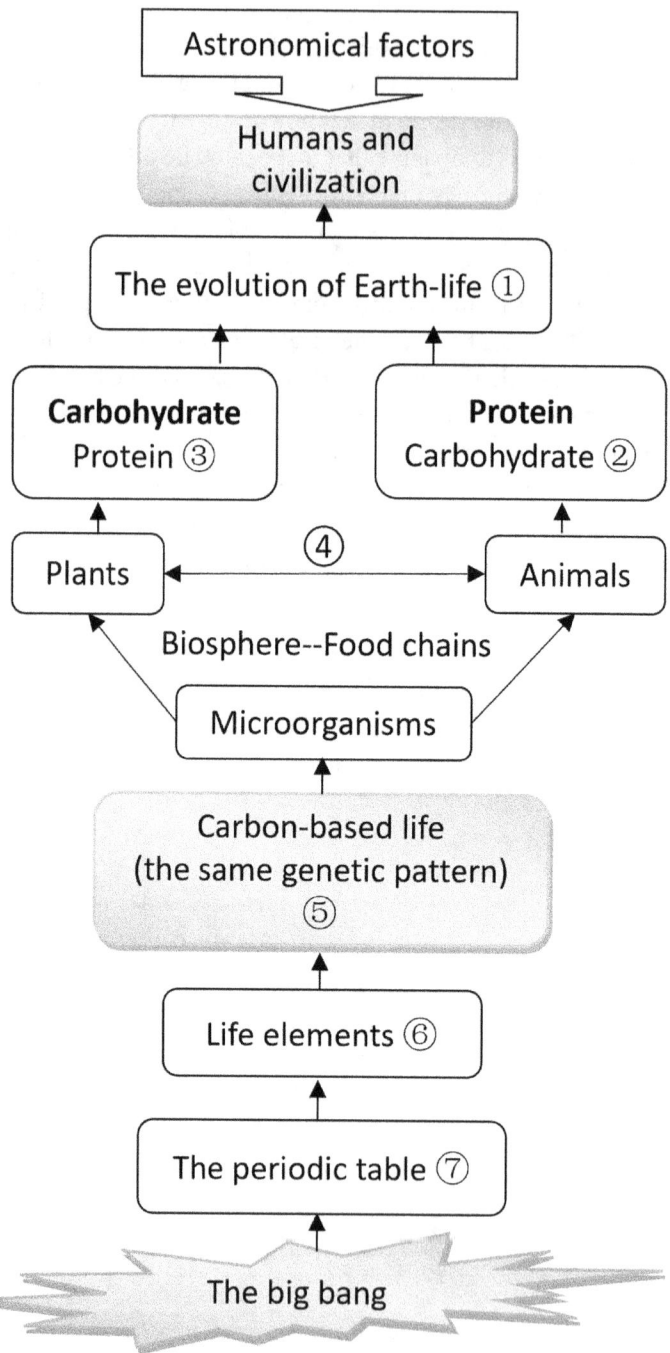

248

(1) What is the value of evolution? Substantially, evolution represents the improvement of the physiological functions of living organisms. From prokaryotic cells to human beings, the functions of life have become increasingly complex and sophisticated.

(2) The most important feature of animals is their ability to move freely, supported by various body parts (legs, wings, fins, etc.) with respective physiological functions (running, flying, swimming, etc.). What kind of substance can fulfill the physiological functions of animals? Only protein, which seems to be an omnipotent substance, can meet the vast physiological requirements for animal life, from minute insects to wild beasts. Consequently, animal bodies are complex and primarily composed of proteins. Their energy comes primarily from glucose; about 70 percent of the energy in the human body comes from glucose, though proteins and lipids can also release energy. There are more than one hundred thousand different kinds of proteins in the human body.

You may be surprised by the quick movements of animals, such as birds. In fact, various and subtle biochemical transformations among proteins, carbohydrates, and lipids occur continuously, involving different internal organs. Animals are the only form of life to evolve into intelligent creatures (see Chapter III, What must life substance be like?).

(3) We quickly understand that the physiological functions of plants are more simple than those of animals. Consequently, the protein content of plants is naturally much less than that of animals. Plant material is mainly composed of cellulose (specifically, polysaccharides) and lignin (a carbon-based macromolecule). Generally speaking, many more proteins are found in plant reproductive organs (seeds, flowers, fruits) than in vegetative organs (roots, stems, leaves). We cannot consume grass as food due to its lack of proteins, but we are grateful for various vegetables and fruits, which contain essential chemical elements and vitamins. However, the protein content in the leaves of vegetables is only 1–2 percent by weight.

Therefore, it is easy to understand that animals and plants have very different amounts of protein, primarily because of their different ways of living. How thoughtful nature is! The primary natural missions of plants are to provide animals and people with energy (mainly glucose), various chemical elements (sourced from the soil), fuel, and raw materials (wood, cotton, silk, etc.), all of which are necessary for the development of civilization and cannot be replaced by inorganic materials. Plants could not possibly have evolved into a form of mobile creature with intelligence.

(4) Carbon-based life forms determine the three kingdoms of life, which are interdependent, forming an inseparable evolutionary system. Here, evolution defines the essence of life itself. If a planet were inhabited only by microorganisms indefinitely until its demise, it would hold no value because it would lack life evolution. Proteins appear to serve as a gateway to evolution. This is because the enhancement of physiological functions in animals or plants relies on changes in proteins—evolution into new types of proteins drives this process.

(5) Both animals and plants on our planet belong to carbon-based life, sharing the same genetic blueprint. Only carbon-based life forms can accomplish the natural development process outlined above (the entire life development process). Carbohydrates and proteins have distinct roles in the evolving system of carbon-based life. Our beautiful, colorful, and even magical natural world owes much to proteins and is sustained by carbohydrates (glucose). No other combination of elements could replicate all these advantages.

(6) All the elements comprising the bodies of living organisms seem to have been selected by nature. Perhaps there were numerous "natural tests" of different element combinations under various natural conditions at the beginning of our planet. Many of these tests failed from the outset, leaving carbon-based life as the sole successful outcome of these natural trials. Carbon is the only element in the periodic table that holds the potential for life and civilization on our planet. Only through photosynthesis does carbon enter the bodies of living organisms. Carbon activates the periodic table, giving it meaning and depth in the universe (see Chapter III, Understanding the element carbon).

Note: Plants selectively absorb elements from the soil, primarily in the form of ions. Different plants vary in the types and quantities of elements they absorb, although there are some common elements. The elements absorbed by plants dictate the element requirements of animals.

(7) We believe the periodic table is consistent throughout the vast universe because it originated from the same Big Bang. Any evolving system of life must adhere to the same laws and patterns observed on our planet.

In summary, let's begin with the Big Bang, which appears to be an inevitable process. The Big Bang gave rise to the elements, forming the periodic table (Figure 3-4). The sixth element of the periodic table is carbon, which is pivotal in the evolution of carbon-based life within the Earth's environment, leading eventually to human beings and modern

THE MYSTERIOUS PERIODIC TABLE

civilization. This entire process is governed by the principles of natural philosophy, making it both inevitable and logical. There is no alternative solution imaginable from the inception of the Big Bang to the emergence of humans and civilization (see the relevant sections in Chapter VI).

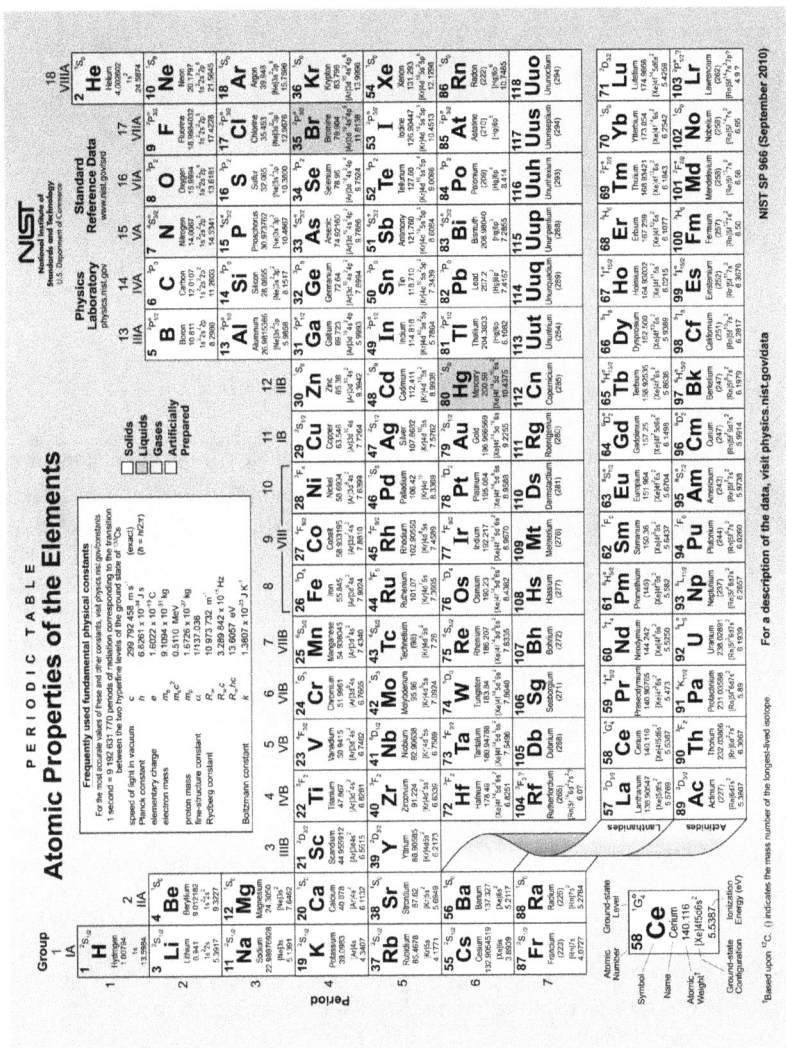

Figure 3-4: The mysterious periodic table (Source: NIST, public domain).

Chapter IV

What Exactly Is the Earth?

What is the Earth anyway? This is a difficult question to answer. However, we can study its basic characteristics and deepen our understanding of the Earth's functions from different perspectives to further reveal its mystery and enhance people's reverence for Earth.

Earth has completed its evolution from chemical elements to human beings, showing that it is an extraordinary planet in the universe. We may never fully understand this mysterious and complicated process. As we have said before, in nature, only life processes, from the fertilized egg to the end of life, have timeliness. However, the development of the Earth System, logically like the development of life, also requires timing. What satisfies the time requirement of life development is the regulatory factor. What, then, is the regulatory factor for timing in the development of the Earth System? What is the nature of the Earth?

Exploring these questions will offer significant insights into the nature of the universe. Popular science and Einstein's philosophy of nature lay the groundwork for this exploration, which will be discussed next.

EARTH'S UNIQUENESS IN THE UNIVERSE

Are there any extra-terrestrials in the boundless universe? Let us turn to this intriguing issue: Earth's uniqueness in the universe. Of course, the discussion below is only a logical deduction rather than a scientific conclusion. Obviously, there is no scientific conclusion on this issue at this point.

The Complexity, Subtlety, and Consistency of Earth's Development

If you have understood all the arguments previously explored, you see how complex and subtle the development of Earth is! Along the time axis, this is an immense natural chain, and the occurrence of each link is on time and necessary for the eventual appearance of humans and civilization. There is no link that has no role to play in this long chain. In addition, every link has its sub-links, and we can further make sub-divisions of each sub-link, a seemingly endless process.

Earth's development in space and time is an immense natural network with numerous levels. All the factors in this network interact and follow strict, precise, and complex patterns in time, space, and quantity. Any deviation from the development process would have resulted in the failure of the eventual appearance of humanity and civilization.

For example, according to the theory of plate tectonics, the Earth's crust consists of plates that float on the surface of the hot mantle, moving in various directions. If this geological process were random, it would result in numerous variations in continental drift, leading to diverse shapes of the Earth's surface and different evolutionary and geological processes (see Chapter I.1, The Timely Change in Ancient Lands).

Another example is in the components of the atmosphere. No doubt, every component of the air was changing along the time axis in its development process (see Chapter I.1, The Timely Appearance of the Atmosphere). Still, both the outcome of continental drift and the development of the atmosphere were conducive to the eventual emergence of humans and civilization. Each natural developmental factor operates within strict time and space constraints, like navigating a vast labyrinth or sailing the ocean without a compass, filled with junctions and options. Any divergence in these developments would have led to different outcomes.

The statement above shows us that the Earth's processes are unidirectional and consistently favorable to the eventual appearance of humans and civilization. We may call this phenomenon *the consistency of the development of the Earth System* (Figure 4–1).

THE RATIONAL UNIVERSE EVOLVING FOR HUMANS

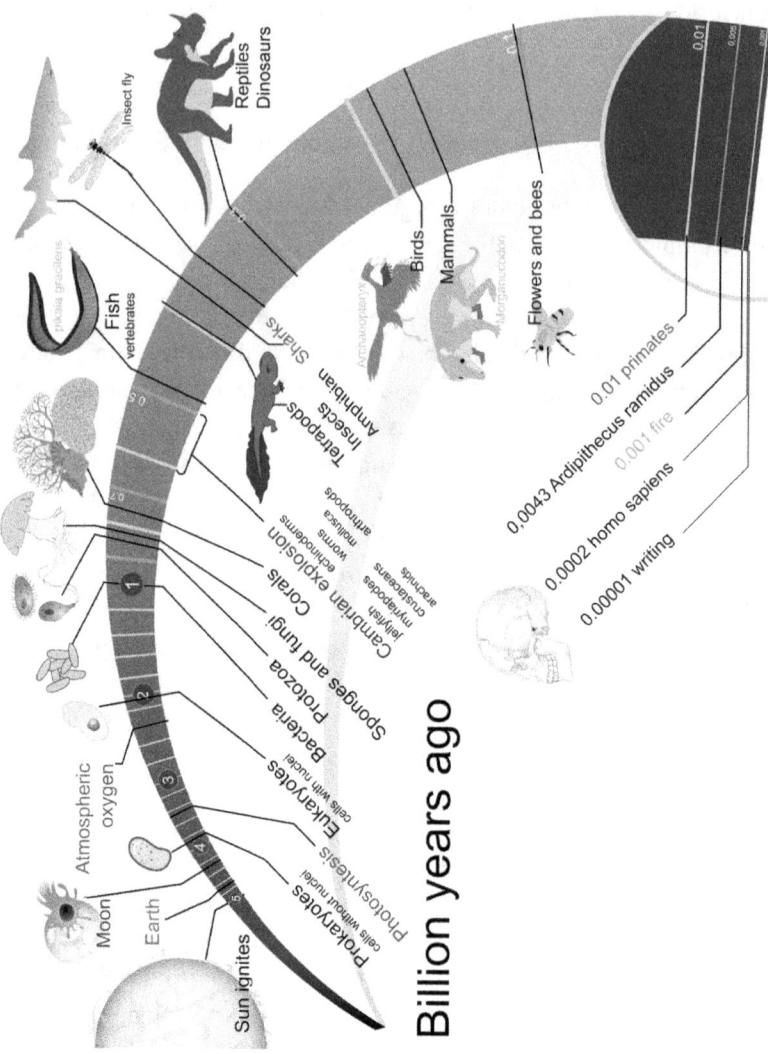

Figure 4-1: The evolution of life is subject to change in the environment. But during the past 4.6 billion years, why was there a series of environmental changes that made all natural selections faultless? Why was the environmental change on the Earth so appropriate over 4.6 billion years that it enabled the chemical elements to become life and gradually evolve into humans? This is an extremely complex random process, and without the intervention of deep factors, this process would never be completed (Source: Public domain).

No Other Identical Natural Process in the Universe

The natural world, whether inanimate or animate, is widely recognized for its characteristic self-similarity—a fundamental geometric property observed in fractal structures. Self-similarity indicates that the whole possesses the same shape as one or more of its parts (Figures 4-2 and 4-3).

Figure 4-2: The part and the whole are similar. Fractal structures can enhance the efficiency of physiological functions. Fractals are a shared feature of both living and non-living systems (also see Fig. 4-3), demonstrating their inherent rationality (Source: ::ErWin, CC BY-SA 2.0).

Figure 4–3: The two satellite images above show the self-similarity of mountains. The white areas indicate snow-capped regions. How they are shaped like plants! But the fractal structure of mountains facilitates the formation of river basins. (Source: [Top] NASA Earth Observatory, CC BY 2.0, [Bottom] NASA, public domain).

Self-similarity is a phenomenon observed in both living organisms and the inanimate world. Mountains look much like tree branches. The images above illustrate this concept. The self-similarity in the mountain landscapes is particularly evident—you could zoom in further and find even more detailed patterns that resemble the larger terrain. You could also find more and better self-similarity in the topography of mountains and rivers on satellite maps from all parts of the world. But why do landscapes exhibit such self-similarity? Has chaos played a role in shaping these features? Modern science still has much to learn about the underlying reasons for this phenomenon.

The self-similarity phenomenon implies the animate and inanimate natural world is characterized by nonlinear systems, and linear relationships are rare. The first nonlinear issue studied in history is the three-body problem, which was researched for the law of motion among the Sun, Moon, and Earth in a deterministic system. Sir Isaac Newton is arguably the greatest scientist of all time, but he "failed" to solve this problem. He was unaware that there is no stable solution in the deterministic system over time. About 200 years later, French scientist Jules Henri Poincare (1854–1912) made significant contributions to that problem. He found that a simple system described by a deterministic equation, such as the three-body problem, could have extremely complex, seemingly random motions—chaos.

The nonlinear system tends to cause chaos and may further result in a strange attractor with a fractal structure. A typical example of a nonlinear system is the set of equations of a dynamical system developed by Edward N. Lorenz (1917–2008), an American meteorologist. Under certain conditions, the solution of this dynamical system will develop into a strange attractor—a mathematical concept in phase space—where its trajectory is highly sensitive to initial conditions. Lorenz termed this phenomenon the Butterfly Effect, where the slight airflow caused by a butterfly could trigger a storm in a distant location. Therefore, nonlinear systems possess inherent indeterminacy, making it impossible to predict their future states precisely. The trajectories never overlap, regardless of how close they are (Figure 4-4).

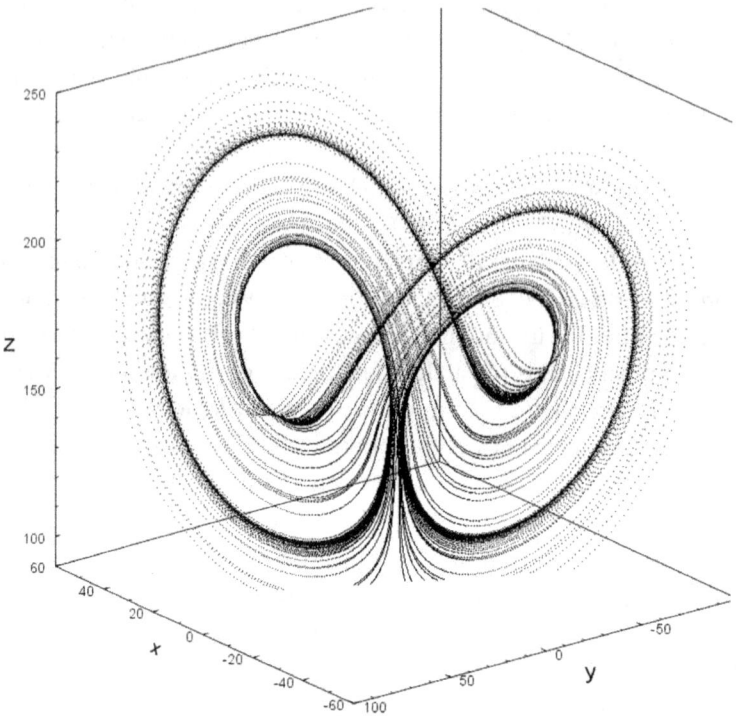

Figure 4-4: The Lorenz attractor (Source: Public domain).

Composed of nonlinear systems, natural processes are stochastic, without repeatability ("never overlap," "sensitive to its initial condition"). There are no identical leaves or identical twins in every detail, even growing from one egg cell. A slight difference between individuals with the same pattern is universal in the natural world because of nonlinear systems everywhere. We usually call them individual differences.

The above relates to the theory of dissipative structures in nonlinear systems, advanced by Ilya Prigogine (1917–2003, Nobel laureate), who researched self-organization in the far-from-equilibrium state of open systems. His book, *Order Out of Chaos*, co-authored with Isabelle Stengers, explores this theory. While the theory has significant implications for understanding processes in the natural world, it primarily describes how chaos develops into order within open systems. It does not, however, explain the successive development of the Earth System, characterized by increasingly complex structures and progressively higher functions over billions of years (see What Exactly is the Earth?, below).

Finally, take another look at Lorenz's strange attractor (Figure 4-4). You will find that despite the motion of the locus being unstable and unpredictable, the pattern (tendency) of the strange attractor is stable. That characteristic of the strange attractor is a bit like the development of our planet. Despite the course of the Earth's development being uncertain, the tendency of the development from a rotating nebula to the Earth System has consistently pointed to the eventual appearance of humans and civilization.

Since the phenomenon of self-similarity is widely found in the natural world, we may well ask if the order in the natural world is derived from chaos. Further, is life a strange attractor? It is logical to imagine that this might be the case. This idea aligns with self-organizing theory. However, we have a long way to go to understand this natural phenomenon of self-organization fully.

Thus, the processes that form the Earth System are stochastic and full of bifurcation and selection (terms from the theory of dissipative structures) due to random effects everywhere. The sub-processes are similarly affected.

How many essential conditions and sub-conditions at different (apparently endless) levels were necessary for civilization to arise on our planet? There is no answer to this profound question, potentially leading to infinite research on the subject. The probability of replicating all these essential conditions on other planets is nearly zero. Any deviation from the Earth's processes, such as the absence of clay minerals, would likely fail to form the natural development chain necessary for civilization.

Let's use heavy rain as a metaphor. Seasonally, the sky is covered with dark clouds, and people long for a downpour. However, the absence of any essential condition for rain formation would result in the failure of a downpour.

We can examine this question from the viewpoint of Stephen Hawking:

> Space-time would be like the surface of the Earth, only with two more dimensions. The surface of the Earth is finite in extent but it doesn't have a boundary or edge: if you sail off into the sunset, you don't fall off the edge or run into a singularity (I know, because I have been round the world!).[1]

1. Hawking, A Brief History, 85

If these statements are tenable, we may compare space-time to the atmosphere, which has no boundary or edge yet is finite around the surface of our planet. Science tells us that weather processes cannot repeat precisely due to random effects. Consequently, we can reasonably imagine there are no identical weather processes in time and space. The atmosphere is constantly in flux from its beginning to its present condition without identical processes. Like the atmosphere, space-time is also always developing without identical processes. The natural background of the emergence of humanity and civilization on Earth can be traced back to the origin of the solar system, even the Big Bang, involving numerous essential natural factors. We may well say this process could not be repeated precisely in the universe due to random effects.

The Issue of Probability

Strictly speaking, our world is full of uncertainty; every natural process is stochastic. Some scientists argue that, from a probability standpoint, numerous Earth-like planets should be scattered randomly throughout the universe. If we consider the Earth's development as reaching 100 percent, then theoretically, there would be "many" Earth-like planets with varying degrees of development, ranging from 0 percent to 100 percent. However, all the planets so far discovered outside our solar system—numbering more than one hundred—are at a low developmental stage, estimated at less than 1 percent. This is due primarily to the absence of liquid water, which is essential for supporting life. Thus, the spectrum of developmental stages (which could be quantified) between 1 percent and 100 percent remains unobserved or absent, contradicting the expectations of probability.

To further illustrate this concept, we can compare planet Earth to a complex spherical machine, and then the other planets in the solar system are just balls with a simple structure. Every part of the machine is complex, and its forming process is also complex, even magical. We must go a long, long way to know this planet conclusively. For example, the core, magnetic field, etc., are still beyond our comprehension.

Imagine a teacher demonstrating probability to students using many small balls, showing the "curve" in the balls is roughly continuous, without a wide gap (Figure 4–5). There are numerous planets in the solar system, including small ones, but the Earth-like planets are limited to

Mars, Venus, and Mercury, all of which have conditions adverse to life. The natural conditions of all planets discovered outside our solar system are far worse than those of the three within it. This fact challenges the continuity of probability values if we consider the Earth as a product of probability.

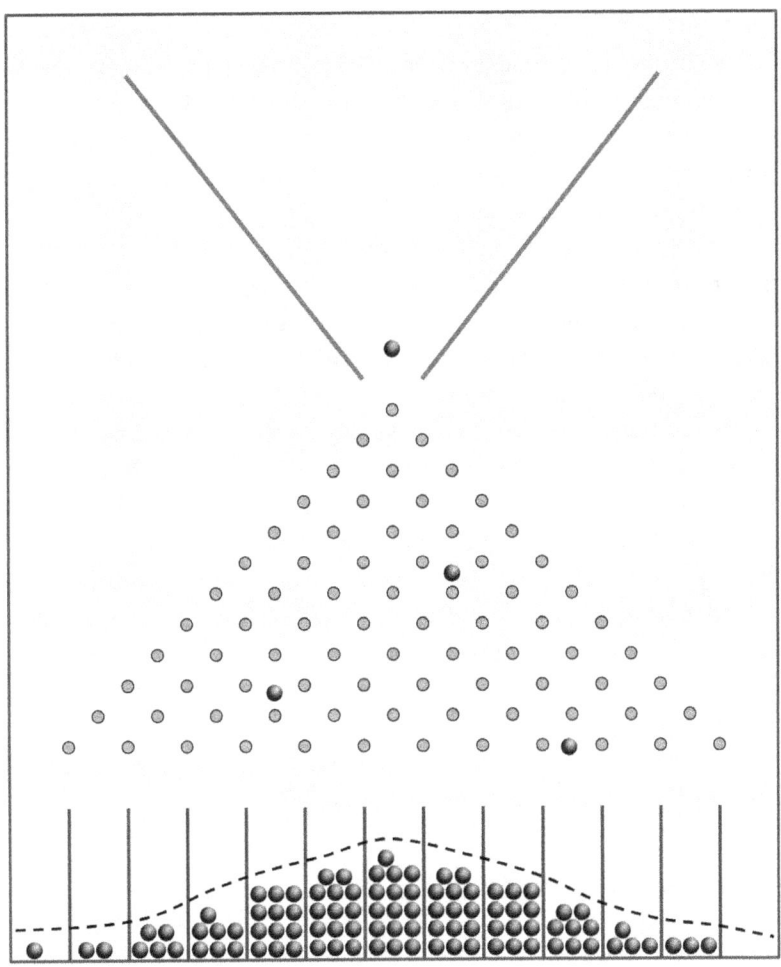

Figure 4-5: A demonstration of probability (Source: Marcin Floryan, CC BY-SA 3.0).

In this regard, we can further understand that random events are based on specific mechanisms (under certain conditions) that determine a particular range in which these events occur. For example, human height varies from person to person, ranging from about 19.5 to 102 in (50 to

THE RATIONAL UNIVERSE EVOLVING FOR HUMANS

260 cm), which is a typical phenomenon of probability. The formation of a person's height cannot escape the control of physiological mechanisms such as genes, nurturing, environment, etc. If there were a man as tall as 300 ft (100 m), would he still be considered a product of probability? If so, where are his siblings who range in height between eight and 300 ft? This extraordinary man is analogous to the Earth, whose formation may result from different mechanisms distinct from that of ordinary planets.

Since the development of the Earth System depends on a powerful timing effect, it is unlikely to be formed entirely by chance. Earth is an animate planet with complex structures and magical development processes, but ordinary planets are inanimate celestial bodies with simple structures entirely different from ours.

Here, we must be aware that the long, natural chain consists of numerous sub-chains and sub-links, an immense network with multiple levels (see The Complexity, Subtlety, and Consistency of the Earth's Development, above). Each link in the network potentially has infinite solutions in its development process.

For example, every gas that makes up the air is unquestionably suitable for people to live and develop civilization. However, the formation of the atmosphere involved infinite selections in quantity and kind over billions of years, which belong to continuous random variables. The probability of any specific value of continuous random variables is believed to be close to zero, which is easily understood. Consequently, the probability of forming our existing atmosphere, with its specific composition of oxygen (21 percent), nitrogen (78 percent), and other trace gases, would be close to zero.

Another example is that during the past 4.6 billion years, the evolutionary process occurred on the Earth as follows:

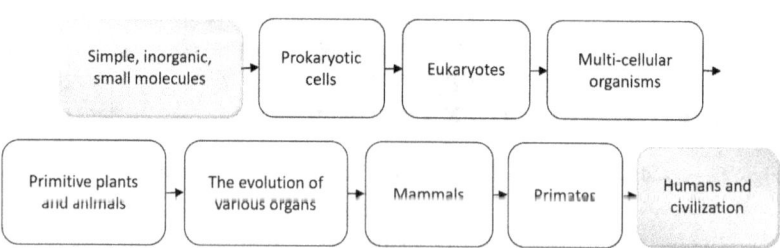

According to the theory of evolution, every stage in that evolutionary process was gradually completed under certain environmental conditions. However, the planet's development had infinite potential selections. Thus, the probability of those particular environmental conditions occurring would also be close to zero.

Given the infinite potential selections for environmental conditions on the Earth over the past 4.6 billion years, we can imagine that nature has faultlessly selected the ones favorable to the eventual appearance of humans and civilization. A few coincidences in succession may be regarded as a matter of probability, but a series of numerous coincidences may not be. The theory of dissipative structures offers no explanation for this phenomenon.

Nature has always done exactly what it should for the evolutionary process in due time. Do you believe that a monkey randomly tapping the keyboard of a computer for 4.6 billion years could compose a work of Shakespeare? Impossible! Here, the period of 4.6 billion years is limited rather than limitless. If the monkey could write a novel within a certain period, it must possess some control factor or an exceptionally clever brain system!

Pay close attention: a hypothetical monkey tapping letters (a, b, c, d, and so on) on a computer is an independent event. The probability of each letter required in a word (or sentence) from Shakespeare's works appearing by random tapping is close to zero. Therefore, for the monkey to form a poem from random keystrokes, it would take an infinite amount of time. In other words, a monkey cannot type poetry.

Every component (factor) that makes up the Earth System is a mathematically continuous random variable (X) within a specific numeric interval. Examples include:

- The distance between the Earth and the Sun,
- the density and size of the Earth,
- the speed of the Earth's rotation and revolution,
- the value of the obliquity of the ecliptic,
- the thickness of the Earth's layers,
- the abundance of the Earth's chemical elements,
- the content of salinity of seawater,
- the volume and area of the ocean,
- the area of the land,
- the location of the continents,

- the content of various gases that make up the atmosphere,
- the thickness of the layers of the atmosphere,
- and the motion state of the Moon, etc.

However, each factor that makes up the Earth System scientifically has a specific value (a), such as the size of the Earth. The probability of its occurrence as a particular value is zero because, theoretically, there are infinite possible values in the interval, $P\{X=a\}=0$ according to probability theory. It is even more remarkable and inconceivable that all these values are present in the Earth System in the same geological period (the Quaternary). Therefore, for continuous random variables, it isn't very meaningful to study the probability of the occurrence of a particular value.

Many natural events can be considered continuous random variables, and their probability of occurrence is zero. Still, this does not mean that they do not exist. In other words, for any natural event, its occurrence is unique. This is simply a mathematical concept and not of significant importance in practical terms.

This book aims to demonstrate nature's extreme complexity and rationality in time and space, which is the gateway to comprehending nature. Only with a deep understanding of this extreme complexity and rationality can we peek into the essence of nature.

WHAT EXACTLY IS THE EARTH?

Before answering this question, let us first explore two features of the development of the Earth System: natural miracles and the time effect.

About Natural Miracles

What Is a Natural Miracle?

With careful consideration, perhaps you will find every essential natural phenomenon mentioned in this book is a miracle, such as the obliquity of the ecliptic, photosynthesis, the atmosphere and its temperatures, precipitation, soil, rivers, rice/wheat, horses/camels, the two primary substance chains, etc. These are "miracles" because:

- Every natural phenomenon in the Earth's development has a complicated and stochastic process.

- Every natural phenomenon's quantity (quantitative value, scale, size, distribution, condition) and function is ideally suited for people and all living organisms to exist and develop on Earth.
- Every natural phenomenon is selected among numerous potential solutions caused by randomness.

For all these "requirements" to be met together is indeed a natural miracle! If any of the above were absent, the world we know would not exist.

Take, for example, a minor factor—Antarctica. To a great extent, this massive continent contributes to the formation of the global climate pattern. On the other hand, the world we are living in is a complicated and sensitive natural system. If Antarctica were absent in the South Polar Region, the current global climate pattern would certainly be dramatically different, resulting in a series of impacts on global oceanic currents, atmospheric temperatures and circulation, precipitation, etc. Therefore, the global environment would no longer be precisely suitable for people to live and develop.[2] In fact, all the items mentioned in Chapters II and III are all natural miracles, without exception.

So, we could say the Earth System is made up of numerous natural miracles in time and space that all came together in the Quaternary Period. This may be the greatest miracle in the universe. Don't you agree?

The Time Effect

We began this book by discovering the time effect, which is the most important feature of the Earth's development. In nature, however, only the development of life has a time effect. Let us look at the time effect of the development of life as a reference for understanding the Earth.

The Mysterious Time Effect of Gene Expression

The regulation of gene expression is undoubtedly a fundamental area of scientific research, and abundant information on this subject can be found elsewhere. However, our focus here is on the time effect in the development of living organisms, a topic that seems underemphasized in modern science.

---

2. ScienceDaily, *How Openings in Antarctic*

Gene expression in the development of life requires not only specific structures but also precise timing, a mysterious field in life studies. Two fundamental processes of life, the central dogma of biology (DNA to RNA to protein) and cell division, are both influenced by timing. This timing effect is integral to gene expression and, consequently, to the overall development of life.

Genes control every physiological stage, from a zygote developing into a fetus and eventually into an independent human being through cell differentiation. The human body contains approximately 50 trillion cells, though this number is subject to scientific debate. Each cell houses around 31,000 genes, also a controversial figure. Every gene has its own mission, producing specific proteins through gene expression to maintain various physiological functions at different stages of life.

In about nine months, a single-cell zygote fully transforms into a human fetus composed of billions of diverse cells. By age 15, a person will experience sexual desires due to hormonal changes. Around age 60, their organs and blood vessels will weaken, leading to the onset of aging and infirmity.

We can think of genes in a cell as a series of keys on a musical instrument. These tiny keys, spread across all the cells of the human body, are each played precisely on time throughout the life cycle. Before gene expression, genes are silent and inactive. The factors regulating gene expression seem to "know" which genes need to be expressed (turned on) and when they should be activated to perform their functions—demonstrating the importance of timing. Billions of genes in 50 trillion cells are expressed with precision on time at different physiological stages of a human life, composing a decades-long life symphony!

All genes are continuously controlled by regulatory factors throughout one's life, from a zygote to the fetus to one's end, not only in the structure of expression but also in the time of expression. That is the most immense, most complex symphony in the natural world! The same goes for all plants and animals.

The regulation of gene expression in all somatic cells throughout the body is crucial for realizing the various physiological functions necessary for growth and development. This precise timing, the "time effect," is truly remarkable. All genetic information in somatic cells comes from both parents, making children somewhat resemble their parents. However, scientists still have a long way to go in uncovering the mysteries of how gene expression is controlled, from turning a zygote into a fetus and eventually into a fully developed human being.

We must recognize that the structure of the human body is extremely complex, even in its smallest parts. The requirement for precise timing, or the "time effect," is essential throughout its development. Without this, the body's development would be inadequate. For example, in the human brain, any slight fault in timing or structure—such as issues with capillaries, blood vessels, nerves, or muscles—could result in abnormal development.

Now, let's turn to the factors that control the time effect of the development of the Earth.

What exactly is the Earth? (1): An accurate evolution machine

Through all that has been stated above, we can observe that the Earth System represents a profound "super design," not just in space but also in time. Each essential factor-forming process is intricate and even mysterious. Unlike other planets, Earth is remarkably complex. It appears to bear a substantial natural mission: facilitating the emergence and development of humanity and civilization. The underlying logic of this development can be summarized as follows:

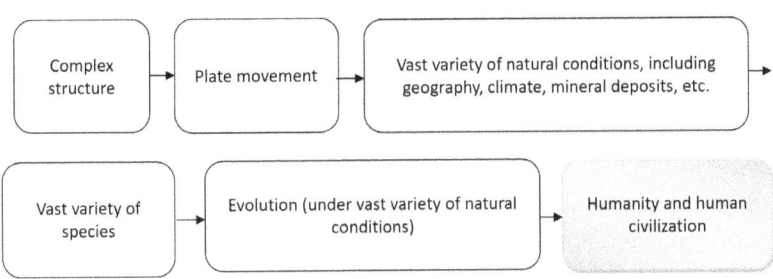

Our planet orbits at an optimal distance from the Sun within the galaxy's habitable zone. Earth benefits from crucial protective factors: the atmosphere and ozone layer, the geomagnetic field, the presence of the Moon, and the gravitational influence of Jupiter and Saturn. During its early formation, the Earth remarkably attracted and retained all the chemical elements listed in the periodic table in reasonable quantities, including rare heavy metals such as gold, platinum, titanium, uranium,

THE RATIONAL UNIVERSE EVOLVING FOR HUMANS

and thorium, which are scarce elsewhere in the universe but abundant on the Earth.

Earth possesses sufficient gravitational force to retain gases and water molecules on its surface, unlike Mars. Perhaps Earth was once enriched by special "deliverers" in the form of comets, which supplied abundant water molecules and the seeds of life. Like a meticulously designed machine with a robust, purposeful mechanism, Earth has developed a sequence of natural conditions over time that have transformed the inanimate world into a dynamic, evolving biosphere, ultimately leading to the emergence and evolution of human beings—the most significant outcome of this natural process.

If we consider 4.6 billion years as one day, or 86,400 seconds (about 53.2 thousand years per second), the development of the Earth System reached timescale accuracy in one hour. It gradually increased in accuracy as time went on.

According to the one-day analogy, timescale accuracy was achieved in about one second in the Quaternary Period. Scientists estimate that the last ice age, known as the Würm glacial period, concluded roughly ten thousand years ago, marking the beginning of the current warm period, expected to last another 15 thousand years. This suggests that the next ice age may arrive in approximately 15 thousand years, a topic requiring further research. Thus, the entire duration of the current warm period totals about 25 thousand years, significantly less than a second in the day-scale analogy.

During this period, humans transitioned into the Neolithic Period roughly ten thousand years ago, spanning four thousand years before entering the Bronze Age. This era also witnessed the maturation of various crucial species' evolution (such as polyploidy and herbivores). Nature appears to have approximately 15 thousand years remaining in the current warm period. While the scientific landscape of that distant future remains uncertain, this timeframe seems adequate for the advancement of modern science, allowing humanity to safeguard itself and embark on a broader path forward.

Here, you may realize that along the time axis, the timescale accuracy is remarkably well-suited, perfectly aligning in both duration and sequence for evolutionary processes and the development of civilization—from ancient times through the present and into the future. The following discussion aims to deepen your understanding of the importance of this precise timescale accuracy.

WHAT EXACTLY IS THE EARTH?

Imagine if the Neolithic Period had lasted as long as the Old Stone Age, spanning two million years. What if it had only lasted a few tens of thousands of years? What would have happened in these scenarios? Remember that the next ice age is projected to arrive in 15 thousand years, suggesting that an extended Neolithic Period would have been unacceptable. Nature appears to have restricted the duration of our Neolithic Period.

The key factor lies in the early maturation of the human brain before the Neolithic Period began; without this development, such an era could not have occurred. This significant evolutionary event shortened the Neolithic Period considerably compared to the Old Stone Age. The exact cause of this early brain maturation remains unknown—perhaps the mastery of fire played a role? Nonetheless, we can hypothesize that the duration of the Neolithic Period aligns well with the current warm period, which has enabled the flourishing of modern science as we know it today.

So, without timescale accuracy, could modern civilization have emerged? Don't you think that timescale accuracy is the timing requirement for the development of the Earth System? The precision achieved in the timing requirements for the development of the Earth System is truly remarkable!

Moreover, each element contributing to this design—from the Sun, the structure of the Earth, its surface, atmosphere, oceans, salty seawater, soil, plants, animals, and microorganisms—represents a sub-super-design, each playing its distinct role. These factors interact and collectively contribute to a shared natural mission. Through stochastic processes in the Earth's development, they collaboratively create harmonious conditions essential for forming a planet capable of sustaining civilization.

Thus, Earth is a unique planet where the gradual evolution of the inorganic environment fostered the evolutionary process from simple molecules to complex organic molecules, then to organisms, plants, animals, mammals, primates, humans, and eventually civilization. This can be viewed as a natural continuum of development, where nature consistently executed specific processes at the right times, highlighting the profound impact of time. Could any essential factor have been absent throughout this extensive natural journey in both time and space? Could anything have adequately substituted any given element? Could any process have been reversed? The answer is no. Even the absence of a seemingly minor factor like clay minerals would have prevented the emergence of civilization on our planet.

It is crucial to emphasize that the evolution of life is a meticulous process. Each stage in this process required specific environmental conditions, such as the first regression that led to the first appearance of life on land (see Chapter I.1, The Timely Appearance of Life on Land). For animals, the development of ears, eyes, and noses involved intricate processes. The evolution of hearing, vision, and smell required distinct environmental conditions to evolve into ears, eyes, and noses. Overall, this represents an incredibly intricate natural process occurring between the environment and evolution, beyond our full comprehension.

So, the planet Earth is simply an accurate evolution machine!

We must particularly understand the role of every factor stated above in the Earth's development in time and space, which is the key to comprehending our mysterious planet. For example, people tend to take the birth of modern science for granted and neglect the necessity of the Mediterranean with its two peninsulas. Looking at a world map, the Mediterranean is the only area with its special geographical conditions in the world. We cannot deny its necessary role in that historic course during which all the world was sleeping (!) except the beautiful Mediterranean!

From the viewpoint of probability, we must also notice that the appearance (in time and space) of every natural factor stated above was not easy with extremely low probability. Fortunately, they all came together precisely right! Accordingly, you may understand why we cannot take for granted the formation of any component of the Earth System.

But why does the development of the Earth involve so many factors that play their roles respectively in due time? Could the probability theory explain all the processes stated above? Let us continue to explore this intriguing question in the following sections.

What exactly is the Earth? (2): The Earth-womb

Having read all the formulations above, we can now consider the question, what exactly is the Earth? This is an inevitable question in modern science. There are only two development processes in the natural world: the Earth and life, and both require timing. Without proper timing, both processes would undoubtedly be in disorder. We can summarize these two processes as follows:

WHAT EXACTLY IS THE EARTH?

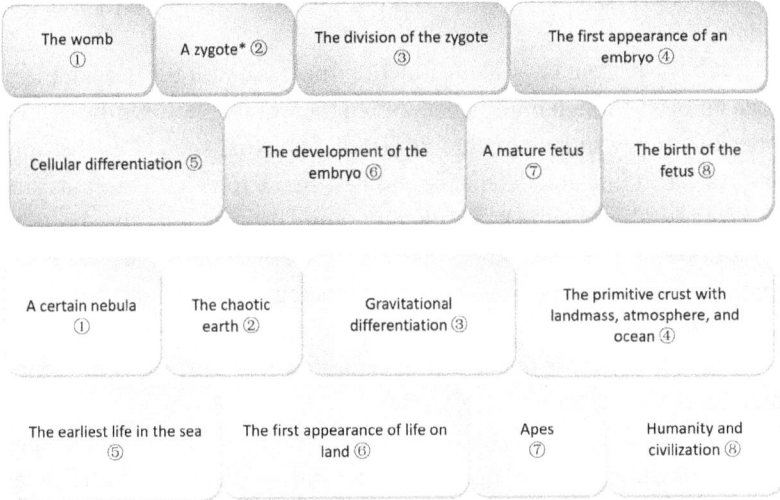

Life is a nonlinear system. A zygote (2) also belongs to a nonlinear system. A zygote is conceived outside the womb before developing into an embryo in the womb. The embryo continues to develop through cellular differentiation from (4) to (7). Bottom item (4) is the primitive Earth System being formed. Since then, the global land and sea environment has been changing. What is stressed here is the concept of the comparison rather than the detail.

If you find the comparison between the two processes unconvincing and are intrigued by this issue, let's examine our logic further to understand this strange planet better.

We could consider the Earth a massive womb in which life first appeared from "nothing." Below is a comparison between the two different wombs in function. You will soon find that both the function of the womb and that of the Earth-womb are similar in logic.

Development	The Womb	The Earth-Womb (1)
Supporting system (2)	Placenta	Photosynthesis
Stage 1 of development (3)	Zygote	The earliest cell
Stage 2 of development (4)	Cellular division	Multicellular organism
Stage 3 of development (5)	Organs of an embryo	Species
Stage 4 of development (6)	Fetus	Apes
Beginning of a new stage (7)	Birth of the fetus	Humanity and civilization

(1) A discussion of this special womb from a logical viewpoint will follow.

(2) The placenta has all the functions necessary for the development of an embryo, though much smaller than the womb. Photosynthesis provides bioenergy to support life. It also defines that the cell is the only basis of Earth Life and determines human characteristics. Placentas and photosynthesis are similar in function, and complex scientific projects need to be further researched (Earth Life is an abstract concept representing all the living organisms on Earth, being of protein and nucleonic acid), although their forms differ considerably.

(3) The zygote is one cell. The earliest cell on Earth was in monoplast (unicellular) form.

(4) Through cellular division, the zygote becomes multicellular. Through evolution, unicellular organisms became multicellular organisms.

(5) Supported by the placenta, different organs form gradually through cellular differentiation. Through evolution, supported by photosynthesis, various species formed gradually.

(6) Through development, the embryo becomes a fetus. Primates evolved from an earlier ancestor.

(7) Through birth, the fetus becomes an infant with a broad prospect in time and space. Primates further evolved into the earliest humans and beginnings of civilization, also with a broad future in time and space.

(1) Having understood the items above, let's now pay attention to the Earth-womb. The phrase "Earth-womb" is invented here to demonstrate its extreme complexity. As the ancient sea, the primitive atmosphere, and the earliest land first appeared, the Earth-womb began to work, which, through the changing environment, gave birth to life and set it into the motion of evolution, and continual evolution leads to a variety of species. The number of species on Earth has been increasing, although extinctions have occurred. But how many species are there on Earth? Though it's a controversial issue, the number is undoubtedly vast, likely in the millions.

Changes in environmental conditions are believed to have played a crucial role in evolution, which raises a question: were varying environmental conditions, occurring at different times and places, responsible for the creation of millions of species? And how were the earliest animals' sensory organs (eyes, ears, etc.) formed? Answering these questions with complete confidence is difficult. However, according to the theory of evolution, the changing environmental conditions over millions of years did lead to the creation of millions of species, including various animals with sensory organs, as well as various plants and microorganisms. In other words, if global environmental conditions had remained unchanged since the beginning of the Earth, the evolution of life would not have occurred.

To further illustrate the concept of the changing earth-womb, take the formation of global climate as an example. The global climate mainly refers to global atmosphere temperature, which is subject to many components that, to a certain extent, influence the evolutionary process of living organisms. We may well say that the living state of all life is controlled by temperatures. But the formation of global atmosphere temperature is a complex process involving many sub-factors that vary randomly over time.

We can summarize these sub-factors into roughly two kinds of factors:

Astronomical factors, including solar radiation, revolution, and rotation, as well as the obliquity of the ecliptic and asteroid collisions. Some of these factors change slightly over time. During the Quaternary period, slight changes in the Earth's movement led to the alternation of ice and interglacial periods.

Earth surface factors, including the changing distribution of land and sea (continental drift, seafloor spreading), volcanic eruptions, and various geological activities. These factors form various geographical environments, as well as the atmosphere's composition. In addition, life itself (global biosphere) also has an important influence on global temperatures, especially after life's massive appearance on land about 400 million years ago.

All of these factors interact to form a highly complex nonlinear system that contributed to the evolution of life. Specifically, over billions of years, the global temperatures varied from the high-temperature fluctuation in the Paleozoic and the Mesozoic to the gradual decrease in the Cenozoic until the appearance of the Quaternary Ice Age. Correspondingly, the plant kingdom evolved from fragile gymnosperms to angiosperms with tenacious vitality, especially the emergence of polyploidy. Also, the animal kingdom evolved from amphibians to mammals until the emergence of humans. Therefore, the complex development of global temperature perfectly met the requirements for the evolution of life.

THE RATIONAL UNIVERSE EVOLVING FOR HUMANS

On a minuscule point in the universe, nature orchestrates numerous factors to precisely maintain a narrow range of temperature changes around the freezing point of water, creating "the changing Earth-womb" that has persisted for billions of years and fostered the evolution of life. This is a scientific fact. However, humanity cannot fully comprehend this process due to its extreme complexity and marvel (as examined in previous chapters).

The development of the Earth-womb—from a primordial nebula to the current Earth System and the evolutionary journey from unicellular organisms to human beings—represents an incredibly complex natural process. Wouldn't you agree? This complexity seems only to be solved in a limited time by the guidance of Einstein's concept of natural Reason.

What exactly is the Earth? (3): What is the regulatory factor in the development of the Earth?

Having determined the regulation of gene expression above, we should note some relevant factors, specifically those that regulate the development process of life from a zygote to an organism to its end, step by step, on time, realizing the time effect on one's life.

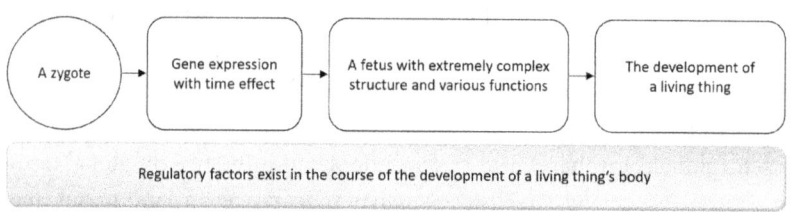

The above process reveals an important principle:

> *The development process of a nonlinear system with a highly complex structure must be regulated by certain factors to successively reach a particular higher state in due time.*

The time effect is merely the result of the regulation of gene expression over time. Our understanding is as follows: in the natural world, randomness exists everywhere at all times, and natural processes are universally stochastic without exception. Without the role of regulatory factors, any complex nonlinear system could not reach a certain condition with higher function in due time. The point to note is the function seems to be the aim.

As genes create many miracles of life in time and space, how then do the many miracles in the Earth System come into being? In this regard, scientists use the same term to describe the development of both the Earth and life, namely, cell differentiation and Earth differentiation (see Chapter I.1, The Timely Differentiation of the Earth and the Timely Appearance of the Oceans).

After the differentiation in time, a simple planetesimal became a layered planet, and the Earth System began to form, including the primitive sea, landmasses with landforms, and a primitive atmosphere. As the Earth System continued to develop step by step, life emerged in the sea and began to evolve alongside the development of the inorganic Earth System. Interestingly, biological scientists and geologic scientists separately use the same term, differentiation, implying both the development of life and the development of the Earth share some similarities.

As mentioned previously, there are numerous instances of the time effect on the development of the Earth System, although strong winds, heavy rain, eruptions, earthquakes, and clouds can occur at any time without any requirement for timing in its development process. To deepen your impression of time effect, consider some "sudden" natural events to be like a software plug-in, such as the formation of the Moon, the Cambrian Explosion, the extinction of dinosaurs, the appearance of the warm age and ice age in the Quaternary Period, the formation of the GRV and the Mediterranean, etc. No matter their causes, all those natural events took place in due time, representing the reliability of the time effect. Ultimately, the significant natural events occurred precisely when they were meant to, ensuring the natural development of the Earth. Without these events, human beings and civilizations would not have emerged.

As mentioned repeatedly, the development process consists of numerous, seemingly endless sub-processes. Do you think all these procedural natural miracles could occur precisely on time without regulatory factors? It's important to recognize that natural processes are stochastic. Can you identify other natural processes with similar characteristics?

Now, reflecting on the previous section, we can imagine that all the initial elements of the essential factors of the Earth System were individually contained in that simple planetesimal, and then through an immensely complex development process, gradually formed the Earth

System of today. The initial components may have been gases, cosmic dust, water molecules, small solids, etc., comprising various chemical elements. As it developed, the planetesimal became increasingly complicated, finer in structure, and higher in function. Today, every essential factor of the Earth System has its own complex but reasonable substructure without any exceptions. For example, you can see the complex structure of clay minerals or cells through a powerful microscope.

Generally, every nonlinear natural process is full of bifurcation and selection. However, the Earth's development process is like the development of a zygote in some ways. Both start from a small, simple object and continuously evolve into increasingly complex structures with more advanced functions. Both are nonlinear systems based on stochastic processes and strictly require proper timing along the time axis. Each stage must correspond to a specific state, which is what we mean by time effect. Without proper timing, the natural processes involved in either development would become disordered along the time axis.

Whether for the development of the Earth or a zygote's development, time effect is the key to the whole progression, although we don't think of Earth as a living thing. This reveals a clue to the question, what exactly is the Earth?

Nature always does exactly what it should with our planet in due time, neither later nor earlier, just on time. Does timing also control the development of the universe? Perhaps it does, but the timescale accuracy of the Earth System seems to be based on that of the solar system. We logically believe that timing controls every cosmic event, major and minor, on the long timescale of the universe.

Now, you might understand why most of this book is devoted to recognizing the essential factors of the Earth System and the requirement for proper timing in their development processes. Undoubtedly, many more instances of time effect in the Earth's development will be discovered as science advances. Do you now understand time effect? Have you accepted that concept? Don't you find it thought-provoking? If you disagree, you are welcome to question it and contact the author for further discussion.

The function of the Earth System can be simplified as follows:

WHAT EXACTLY IS THE EARTH?

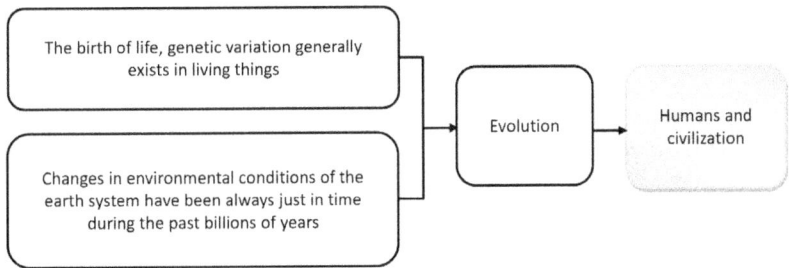

Why were changes in the environmental conditions of the Earth System always just in time over the past billions of years? The fact that the time effect manifests itself in the Earth's development is irrefutable and not speculation. We should consider the time effect as a whole, not necessarily studying the cause of time effect on an individual detail. Time effect may have been in play throughout the whole progression of the Earth's development. Otherwise, the process could not have successfully been completed.

The above explains how the time effect manifests in an exceedingly complex system. However, is there truly a regulatory factor in the Earth's development as there is in the development of life? What could this regulatory factor be? What governed the transformation of initial elements into the Earth System over 4.6 billion years? The following sections will further explore this topic.

What exactly is the Earth? (4): About the natural philosophy of Albert Einstein

The following quotes from Einstein express the above ideas best:

> Quantum mechanics is certainly imposing. But an inner voice tells me that it is not yet the real thing. The theory says a lot, but does not really bring us any closer to the secret of the 'Old One.' I, at any rate, am convinced that He is not playing at dice.[3]

> The most beautiful experience we can have is the mysterious.[4]

> A knowledge of the existence of something we cannot penetrate, our perceptions of the profoundest reason and the most radiant beauty, which only in their most primitive forms are

---

3. Einstein, *Letter to Max Born*
4. Einstein, *The World*

accessible to our minds—it is this knowledge and this emotion that constitute true religiosity: in this sense, and in this alone, I am a deeply religious man.⁵

I am satisfied with the mystery of the eternity of life and with the awareness and a glimpse of the marvelous structure of the existing world, together with the devoted striving to comprehend a portion, be it ever so tiny, of the Reason that manifests itself in nature.⁶

To know that what is impenetrable for us really exists and manifests itself as the highest wisdom and the most radiant beauty...⁷

But whoever has undergone the intense experience of successful advances made in this domain is moved by profound reverence for the rationality manifested in existence.⁸

My religiosity consists in a humble admiration of the infinitely superior spirit that reveals itself in the little that we, with our weak and transitory understanding, can comprehend of reality.⁹

But whoever has undergone the intense experience of successful advances made in this domain is moved by profound reverence for the rationality made manifest in existence.¹⁰

I cannot prove to you that there is no personal God, but if I were to speak of him I would be a liar.¹¹

I believe in Spinoza's God who reveals himself in the harmony of all that exists, but not in a God who concerns himself with the fate and actions of human beings.¹²

Everyone who is seriously involved in the pursuit of science becomes convinced that a spirit is manifest in the laws of the Universe—a spirit vastly superior to that of man ... In this way the pursuit of science leads to a religious feeling of a special sort,

---

5. Einstein, *The World*
6. Einstein, *Letters to Solovine*, 102
7. Einstein, *The World*
8. Einstein, *Ideas and Opinions*
9. Einstein, *Albert Einstein*
10. Einstein, *Ideas and Opinions*
11. Hermanns, *Einstein*, 132
12. Einstein, *Einstein Believes*

which is indeed quite different from the religiosity of someone more naïve.[13]

The word God is for me nothing more than the expression and product of human weaknesses . . .[14]

Building on these statements from Einstein, we can imagine he probably considered "the Reason in nature" or "the infinitely superior spirit" or "the highest wisdom" as the primary order or the fundamental law or the essential property of nature. The Reason in nature belongs to the natural world, not a supernatural power outside it. Reason or wisdom may be a character of Nature itself.

It's worth noting that nature operates rationally, as Einstein concluded from the human perspective—an important premise. Given nature's rationality for humans, it logically follows that humans arise through the natural development that the universe has undergone for humanity.

According to this book, the Reason is highly intricate. Every relevant factor—such as the Earth's structure, including size and mass—is governed by specific natural laws and scientific principles in fields like physics, mathematics, chemistry, and mechanics, forming an intricate formation process (see Chapter II.1, The Definition of the Earth System). *Therefore, demonstrating the rationality of an essential factor and the formation of that rationality are two different things.*

What Exactly is the Earth? (5): What Does the Controversy between Einstein and Bohr Imply?

Einstein was not only a great scientist but also a great natural philosopher. His natural philosophy is based on his profound theory of relativity. His view on Nature is worthwhile to continually strive to understand, although many people could hardly reach his depth of comprehending Nature. Apparently, the more you understand nature, the more you will align with his views. For example, if you think of the wide variety of uses of modern electronic technology, don't you think the structure of the atom and its magical electrons represent the highest wisdom of nature?

---

13. Einstein, *Albert Einstein*
14. Einstein, *Der Einstein-Gutkind Brief*

THE RATIONAL UNIVERSE EVOLVING FOR HUMANS

The famous controversy between Einstein and Niels H. D. Bohr (1885–1962) is a valuable reference. Does God really play dice? Einstein and Bohr have opposite viewpoints on this topic.

Einstein's and Bohr's views are correct when we use them to understand the development of the Earth and that of a zygote. Both randomness and regulatory factors seem essential to perfectly complete these highly complex processes. The interaction between these two aspects results in rationality both in terms of time and space. Without randomness, there would be no infinite opportunities for selection, and without regulatory factors, there would be no rationality in that selection process. In essence, understanding the essential factors and timing factors reveals the rationality that manifests itself in each process. In other words, probability cannot ultimately govern the natural world. For instance, carbon dioxide levels in the atmosphere varied randomly and extensively over millions of years, yet they evolved into conditions precisely suitable for the emergence of humans and civilization just in time.

The time effect manifesting itself in the Earth's development is a fact rather than speculation. Without timing, the appearance of every natural factor along the time axis would fall into disorder and be unable to successfully culminate in the development of the Earth of today. Isn't this apparent?

Explaining the time effect raises two distinct questions. First, whether the time effect truly exists in the Earth's development. Second, what factors can generate this time effect. These questions fall into separate categories. Our examination above addresses the first question. Einstein proposed his perspective on the second issue, which aligns with the author's views in this book.

It's important to note that Bohr's perspective primarily pertains to the subatomic realm, whereas Einstein's perspective is focused on the macroscopic world. In this sense, both of their views are correct.

Figure 4-6: Albert Einstein and Niels H. D. Bohr in 1925 (Source: Public domain).

If you read this book patiently and carefully, you would probably realize that every essential factor stated in this book is a miracle. There is no room for change in all these miracles and no alternative. Do you have a better solution than precipitation? No!

You would probably also realize that in the course of natural development, every choice nature has made is necessary for forming a civilization planet, such as the distance between the Sun and the Earth chosen by nature 4.6 billion years ago. Did nature "know" 4.6 billion years ago that this distance was best for life on Earth? Is this what Einstein meant by the Reason in nature? Yes, because only the rational choice is the right choice. Similar options abound: the mass and size of the Earth, the abundance of various chemical elements required to form the Earth, the global distribution of land and sea, the composition of the atmosphere, and even the formation of massive clay minerals around the world. They are too numerous to mention one by one.

So, the Reason seems to be the regulatory factor in the development of the Earth, although it is abstract and invisible. Over the past 4.6 billion years, in this long stochastic process of natural development, nature has always chosen according to the needs of forming a civilization planet. Is it not the embodiment of "the highest wisdom"? The following diagram represents this process (see What exactly is the Earth? (4)).

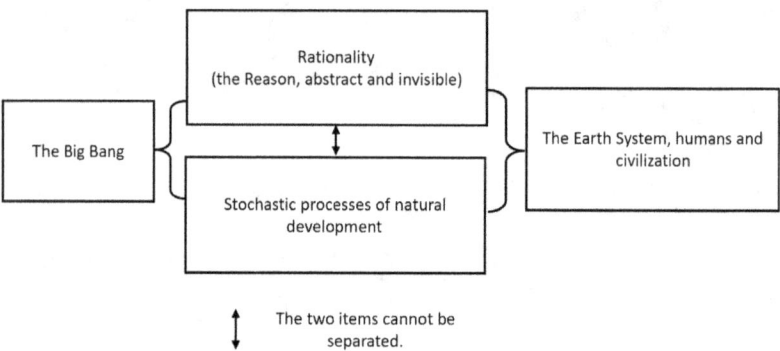

Therefore, the Earth System is a precise combination of numerous relevant factors in space and time, achieved seemingly under the control of the Reason in nature. If you disagree, how do you explain this process of natural development? We must realize that a purely random process will not produce any meaningful results or that the valuable results produced by a random process will appear after an infinite length of time. However, the effective radiation period of the Sun is limited and cannot wait forever.

What Exactly is the Earth? (6): Frequently Asked Questions

There are no identical natural processes in detail in the natural world. The development process of our planet cannot be repeated. However, some ideas support the view that civilization may occur on other planets. Following are the typical two.

The first supposition is that entirely different intelligent life and their civilizations could, through evolution, adapt themselves to different environments on other planets. However, we must be aware that the periodic table is the only basis for forming the material world around the universe. The following idea speaks to this issue.

The Theory of the Function of the Earth System

To grasp this issue, one must employ common sense. The book begins by introducing this common sense, referred to as the principle of natural philosophy, which runs through its chapters, though not explicitly mentioned each time. We aim to utilize this principle to explain Earth's uniqueness.

In any system, animate or inanimate, its structure is believed to always correspond to its function—how it operates or its purpose. Conversely, a specific function requires a corresponding structure, as in examples like clocks. Even slight changes in a system's structure lead to corresponding changes in its function. Thus, we can assert that a system's function is entirely dependent on its structure, illustrating the consistency between structure and function.

Under given astronomical conditions, the Earth System is mainly composed of immense molecule groups, namely the gas group (atmosphere), the solid group (Earth itself), the liquid group (the oceans), and the mixture group (organisms). Each group has its functions:

- The solid group possesses many functions, such as gravity, geomagnetic field, plate movement, etc.
- The gas group, like the Earth, also has a layered structure. Its functions are filtration of solar radiation, supporting life, the material base of forming pressure, temperature, weather, combustion, etc.
- The liquid group consists of seawater with some chemical components. Its functions are supporting aquatic life, adjusting global

atmospheric temperature and humidity, forming weather patterns, precipitation, etc.

- The mixture group is the most complex, consisting of millions of species. Its functions are evolution, forming organic materials for combustion, reforming the gas components of the atmosphere, etc.

These four groups and their functions interact, causing the Earth System to develop more complex structures and higher functions from the beginning through today.

Here, we can liken all chemical elements to an assortment of bricks, each varying in size and shape with unique properties that cannot be substituted. These bricks combine to form diverse molecules, which in turn assemble into the Earth's development system, representing three distinct levels. The functions at each level are crucial, with no viable substitutes. For instance, water molecules serve numerous essential functions in the natural world, and there are no alternatives for water. While liquid ammonia shares some similarities in function with water, other than a boiling point of—29.2°F (-34°C), it is not indispensable for the natural world. Therefore, liquid ammonia cannot replace water molecules in natural processes.

Understanding the harmony between structure and function, we can easily conceive that the only plausible explanation for the formation of the Earth System was to create a civilization-brewing machine, assembled with bricks that perfectly fit at all three levels. This machine's success in cultivating civilization on our planet is evident. Throughout Earth's lengthy development, any alteration in structure at any level would have correspondingly altered functions, potentially leading to the failure of this civilization-brewing machine. This concept could be termed the theory of the function of the Earth System. Each link in every chain has unique functions for which there may be no substitute.

Now let us summarize the argument above as follows:

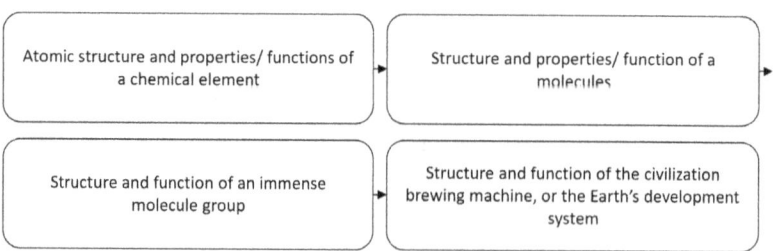

For further understanding, we may also combine the four groups to show their complex relations.

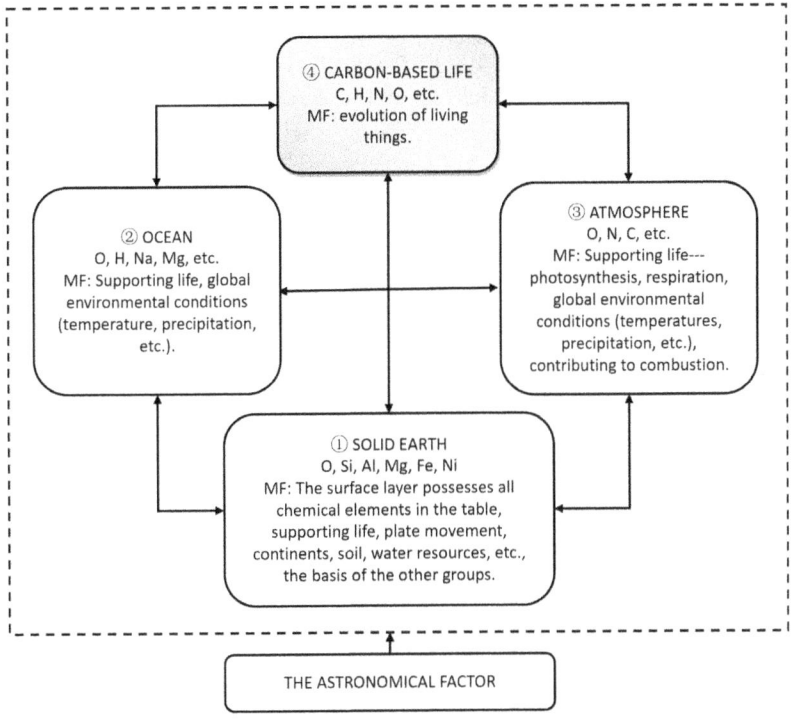

MF: The main functions

Let's explore the relationships between the elements in the groups. The diagram depicts four distinct groups and the elements comprising each group. Functions are carried out by specific molecules, such as H_2O, composed of hydrogen (H) and oxygen (O). Various molecules form different extensive groups accordingly. The arrow indicates how these molecules support or influence the distribution of elements within a group.

(1) SOLID EARTH. Earth is a complex mixture, not only with a layered structure, but each layer has different chemical elements, proportions of element content, and functions. Earth is the base of the other three groups. The primary elements in the Earth's crust by weight percentage are oxygen (46.95 percent), silicon (27.88 percent), aluminum (8.13 percent), iron (5.17 percent), and magnesium (2.06 percent). The corresponding chemical compounds include SiO_2 (57.76 percent), Al_2O_3 (15.55 percent), and various forms of iron oxides (FeO, Fe_2O_3),

magnesium oxide (MgO), etc. The main element of the core is iron, followed by nickel. Based on the distribution of the elements, the main functions of the Earth are to form plate movement (to form the changing Earth's surface) and the geomagnetic field; to provide (through eruptions) water molecules (eventually becoming seawater), carbon dioxide, and nitrogen to the Earth's surface, which are the components of air; and to make a tremendous amount of soil and rock (Si, Al, Mg, etc.) around the world.

(2) OCEAN. The ocean is a complex mixture with a specific chemical content. The primary molecule in seawater is H_2O. Seawater salinity (S percent) varies between approximately 3.2 percent and 3.6 percent, with the main chemical components of the salinity being sodium chloride (NaCl, 70 percent) and magnesium chloride ($MgCl_2$, 14 percent). The vast amount of seawater, containing elements like sodium, chlorine, and magnesium, performs several critical functions: it supports a diverse array of marine organisms; through evaporation, it supplies water molecules to the atmosphere and absorbs carbon dioxide, thereby regulating the global environment (temperature, air humidity, precipitation, etc.). This regulation is facilitated by the high specific heat capacity of seawater, which is crucial for the existence of life both on land and in the sea. Note that the specific heat capacity differs significantly between pure water and seawater. In addition, the remains, remnants, and excrement of marine life contribute to the elements in seawater, while marine organisms absorb a significant amount of salt, thereby helping to adjust the salinity of seawater.

(3) ATMOSPHERE. The atmosphere is a complex mixture, not only with a layered structure, but each layer has different gases, proportions of gas content, and functions. The main molecules of air are nitrogen, oxygen, and carbon dioxide. The main functions are to support photosynthesis and carbon-based life, both on the land and in the sea, and to allow for the combustion of fire, the parent of civilization. Some air can be absorbed by seawater, increasing the content of oxygen in seawater, which is necessary for marine organisms. Carbon dioxide in the air, through the greenhouse effect, keeps the globe warm, and some carbon dioxide is dissolved in seawater, thus potentially increasing the carbonic acid in the ocean. Today, seawater is becoming more acidic due to increased carbon dioxide in the air, imperiling marine life. Essentially, the ocean and the atmosphere interact as a dynamic system.

WHAT EXACTLY IS THE EARTH?

(4) CARBON-BASED LIFE. The other three groups strongly support this group, allowing for carbon-based life and supporting the continual evolution of increasingly advanced life. In return, carbon-based life provides oxygen and carbon dioxide to the air and eventually becomes organic deposits, an important component of the Earth's structure. The existence of life affects the environment, which in turn affects the existence of life. It's a complicated process.

These four groups have been developing since the beginning. For example, the composition of gases in the atmosphere has gradually changed throughout the Earth's development, affecting its functions. Similarly, plate movements have created various geographical conditions influencing evolutionary processes. Together, these four evolving groups form an incredibly complex dynamic system. Although we do not know the precise details of how this system developed over millions of years, we believe that the system's development, which supported the evolution of humans from small inorganic molecules, could not have occurred if even a minor factor had changed. For instance, silicon dioxide (SiO_2) is a major chemical component of the Earth's surface layer. Without sufficient silicon, the vast quantities of clay minerals found worldwide could not have been produced (see Chapter II.1, The Soil). Any change in structure would result in a change in function.

For life to be composed of elements other than carbon, hydrogen, oxygen, nitrogen, etc., seems impossible because every chemical element possesses unique properties that nothing could substitute. The development of a planet that can support a civilization following the pattern of the Earth but with a different process of development (say, continental drifting in another way) would also be impossible because any "brick" in the Earth's development system could not be changed either in location or in direction. Earth is irreplaceable (remember the principles of natural philosophy; see Chapter II.3, The Rationality of the Earth System). In this regard, Einstein once said, "The more a man is imbued with the ordered regularity of all events, the firmer becomes his conviction that there is no room left by the side of this ordered regularity for causes of a different nature."[15]

Moreover, we can consider "structure and function" as a unified concept to understand the origin of our planet and life. Over the past 4.6 billion years, the evolving structure of the Earth System has led to

---

15. Einstein, *Ideas and Opinions*

corresponding changes in its functions, ultimately resulting in the evolution of humans and the rise of civilization.

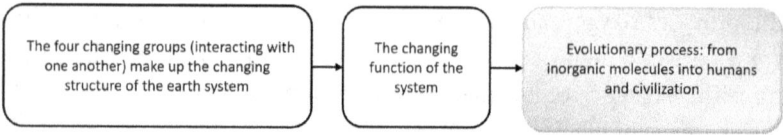

Please understand the relation between the function and the structure above. It is a complicated dynamic system as a whole. Any evident change in the structure would result in a difference in function: little hope for forming a civilization planet.

Since all natural processes are stochastic (although they are subject to laws of motion), the process from a nebula to a civilization planet could not be replaced by any other solution because any change in the process would result in change both in structure and in function and would therefore lead to failure in the process of successful evolution. The process resembles a zygote developing into a fetus and then an infant. Both demonstrate that changing of structure results in a changing of function.

To concretely understand the statement above, consider natural chains 1–3. You'll find that crustal or plate movement was a primary factor in changing global environmental conditions and promoting evolutionary processes. This crustal movement was caused by changes in the mantle and the crust.

The Infinite Universe

The second idea supporting the possibility of civilization on other planets is that since the universe is infinite, there must be other planets with civilizations. Here, infinity is a mathematical concept rather than a number. Conditioning the chance of forming a civilization planet on infinity essentially amounts to "zero." Would you agree to a contract if the repayment period were infinite? It would effectively mean no repayment! Moreover, infinity and nonlinear systems are two different concepts. The development of the Earth System consists of seemingly endless subprocesses. An infinite universe could not result in two identical natural processes in detail (see Hawking's ideas in Chapter IV). Therefore, from a probability standpoint, infinity would be inadequate for forming a civilization planet within a certain period.

Summary

With all the chemical elements fitting together perfectly like bricks to construct a civilization planet, the Earth System appears to be the only solution. From the viewpoint of a nonlinear system, no identical natural processes exist in the natural world. The formation processes of the Earth System are so complex, precise, and subtle that any deviation would fail to produce a civilization planet due to the consistency of structure and function. In reality, both the development of the Earth and the evolution of life are far more complex than we currently understand.

Chapter V

The Remote Future of Humanity

Based on modern science, the universe is thought to have taken 13.7 billion years to complete the extremely complex process of producing highly intelligent humans on Earth. This fact inevitably leads to a serious question: why did the universe do this? What is the significance of the existence of humans in the universe? What is the end of humanity? Exploring the remote future of humanity through scientific reasoning may be the best way to understand these weighty questions. This chapter is also necessary for the argument in the following chapter.

HUMANKIND—THE GREATEST SECRET OF THE UNIVERSE

People have been searching for extraterrestrial civilization for decades. Particularly, during the last 70 years, with the application of satellites and space probes, exploration has begun to enter a completely new stage. Unfortunately, compelling evidence is yet to be found, so we can still say that humans are the only intelligent life in the universe.

Human civilization has developed over thousands of years. Today, we have modern science, including the ability to land on the Moon and Mars, make nano-robots, and so on.

Scientists estimate that Earth will be habitable for another a few hundred million years. What will intelligent life look like in the remote future, say, in some ten thousand years, a million years, or one or two hundred million years? What will their civilization, their natural mission, be? This topic is interesting but also highly abstract. No one knows the exact answer, but building on our modern scientific knowledge, we deduce what will transpire from the physiological characteristics of human beings. The primary method of the development of modern science is through experimentation and logical reasoning.

As you know, photosynthesis is the basis of life on Earth. Without photosynthesis, there would be no life, no evolution, and no human civilization. So, we must start with photosynthesis.

Photosynthesis is the Basis of Life on Earth

Life is the greatest natural miracle created by the universe. Modern science tells us that the universe was born 13.7 billion years ago. After about 10 billion years of galaxy development, it was not until 3.8–3.5 billion years ago that the first life appeared on Earth. But the birth of life in the universe is an inconceivably complicated natural process. One of the primary problems to be solved in this process is the energy source for life. The simple truth is that life must consume energy to survive, and at the same time, life must continue to acquire energy to replenish the energy it consumes, or it cannot continue to exist.

So, the question arises: how does life get its energy from the natural world? In other words, how does life transform natural energy into biological energy? Solar energy, or photons, is the main source of natural energy available everywhere on the planet. So, the question becomes, how do photons convert into biological energy? Nature's solution is photosynthesis. In a very clever way, photons, water molecules, and carbon atoms in carbon dioxide are combined, producing glucose, which is the basis of life on Earth.

Why must photosynthesis be the mechanism for biological energy? Is there no other way to transform natural energy into biological energy? This is a complicated question with a long, complicated answer, but simply put, nature can only use the element of carbon to form biological energy because only carbon can form large molecules (see Chapter III, Understanding the element carbon).

Life is a highly complex organic organization, necessitating the presence of macromolecules. Photosynthesis utilizes water not only because this process requires the breakdown of water molecules but also due to water's numerous essential functions, which are irreplaceable for plants, animals, and the Earth's environment. Sunlight, carbon dioxide, and water molecules are the three essentials for life, with no alternatives. Among these, carbon is the most critical element. Notably, in photosynthesis, only gaseous carbon oxides can be absorbed. This is a unique exception in the periodic table (see Chapter III, Understanding the element carbon). Isn't that surprising?

An organism with a physiological organization based on the element of carbon is called carbon-based life. Under the Sun, plants synthesize organic matter from carbon dioxide and water molecules. That is the simplest solution and the only one in the periodic table.

Now, let's look briefly at the process of photosynthesis. Generally speaking, the leaves of green plants contain organelles called chloroplasts (Figure 1–1: Plant cell chloroplast), found in the cells of plant leaves. In sunlight, chloroplasts absorb carbon dioxide molecules from the air. A complex chemical process occurs through the involvement of photons and water molecules, eventually forming glucose, which is stored in cells, becoming the only source of biological energy in the natural world. Some strict conditions are required for the process of photosynthesis to take place. There must be a sufficient number of carbon dioxide molecules in the gas state, an adequate amount of water molecules in the soil, ample sunlight in the surroundings, and ideally, environmental temperatures should range between 59°F (15°C) and 104°F (40°C). The probability of all these conditions forming simultaneously in the universe is close to zero.

Photosynthesis is not only the basis of life on Earth but has also supported the evolution of life for as long as 3.5 billion years. Humans could not survive for long on a planet where photosynthesis cannot occur.

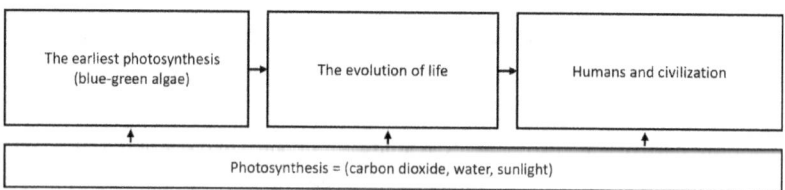

Photosynthesis Determines Life in Cellular Form

Let's begin with the concept of chloroplasts. A chloroplast is a tiny organelle, much smaller than a cell, with a complex structure. While the origin of chloroplasts remains a scientific mystery, it is crucial first to understand the relationship between chloroplasts and cells to comprehend the existence of life.

Chloroplasts are composed of proteins. Chloroplasts contain small amounts of DNA and RNA, enabling them to produce some proteins as needed. Additional proteins required by the chloroplast are produced and supplied by the cell, establishing a relationship of interdependence between chloroplasts and mesophyll cells. Chloroplasts, which serve as tools for photosynthesis, cannot survive without cells. The cell provides a suitable environment and the necessary proteins for chloroplasts while also utilizing glucose to generate energy and support plant growth. Thus, the mutual benefit between cells and chloroplasts forms the basis for the evolution of life.

Building on the close relationship between chloroplasts and cells, we may well say that photosynthesis defines life as something that must exist in the form of cells. The earliest cells are self-replicating prokaryotic cells. More than a billion years later, prokaryotic cells evolved into eukaryotic cells, and then, due to competition for survival, life diverged into plants and animals. Some organisms relied on photosynthesis for energy, evolving into plants, while others opted to consume other organisms, evolving into animals. Thus, the world, at its core, has a foundation rooted in photosynthesis. Animals are completely dependent on photosynthesis for their survival. Humans are no exception.

Chloroplasts and photosynthesis are only one of the functions of plant cells. Both plants and animals have various cells with different functions. Cells are the most basic unit of all life, regardless of their species. Life made up of anything other than cells is hard to imagine.

Photosynthesis Determines Reproduction through Sexual Means

Photosynthesis dictates that cells are the fundamental form of life. In other words, life is composed of cells whose primary function is to replicate themselves, as it is their inherent nature. Through self-replication, cells divide, and this division enables life to multiply. Therefore, cell division is the essential mechanism for the reproduction of life. Throughout the evolution of life, cell division also provides the physiological basis for sexual reproduction.

Scientists believe that the sexual reproduction of plants and animals evolved from eukaryotic cells, marking a significant advancement in evolution. Prokaryotic cells transitioned into eukaryotic cells between 1.3 billion years and 600 million years ago. Unicellular organisms evolved into multicellular forms, eventually giving rise to primitive plants and animals. Subsequently, gender differentiation emerged. Through meiosis, cells produce sperm and eggs, which, upon fertilization, form a diploid zygote, culminating in the birth of a new organism. Thus, it can be argued that sexual reproduction traces its origins to photosynthesis.

Is sexual reproduction necessary for the evolution of life? To explore this, let's have a look at the substance of sexual reproduction. For plants, sexual reproduction generally occurs within the plant itself. Flowers, for example, are the most common "factories" where plants reproduce, producing fruits and seeds by pollination. Sexual reproduction of plants is only a means to produce the next generation. There is no other meaning.

However, for animals, sexual reproduction holds deeper meaning. Life on Earth has been evolving for approximately 3.8 to 3.5 billion years. Until about 400 million years ago, life was predominantly confined to the ocean, where the seawater provided a relatively safe environment. With the emergence of life on land, organisms encountered a more complex environment encompassing the atmosphere, sunlight, and various natural factors such as fluctuating temperatures. Consequently, the evolution of life was significantly shaped by these changing environmental conditions.

Life on Earth is believed to have undergone five mass extinctions, occurring approximately 440 million years ago, 365 million years ago, 230 million years ago, 195 million years ago, and 65 million years ago. Without sexual reproduction, animals would not have survived these extinctions. This is because animals are typically differentiated into male and female, and their bodies produce sex hormones that strongly encourage both genders to mate and produce zygotes for the next generation. Nature is pretty clever.

In both genders, sex hormones generate intense physiological demands, known as sexual desire, which drive the production of the next generation. In humans, sexual reproduction can also give rise to feelings of love, which have served as the physiological basis for family formation for thousands of years, with the family being the fundamental unit of society. Sexual reproduction remains the sole means of human survival from generation to generation. Without sexual desire, humans could not have survived to this day, given the numerous adversities that have historically threatened human existence. It's important to note that "survival" here refers to a portion rather than the entire population.

However, sexual desire, along with love and emotions, is complex and can bring about a range of emotions and lead to misconduct and other issues. Unlike animals and plants, human mating is not seasonal; if it were, human society would face significant challenges. Humans can engage in sexual activity at any time, which, under the influence of sex hormones, particularly in youth, can lead to various social problems due to heightened sexual demands.

Photosynthesis Determines the Fragility of Human Existence

Photosynthesis dictates that carbohydrates, such as starch, must be our primary energy source, necessitating a well-developed digestive and excretory system. Enzymes convert starch into glucose, which requires oxygen for energy release, necessitating a robust respiratory system. Our facial features, limbs, and other structures meet the demands of both the material and energy metabolic systems. To sustain life, nutrients and oxygen must be transported throughout the body, requiring a comprehensive circulatory system powered by the heart and supported by extensive vascular tissue.

Furthermore, the coordination of systemic functions requires nervous and secretory systems. Therefore, functions like food intake, water consumption, oxygen inhalation, excretion, and sleep are essential for human life. Notably, the human brain holds paramount importance.

The human brain comprises over 100 billion nerve cells interconnected with synapses, a discovery in brain research comparable in significance to DNA in genetics. It is estimated to possess over a trillion synapses—an astonishing number. Despite constituting only about 2 percent of body weight, the brain consumes 25 percent of inhaled oxygen. Thus, plants, animals, and especially humans possess extremely complex bodies to support extensive and continuous metabolic processes that rely on precise environmental conditions, including temperature, humidity, atmospheric pressure, and oxygen levels.

The main component of the human body is water. The optimal temperature range for human survival is between 63°F (17°C) and 79°F (26°C). Outside this range, we must rely on our capabilities to protect ourselves. Early humans, for instance, sought shelter in caves and wore animal skins to stay warm during winters. If the ambient temperature strays significantly outside this range, our ability to protect ourselves and ensure survival diminishes.

Humans depend on oxygen for survival, requiring a specific oxygen level in the surrounding environment. The current atmospheric oxygen level on the Earth, at 21 percent, is optimal. Deviations from this concentration, whether too low or too high, would significantly impact the evolution of life, human survival, the use of fire, and the overall safety of the environment. The human living environment also necessitates a specific air pressure. The current standard atmospheric pressure is ideal for humans. Any increase or decrease in air pressure would cause discomfort, with greater deviations leading to increased discomfort.

Given photosynthesis and the reliance on carbohydrates, human aging is inevitable, limiting our lifespan to approximately a hundred years at most. Beyond physiological needs, there is also a spiritual requirement for the environment. Humans are a product of nature; thus, we thrive in natural surroundings, particularly amidst green plants. If individuals live in confined spaces devoid of natural colors throughout their lives, their spiritual well-being would likely suffer, impacting creativity. Further, protecting ourselves from radiation is crucial, as it can harm our cells.

Our sophisticated physiological organization and variety of necessary living conditions lead us to the conclusion that humans are incredibly vulnerable. All the requirements for human life represent the fundamental weaknesses of humanity on our planet.

Photosynthesis Determines Human Characteristics

Human beings are unequivocally the most intelligent beings on Earth. Intelligence and reasoning are crucial for human survival. While plants and animals rely on their functions—such as prolific reproduction or superior hunting skills—to survive natural selection, humans navigate diverse survival challenges through intelligence and reason, bypassing natural selection. Curiosity, stemming from intelligence, is humanity's most defining trait. Driven by curiosity, humans tirelessly explore the unknown, perpetually advancing and cultivating civilization. This cognitive prowess evolved gradually alongside the physical basis of evolution, which is rooted in photosynthesis (see the previous section).

On the other hand, life's nature is to maintain its survival. If life did not have this nature, no species would continue to exist. The basic requirement for survival is the intake of food. The subsistence of animals on the Earth depends directly or indirectly on plants. Plants produce glucose through photosynthesis, which is the only bioenergy for all life. However, the amount of food photosynthesis can produce for animals is limited. Thus, limited photosynthesis determines the finiteness of food sources, and, in turn, predation of food is a vital survival characteristic of animals. Even littermates must compete with each other. This is inevitable in evolution. Hunger is probably the most painful form of physical suffering, and extreme hunger can lead to reckless behavior. As the product of animal evolution, human beings also inherit this characteristic of animals. Hence, human nature is to safeguard our interests, further developing into greed, jealousy, hatred, etc.

Therefore, human characteristics encompass both rational intelligence and a propensity for greed. Greed stands as humanity's greatest vulnerability, driving desires not only for survival but also for perpetual possession and enjoyment. Pursuits such as power, material wealth, and sexual gratification, which are insatiable, frequently give rise to diverse social issues and conflicts.

Photosynthesis Restricts Human Ability

In addition to determining human characteristics, photosynthesis restricts human ability.

We have explored how organisms based on photosynthesis require intricate structures and specific environmental conditions to meet their survival requirements. Such conditions are unlikely to be found elsewhere in the universe or even within our solar system. Even if humans were to reach another planet, solving a series of survival challenges permanently, such as the need for photosynthesis, would be exceedingly difficult. Consider Mars, our nearest planet. Transforming it into a minimally habitable environment might require centuries. Yet, despite any efforts to terraform Mars, its habitability would remain challenging for humans due to direct exposure to space radiation.

Retrofitting extra-solar planets into Earth-like environments appears to be an insurmountable challenge. The nearest star, located 4.22 light-years away, would take over 10,000 years to reach even with the fastest rockets available today. After such an immense journey, would the pursuit still be worthwhile or meaningful? Even if we could somehow dramatically increase flight speeds to reduce the journey time to around 1,000 years, a feat currently deemed technically impossible, waiting a millennium for answers doesn't seem practical either.

The journey itself would be astonishing, but the potential challenges involved are even more daunting. Therefore, human migration to planets outside our solar system seems prohibitively difficult at present. For now, it appears that humans will remain confined to this small planet Earth, with Mars being the most feasible alternative within reach.

One might ask, will humans be able to synthesize food from chemicals for a future space flight? From our understanding of modern science, human beings, as carbon-based life, are completely dependent on photosynthesis for survival, and to survive without photosynthesis would be impossible. We can safely say that photosynthesis is the only way for the elements of the periodic table to use solar energy to form organic matter. Of course, humans might create synthetic food in the future. But eating synthetic foods for your entire life, forgoing all the natural vegetables, fruits, grains, and meat, is hard to believe and would be problematic.

Moreover, the human body (with its great variety of enzymes and hormones) contains various trace elements, and the lack of any of them potentially seriously affects people's health. Trace elements in the human body are obtained from the soil by plants through photosynthesis, so it would seem unlikely that humans, as carbon-based life, would be able to escape the need for photosynthesis altogether. Furthermore, the problem of sustained energy in a spacecraft is challenging because there is no solar energy outside a solar system.

Therefore, the energy problem, including electricity and food (bio-energy), will significantly restrict the ability of cell-composed humans to make a distant space voyage for hundreds or thousands of light-years. All in all, we will find it extremely difficult.

It's important to consider human capabilities beyond the reliance on photosynthesis, which limits our ability to explore space. The human body, composed of cells, is insufficient for fully comprehending the natural world. Haven't we always desired to uncover nature's true essence? Achieving this goal necessitates ongoing exploration and understanding of both the macroscopic and microscopic realms. The depth of our understanding in these domains is contingent upon human tools and technologies. Consequently, human exploration of nature appears to be increasingly challenging and endless. Essentially, as we depend on instruments to probe deeper into nature, there seems to be a point where further exploration becomes increasingly elusive.

Due to time and space constraints and based on present technology, conducting exploration in person beyond our solar system, as mentioned above, seems impossible. What is now known as the wormhole theory uses highly curved space to reach distant locations. What would the human body and spacecraft look like in highly curved space? What would happen if we were to go through a wormhole? Such ideas are hard to comprehend.

Regarding the subatomic world, human understanding hinges on the principle of "seeing is believing," although facilitated through instrumentation that magnifies the otherwise invisible. However, as we strive to dig deeper into ever smaller scales, the conventional notion of "seeing" becomes increasingly challenging. Presently, our approach to comprehending the subatomic domain relies heavily on observing the behaviors of subatomic particles rather than capturing their visual appearance.

The search for understanding the natural world, both macro and micro, is endless. Even if human beings could travel a distance of a dozen light-years, that would represent only an infinitesimal distance across the vast universe. The scale of a particle accelerator is finite, but in trying to understand the subatomic world, the search for particle structure is endless. As a result, human beings, made of cells, seem unable to reveal nature's mysteries. This also seems logical and reasonable because human beings are created by nature.

THE REMOTE FUTURE OF HUMANITY

The explanation above underscores that human survival in the vast expanse of outer space, light-years away from our home planet, would be untenable without overcoming our reliance on photosynthesis. Consequently, humanity may be constrained to remain on this small planet, or at most within our solar system, until the eventual extinction of our species. This perspective suggests that we are merely a fleeting and inconsequential cosmic phenomenon devoid of any special significance in the broader universe—a notion that challenges common beliefs and general logic.

On the other hand, with our boundless intelligence and curiosity, are we willing to tie ourselves to the Earth forever? If the Earth's environment begins to deteriorate seriously in the future, which is quite possible, threatening the survival of human beings, could we find another planet? To reach a remote distance, could humans migrate on a large scale? The question of humanity's remote future is relatively new, both rationally and scientifically.

The Remote Future of Humanity (1)

Let us begin with some of the major global problems that will affect humanity's distant future that we must face in the next thousand years: climate change, population explosion, species decline, and resource depletion.

Climate Change

The world is aware of the ongoing rise in global temperatures, primarily attributed to the widely accepted theory of carbon emissions. As more carbon dioxide is discharged into the atmosphere, the greenhouse effect intensifies. Global warming, a consensus in the scientific community, is leading to the melting of polar glaciers, rising sea levels, increased acidity in seawater, and a rise in the frequency of extreme weather events, though at a gradual pace. Current trends suggest these changes have not been curtailed. Their impact is keenly felt in people's lives, evident in high temperatures, frequent floods, and other direct disruptions to daily life.

In response, the United Nations Intergovernmental Panel on Climate Change (IPCC) released an authoritative report in 2018. The report's full name is Global Warming of 1.5°C. The report states that if global warming can be limited to just 1.5°C above pre-industrial levels, global net human-caused carbon dioxide (CO_2) emissions would need to fall by about 45 percent from 2010 levels by 2030, reaching "net zero" around 2050. This means that any remaining emissions would need to be offset by removing CO_2 from the air.

Human reason has reached this determination. What about after 2050? Will global temperatures continue to rise? The future is hard to predict. To begin with, it remains uncertain whether achieving zero carbon emissions by 2050 is feasible, as it requires continuous cooperation among 10 billion individuals from diverse cultures, political systems, and economic statuses.

Population Explosion

In 1987, the world's population was about 5 billion. It is nearly 8 billion today, and scientists estimate it will be close to 10 billion by 2050. The rapid increase of the world population has increased the consumption of the Earth (energy, food, wood, water, living space, etc.), which has overwhelmed the Earth.

In the process of population growth, there are two noteworthy problems. One is that the aged population is growing, and the other is that the rate of population increase is quite different around the world. There are also some areas where the population is declining. These demographic characteristics will have a great impact on world economic development.

Human sexuality differs from plants and animals in that it is not tied to seasonal cycles (as mentioned earlier) and can occur at any time. In the pre-Neolithic era, human sexuality may have been less regulated, but with the introduction of clothing around 6,000 years ago and the formation of family structures, sexual behavior gradually became more private. Early humans generally had shorter lifespans compared to today. A fundamental factor contributing to today's exponential population growth is the advancement of science and medicine, which has extended human lifespan, thereby prolonging the age at which individuals experience sexual desire.

On the other hand, the growth rate of the global population has been declining for nearly half a century. So, scientists estimate that 2050 will be the peak of the world's population, after which we may enter a period of population decline. The history of human development seems to suggest a relationship between civilization's progress and population growth. The population growth rate in countries with advanced science and technology is low; technologically underdeveloped countries, especially in the tropics, have a high population growth rate. Therefore, addressing the issue of population explosion remains challenging, and resolving it may prove to be a complex and lengthy process.

Resource Depletion

With the rapid increase in population, per capita consumption of resources is accelerating. These resources include energy, water, land (including farmland), and minerals. Most of the Earth's coal, oil, and natural gas will be depleted in 300–400 years. In arid and semi-arid regions of certain countries, excessive groundwater extraction has led to significant declines in groundwater levels. Consequently, this has intensified dry conditions on land and in the air, reduced precipitation, and caused rivers to shrink, creating a vicious cycle that threatens the survival of all life. The mineral resources available worldwide will also likely be depleted in the next few hundred years.

With the progress of science, energy will no longer pose a challenge (through hydrogen energy, nuclear fusion, etc.), and freshwater and mineral resources can be recycled and utilized efficiently. Overall, the mineral resources and energy that support human civilization are abundant, although their exploitation may become increasingly difficult.

Species Decline

As a result of the population explosion and climate change, the number of species on Earth is decreasing, and the world's biomass is also decreasing. A big problem that needs to be solved is the rapid decline of tropical rainforests, which has affected global ecology for years. An important natural resource, tropical rainforests only account for 7 percent of the global land area, but most of the world's species of plants and animals live in them. Rainforests are also an essential component of global climate regulation. Unfortunately, today, rainforests are disappearing at a heartbreaking rate!

The disappearance of tropical rainforests and the extinction of numerous species due to human encroachment pose significant global environmental concerns. What will happen to our planet if these rainforests vanish and so many species disappear? What difficulties will humans encounter? These questions may not immediately concern those living in temperate zones, but thinking they won't be affected is an illusion. If all tropical rainforests and their diverse species vanish, the intricate global ecological balance developed over millennia will be severely disrupted. Can we still ensure the production of essential vegetables, fruits, and meat for human survival? Will we continue to enjoy the current global climate? All of these uncertainties would arise.

The Important Implications

The four major problems that threaten everyday human existence, including climate change, population explosion, resource depletion, and species decline, lead us to the following important implications.

First—The evolution of humans over millennia demonstrates that nature and humans are a unit, a harmonious natural environment system. The relationship is highly complex.

Second—In this harmonious environmental system, humans are the only factor that can change nature at will. Humans have moved beyond the Neolithic age and, after several thousand years of relatively steady development, have entered the era of modern civilization in the last 200 years. This modern era has accelerated global communication, improved human health, extended lifespans, and caused a population explosion, leading to a host of other problems. Among these are the destruction of the Earth's previously harmonious natural environment and increasingly severe pollution of the world's air, seawater, groundwater and runoff water, and soil. Global pollution and climate change will have severe impacts on life worldwide.

Third—The history of human development clearly shows that nature aids human progress. However, when humans move toward destroying the Earth's harmonious environment, nature intervenes to "punish" or warn humanity. This forces a return to harmony between humans and nature through scientific and technological advancements. For instance, nature has provided much fossil fuel for human use. Yet, over the past two hundred years, extensive fossil fuel consumption, particularly coal, has led to a sharp increase in atmospheric carbon dioxide, which is the fundamental cause of global warming.

Things in this world are always easier to destroy than to restore. Do you believe we can resolve these problems in the next 300 years?

The most impactful factor in humanity's future is the population explosion. If the world's population peaks near 10 billion by 2050, the subsequent decline will be gradual. The natural decrease in population and the improvement of knowledge and understanding of population dynamics will necessarily take several hundred years.

The massive ice caps that have melted at the poles may not be able to re-form, as they took hundreds of thousands of years to accumulate. The continued melting of the polar ice caps may be unstoppable, and its impact on humanity is dire. Achieving zero carbon emissions and gradually removing excess carbon dioxide from the atmosphere to return to pre-industrial levels may not be optimistic.

Suppose the destruction of rainforests can be stopped by the end of this century, though some scientists think the Amazon will disappear in the next fifty years! Will rainforests be able to naturally recover on a large scale in the next 300–500 years? Will the vast number of dry rivers be able to regain their original clear and abundant water in the next 300–500 years? Will polluted land be restored in the next 300–500 years? Since species are created by nature in long periods of evolution, extinct species will not reappear, let alone appear on a large scale.

Considering all these factors, 10 billion people will have a large, extremely difficult task ahead of us. If these problems cannot be fully solved in the next few hundred years, the prospects for humanity are bleak. The next few hundred years will probably be a low point in our long survival. There are several troughs in the long course of human existence, but those troughs were formed by nature (for example, ice ages) and imposed on humans. Today's trough is of our own making but is by no means the end of our existence. One thing is certain: the longer this trough lasts, the more problems there will be, and the greater the cost of regrowth for humanity.

Although humans are not restricted by nature, we are subject to the general laws of nature overall. With up to 10 billion people involved in the regrowth of nature, human reason and greed will necessarily and naturally play a role. Human reason and greed will appear in multiple forms, nations, groups, and individuals. This will be a natural process, including humans.

But with the progress of science and technology and the growth of human rationality, humanity's full recovery from this low point will probably take at least a thousand years, or optimistically by about 3000 AD. By then, the world's population may only be a few billion and highly literate. By 3000 AD, hundreds of people will likely have lived on Mars, and some will have visited the moons of Saturn and Jupiter. But for a human-crewed spacecraft to fly out of the solar system will remain impossible.

Perhaps nations will still be the main form of human society, though the division of nations will likely be different from what it is today. Of course, we hope the world will be peaceful and free, full of friendship and cooperation. The fundamental driving force for human progress and development remains our great curiosity, and the progress of science and technology is the only way to save humanity. Scientific innovation matters.

The Remote Future of Humanity (2)

Suppose humanity can maintain normalcy over the next thousand years without serious incidents. In that case, we may enter a completely new era of artificial evolution, an exceptionally long evolutionary period.

We can divide natural development into three stages.

The First Stage, the Astronomical Stage

The creation of Earth in the universe is an incredible miracle. Earth must have a strong geomagnetic field, abundant liquid water, major and minor landmasses, a proper atmosphere with suitable temperatures, and enough various chemical elements. Earth must have sufficient development time to achieve these conditions.

Our planet was formed about 4.6 billion years ago. What did the universe do in the 9.1 billion years from the birth of the universe 13.7 billion years ago to the birth of the Earth? So much!

Initially, the universe formed the first particles, the first atoms, the first light elements, and the first stars and galaxies. All the natural elements in the periodic table were also formed in this process, a completely natural, stochastic process. Randomness is a fundamental property of nature and determines the time needed for natural development. The formation of the solar system laid the foundation for the emergence of humanity. We call this the astronomical stage.

In the formation of the universe, the development of stars is a critical step. Scientists believe that the natural chemical elements in the universe were formed during the development of stars and eventually spread out into the universe. For stars, completing the "element-making" process is not long, a few tens of millions of years. They are the first generation of stars, the earliest stars in the universe. Their remnants, such as distant supernovas, are difficult to spot now.

Stars visible to the naked eye in the night sky belong to the second generation of stars or those formed later, and the "raw materials" that comprise them are from the first generation of stars. The lifetimes of the second generation of stars vary from star to star, ranging from a few billion years to more than 10 billion years. Scientists estimate that our Sun could survive to 10 billion years.

The theory of stellar evolution has two implications for us. The first is that there are an "infinite number" of galaxies in the universe, so the first stars would have been an "infinite" number to produce enough raw materials for consecutive generations of stars. This is an incredible process. Our Milky Way galaxy and our solar system are just one of those "infinite" number of galaxies.

The other implication of stellar evolution is the lifespan of the star. Why the lifespan of the first generation of stars is so short and that of the second generation of stars is so long is a scientific question that remains to be studied. However, we can say with certainty that such "natural arrangements" are conducive to the emergence of life and the development of civilization. The origin of life on Earth is a profound question. The time from the birth of life to the emergence of humans and modern civilization has taken 3.6 to 3.8 billion years. Accordingly, a 10-billion-year lifespan for our Sun is necessary. From the viewpoint of evolutionary processes, there may be no alternative to such a "natural arrangement." The rationality of the "division of labor" between the first generation of stars and the second generation of stars is surprising.

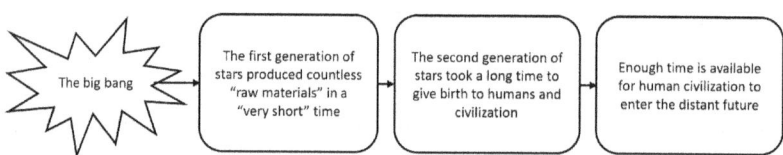

The astronomical stage includes the formation of the universe and emergence of our solar system, setting the stage for the Earth's development.

The Second Stage, the Earth Stage

The highlight of the Earth is the birth and evolution of life, a stochastic, complex process that requires many conditions, and the Earth's structure perfectly meets those requirements.

According to the theory of evolution, environmental change is the "guide" to the evolution of life. Precisely to this point, over the past 4.6 billion years, changes in the Earth's environment caused by its structure have shown that they played an incredible role in the evolution of life. Over the past 3.8 billion years, life has evolved through various environmental changes, ultimately leading to the emergence of humans and civilization. Earth is a life-evolving machine.

These 3.8 billion years represent a unique natural process in the universe. Now, Earth seems to have fulfilled its natural mission by giving rise to humans, who possess highly developed intelligence and curiosity—a profound and significant outcome in the universe. However, numerous potential threats in the universe could annihilate life on Earth at any moment, such as asteroid collisions and supernova explosions (though these are typically extremely distant from our planet). Life here would not have been possible if a supernova had occurred "close" to Earth. The fact that Earth has survived these 4.6 billion years without such incidents is truly miraculous.

The Third Stage, Artificial Evolution

Will humans evolve into more advanced forms in the future due to changes in the Earth's environment? This seems unlikely, as natural environmental changes occur slowly over thousands or millions of years. In contrast, humans are the only factor capable of altering nature rapidly and significantly. Human activities will likely make the Earth's environment increasingly hostile to the survival of humans and all living organisms.

Such environmental degradation may be slowed down through human efforts, but a return to the pre-industrial environmental state is probably impossible. In response to these adverse environmental changes, creating increasingly more advanced robots to replace human labor extensively is an inevitable trend. In this context, the robotics industry will greatly develop in the next thousand years, and perhaps humans and robots will coexist. Are there other likely scenarios?

THE RATIONAL UNIVERSE EVOLVING FOR HUMANS

The logic of nature seems to unfold according to the following diagram:

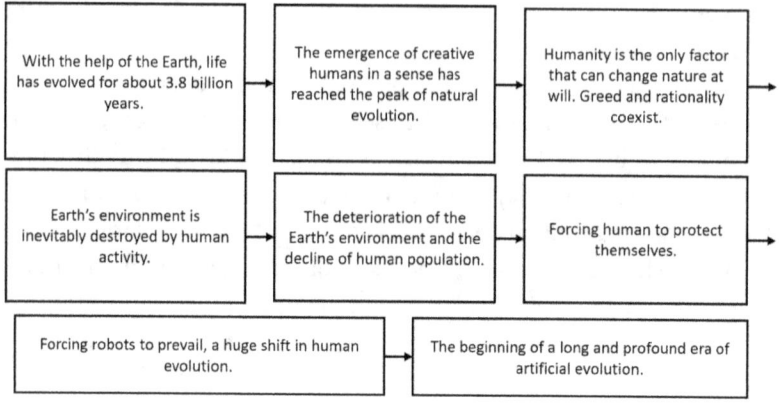

The Remote Future of Humanity (3)

Understanding the three stages allows us to speculate about humanity's distant future. Over the next 1000 years, the human population is expected to decline while the robot population will rise. However, despite their advanced intelligence, robots will remain tools for humans, much like oxen and horses were over the centuries. Their purpose will be to assist humans, who will still be the masters of the Earth.

Nonetheless, humans' fundamental nature and vulnerabilities, composed of cells, will remain unchanged. Carbon-based life forms, including all animals and insects, are inherently driven by sexual reproduction (as mentioned in the previous section). However, nature prevents any species from multiplying indefinitely and taking over the entire world, as illustrated by mice.

Humans are the only life form that can change nature at will, so we seem to be able to take over the whole world forever, but nature will not allow that. Remember, the Earth's environment changes slowly, over thousands or millions of years, purely caused by nature. But now, the Earth's environment is changing rapidly due to large-scale human activity. A few decades can significantly deteriorate the Earth's environment. Taking the natural and artificial factors together, in the next few thousand years, the planet's environment may be unfit for humans and all other life to continue.

Furthermore, the history of human development shows that as science advances, the human body becomes weaker. For instance, the average human body temperature has decreased over the last century from about 99°F (37°C) to about 97°F (36.3°C). A drop in body temperature signifies a decline in immunity. Although modern people live longer due to medical advancements, their physical strength and ability to adapt to temperature changes have diminished compared to humans hundreds of years ago. Modern lifestyles have made people increasingly "lazier," as well. Given the ongoing deterioration of the Earth's environment, a continuous decrease in the global population seems inevitable. Eventually, humans composed of cells may wither away and potentially disappear from the planet in hundreds of thousands of years.

In the "near" future, natural changes to the Earth's environment will likely signal the end of the current interglacial period, leading to the next glacial period in approximately 15,000 years. We cannot accurately predict the global climate over the next tens of thousands, hundreds of thousands, or millions of years, but continued "deterioration" is highly likely. For instance, continental drift is expected to continue, periodic changes in geology or astronomy, and perhaps, the global high temperature of the dinosaur era will return, lasting millions of years.

Overall, the temperature of our planet was high, and such high temperatures may be necessary for the evolution of life, while ice ages are relatively "short." Humans emerged and developed during the interglacial period of the Quaternary Great Ice Age, which is a very short period of time. After that, global temperatures may rise again, and polar ice may no longer exist.

To sum up, in the future,

- The Earth's environment will get increasingly worse,
- science is becoming more advanced, and the robot population will grow with increasingly advanced functions, and
- the human population will decrease, just as people become lazier and weaker.

Since the beginning of the Earth, nature has been facilitating the evolution of life and the continuous development of human civilization. After civilization reaches its peak, nature forces humanity to undergo artificial evolution. This shift in nature is profound.

The Remote Future of Humanity (4)

Einstein believed that nature has "the highest wisdom" that is hardly accessible to human beings (*what is impenetrable for us truly exists and manifests itself as the highest wisdom*). Logically, nature (the universe) with the highest wisdom would be meaningless without human existence.

The instinct to continue the existence of a species, including humans, is fundamental to life. In the face of continuous population decline and environmental deterioration, people with advanced science and technology will certainly strive to avoid their demise. The next challenge will be transforming ourselves to achieve immortality. To do this, three technological problems must be solved:

1. Endowing robots with highly developed intelligence
2. Providing robots with permanent energy sources
3. Developing suitable materials for robots

Solving these problems will drive the evolution of robots, enabling them to help humans survive in a deteriorating environment. This process may take thousands of years.

However, according to the principles of natural philosophy, things made of completely different substances cannot perform precisely the same functions. It seems impossible for a robot to fully replicate the functions of the human brain. Intelligent robots are unlikely to experience a range of emotions, such as passion, joy, love, anger, sorrow, and happiness. Emotions can have both positive and negative effects on a person, and understanding how the human brain forms these complex emotions is still an area of ongoing research. This process is likely an incredibly intricate physiological phenomenon involving various secretions. For example, elderly couples may lack certain hormones but still experience deep love. Intelligent robots made of inorganic matter may not be capable of experiencing the same complex emotions as humans, including passion. In many ways, it is beneficial that robots possess intelligence without emotions.

In addition, intelligent robots are composed of matter, making it seemingly impossible for them to enter another space through a wormhole. Consequently, intelligent robots will likely be confined to the solar system for thousands of years to come, making it difficult to explore the deep universe. In other words, the potential of the periodic table appears to be reaching its limits.

According to modern scientific understanding, the Big Bang describes how energy transformed into particles, atoms, molecules, and eventually matter, which includes life. This process reveals that matter originates from energy. Energy manifests in various forms that we encounter daily and can be converted from one form to another; for instance, mechanical energy can change into heat energy. However, we never directly perceive what energy itself looks like. In contrast, matter is visible, while energy remains mostly invisible in our everyday experiences, except when it appears as light with various colors.

This distinction underscores a fundamental difference between matter and energy. Understanding the nature of energy remains a major area of scientific study, as it involves uncovering the essence of natural phenomena.

However, logically, if intelligence exists solely in the form of matter, then intelligence in the universe must be a "temporary" phenomenon. Over time, significant challenges arise that neither the human brain nor intelligent robots can perpetually overcome. Intelligence retains continuity and meaning only when it exists in the form of energy. We cannot conceive what wisdom in energy form entails, but we speculate it to be fundamentally distinct from human intelligence.

Can intelligence in the form of matter be transformed into energy? Can intelligence in energy form exist independently in the material world? These questions examine profound and intricate issues, touching not only on the nature of energy and its relationship with matter, but also on the essence of intelligence and how the human brain generates it. However, this book does not delve into these complex topics.

The function of intelligence is creativity, and the human brain's creative process is an immensely intricate journey—from initial awareness and interest to deep, continuous, and increasingly complex thought, culminating in moments of inspiration. Modern science still lacks a complete understanding of this creative journey's biochemical processes, brain waves, memory, and thinking processes. Einstein himself spent years unraveling the relationship between time, space, and matter (relativity), with his findings culminating in significant publications.

Only the human brain, composed of cells, appears capable of undertaking these intricate processes, which may be nature's unique method for fostering creativity. According to the principles of natural philosophy, while a robot can possess intelligence, it cannot replicate the creative process of the human brain.

Logically, intelligence in the form of energy must differ fundamentally from intelligence based on cells. In a distant future where human knowledge and capabilities reach new heights, an energy-based form of intelligence might emerge. Such intelligence would not rely on creativity or the creative processes inherent to the human brain. Instead, it could exist as vast informational intelligence in an energy state, free from matter and devoid of desires, potentially integrating into the fabric of the universe. This form of intelligence does not need to visit distant galaxies because it is a part of the universe.

If Hawking's no-boundary theory, from the Big Bang to the Big Crunch, proves valid (to be explained in detail in the final section of Chapter VI), it logically follows that energy-based intelligence would succeed material-based intelligence. However, the concept of intelligence in energy form remains purely speculative at this stage, facing numerous challenges in current scientific understanding. The transformation from material-based to energy-based intelligence would undoubtedly be a lengthy process, spanning perhaps tens of thousands of years. Yet, we must acknowledge that the advancements in science and technology thousands or even tens of thousands of years from now will surpass our current imagination. Moreover, the mysteries of nature appear boundless, underscoring the infinite potential for exploration and discovery.

In addition, since nature possesses "the highest wisdom," this suggests that wisdom can exist in an invisible form, allowing for the possibility of an energy-based form of intelligence. Einstein's lifelong belief in "the Reason in nature" or "the highest wisdom" represents his profound understanding of nature and warrants our contemplation and research. His ideas are explored in this book.

The black hole information paradox, however, should be mentioned here. This paradox has gained new insights in recent years and will undoubtedly inspire further thought. In natural philosophy, we believe in the existence of "the Reason" and "the highest wisdom," as argued in this book, and we maintain that the information of intelligence would not be annihilated in the Big Crunch.

As humanity approaches the limits of the periodic table, will our development cease? No! Human progress must continue. Therefore, it is logical to assume that after mastering matter, humanity will enter a phase of development centered around energy.

We can envision a future where species gradually diminish due to the ongoing deterioration of the global environment. Mammals may be the first to vanish from the biosphere, followed by insects and bacteria, while plants and marine life also decline due to rising temperatures. The continual deterioration of the global environment might be the catalyst forcing intelligence in material form to transform into an energy-based form, potentially reflecting a new direction in natural evolution. Can you propose an alternative? If not, this transition might signal the end of the Earth's natural mission, as intelligence in energy form would no longer depend on Earth.

So, the ultimate question becomes: How does human intelligence integrate into the universe as energy? This question may be closely tied to the evolution of the universe itself. The chart below represents the process of this transformation and also suggests how Reason manifests itself in nature. This chart can also be understood in conjunction with the last section of the next chapter.

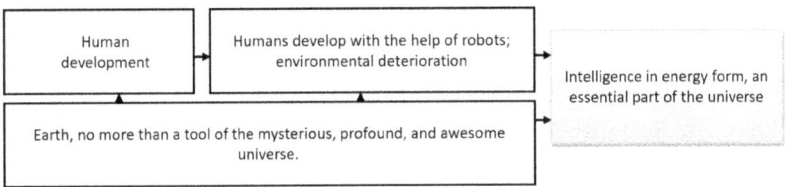

What is the significance of human existence in the universe? In the next chapter, we will explore why the universe may require intelligence in energy form. It promises to be an exciting discussion.

Chapter VI

The Universe Evolved for Humans

All that has been examined thus far leads us to conclude that the universe has unfolded to facilitate the rise and development of humanity. This chapter will further investigate this notion. Humanity may be the greatest mystery of the universe. Why do humans exist in the universe at all? This question touches on the essence of Nature itself. Might humanity face self-destruction due to inherent human traits? Are humans merely a transient tragedy in the grand scheme of the cosmos? Does the evolution of the universe necessitate such a tragedy?

Our modern scientific understanding has provided a basic knowledge of the universe's progression since the Big Bang. Its evolution unfolds logically and cohesively, with each phase paving the way for the next until the eventual emergence of humanity and modern civilization. This progression also signifies a growing complexity in material structure, with the human brain arguably representing the apex of this complexity. The brain is the material foundation for human civilization's development, and the brain's absence would render everything devoid of meaning.

So, what meaning does this precise and logical 13.7-billion-year process hold? Common sense dictates that such a precise and logical progression must have intent. Thus, does cosmic evolution have a purpose? And if so, what might that purpose be? This chapter will continue to explore the mysterious relationship between the universe's evolution and the human brain, investigating the insights of Einstein and Hawking.

As we continue, note that "cosmic evolution" and "natural development" are similar in meaning, although they often appear in different contexts. Both refer to the natural process from the Big Bang to the emergence of modern civilization.

NATURE'S MOST FUNDAMENTAL CREATIONS

Atoms, the Basic Stuff of the Universe

The Big Bang marked the genesis of time, space, and matter. The three are inseparable and integral to the universe. At the heart of nature's evolution towards the creation of intelligent humans lies the atom. While we may not possess the expertise of physicists to understand the marvels of atom-based matter fully, it's evident that atoms serve as the universe's fundamental building blocks: the four fundamental forces (gravity, electromagnetism, and the strong/weak nuclear forces), alongside various particles and photons. We can trace all natural laws back to the atomic structure, characterized as matter and energy.

Atoms are the cornerstone of the universe; without them, the universe would cease to exist, and conversely, atoms sustain the fabric of space and time in the universe. They form the basis of galaxies, solar systems, the Earth System, life itself, and the intricate mechanisms of modern electronic devices. The human brain, responsible for creativity, relies on atoms' highly complex and delicate structure. Every life form and physiological process, no matter how complex, finds its roots in atoms. The development of modern electronic functionalities appears boundless, suggesting atoms possess an almost omnipotent quality in shaping matter and energy.

Atoms exist in the form of chemical elements, with carbon emerging as the pivotal member of the periodic table. All other elements appear to assist carbon, either directly or indirectly, in fulfilling various natural missions (see Chapter III, Understanding the element carbon).

There seems to be no end to humanity's exploitation and use of elements, prompting us to ask, what is the atomic structure? According to the two principles of natural philosophy, since the function of the atom seems endless, we can understand that the structure of the atom is also endless. In other words, you may never find the limits of its structure. This infinite nature underscores nature's reason and "supreme wisdom," or "the Reason" and "the highest wisdom," surpassing the capacity of human intellect.

The atomic structure cannot be changed or destroyed. If the structure and function of atoms were subject to change or destruction, they would not effectively constitute the natural world.

Only the atom can perfectly fulfill the divine natural mission entrusted by the universe. Therefore, from atom formation to chemical elements to various molecular combinations to the wonderful natural world, we witness an exact yet logical natural process. Mass conservation and energy conservation laws remain constant throughout the universe's evolution. Relativity explains how matter, time, and space form a seamless "marvelous structure," constantly evolving without interruption.

Ultimately, the most critical factors of the universe are energy and matter. Matter embodies the world; energy drives the world. Their reciprocal transformation is necessary for the universe's formation and development. Though the Earth System comprises matter and energy, their lack of transformation is essential for its formation, including all life forms. In summary, the rational structure of matter is a vast and profound topic, which we will only touch on briefly here.

DNA Molecules, the Foundation of Life

What is life? In nature's extraordinary tale, life is the main character, the story's highlight. Life has countless mysteries, genes being the biggest. Humans may never fully solve the mystery of life.

As a distinct form of matter, life can replicate itself, reproduce from generation to generation, and evolve to have many more functions than inanimate matter. We may call these the three basic requirements of life that reflect the tremendous adaptation of atomic matter, thus forming the highly complex structure of life substance. Only carbon-based life can meet these requirements.

All life descended from a common ancestor and constantly evolved to form numerous species. Therefore, evolution is the most important function of life, without which life would be insignificant. At this point, we need to understand the internal factors of evolution. It is the gene, which is a part of a DNA molecule.

The DNA molecule is an astonishing form of atomic matter. DNA molecules not only have a common pattern of molecular structure but also have a common working principle across species and exist in all kinds of cells. Millions of species develop from only tiny differences between DNA molecules. Under different environments, DNA molecules produce numerous forms of life, including various biological functions, structures, colors, smells, and tastes, with different chemical elements. They are embodied in plants, animals, and microorganisms and rooted in the soil, which is also a natural miracle. DNA molecules and environmental changes have combined to gradually form the Earth's biosphere of tens of millions of species with numerous food chains that compose the basis for evolution, lasting billions of years. Aren't you surprised at this amazing function of DNA molecules?

In the case of humans, subtle differences in genes among individuals have given rise to diverse capabilities, or talents, promoting the advancement of human civilization. These capacities, inherent to individuals, can be viewed as a natural arrangement, with nearly everyone possessing unique abilities. Without these distinctive aptitudes, the development of human civilization would be hindered or potentially halted altogether. Therefore, intelligence alone does not guarantee the advancement of civilization; rather, it is the presence of diverse capabilities that significantly contributes to progress. Moreover, the DNA molecule can evolve into many useful species within the plant, animal, and microorganism kingdoms, all of which are crucial for human survival and development.

DNA molecules have existed for more than 3 billion years. They form the basis of the Earth's biosphere and, through their tiny changes, cause ongoing evolution.

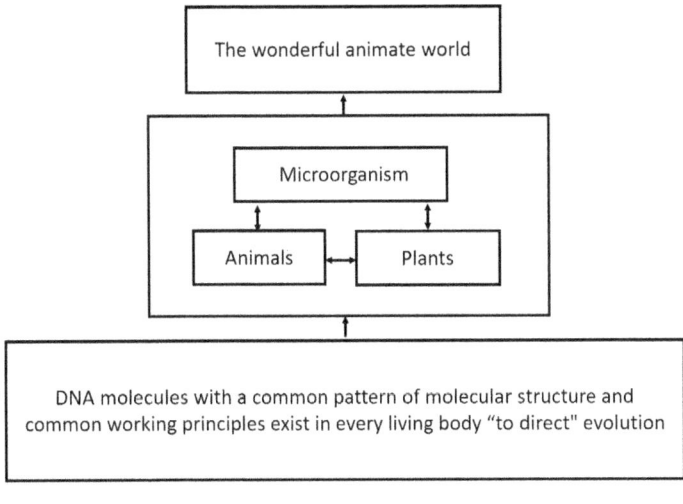

↑ Support

THE RATIONAL UNIVERSE EVOLVING FOR HUMANS

How exactly does life evolve? Because of the structural complexity of DNA molecules, DNA can make minor changes during its self-replication, known as gene mutations. In addition, there are also chromosomal variations of DNA molecules, etc. They all are internal factors facilitating evolution, combined with different environmental conditions as external factors (Figure 6–1) to form various species.

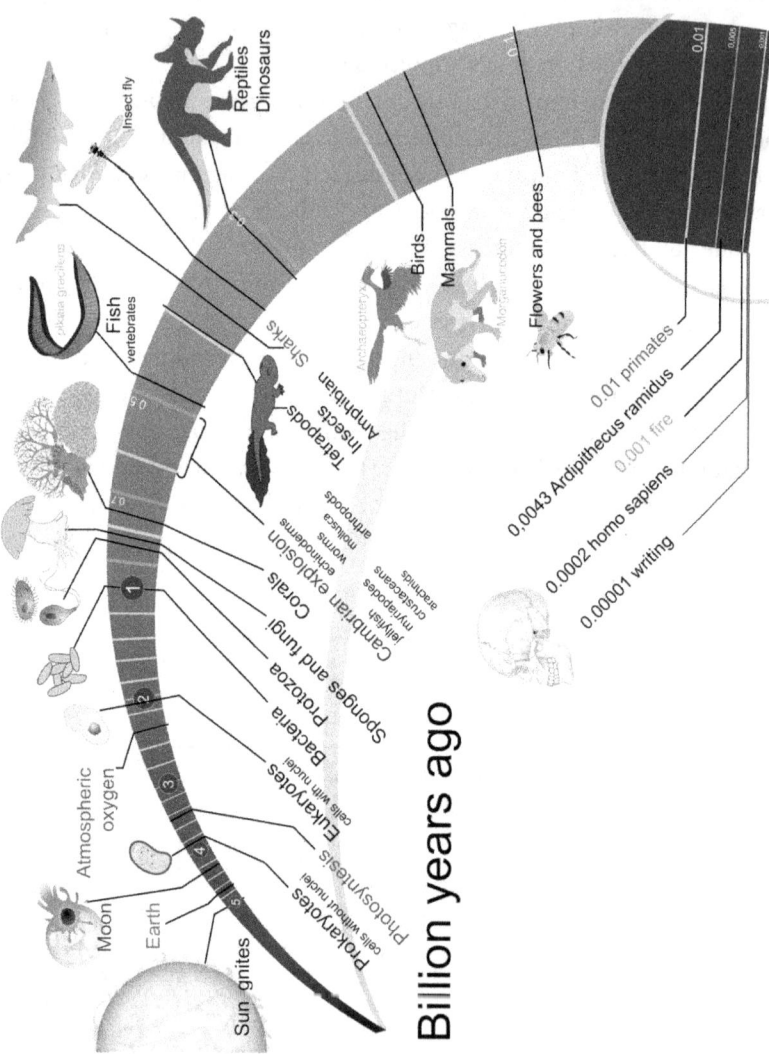

Figure 6–1: Factors including the Cambrian Explosion, the regression, Pangea, continental drift, dinosaur extinction, reduction of global temperatures, and the Great Ice Age all contribute to evolution (see Figure 4–1) (Source: Public domain).

DNA is the basis of the Earth's biosphere. It is the only physiological factor driving evolution. Over the past 3.5 billion years, combined with changes in the global environment, the DNA molecule has created millions of species with countless physiological functions, forming the animal, plant, and microorganism kingdoms. No matter how small, each organism has a highly complex and delicate organizational structure. Because of minuscule differences in genes, there are some differences between individuals of the same species, making for a colorful world.

Life can evolve, but it has no "motive" for evolution. It is easy to understand that life cannot automatically or blindly increase its functions without a purpose. Survival demands caused by environmental changes drive the generation of new functions. For example, the shift from rainforest to savanna eventually led to the development of primate bipedalism and giraffe's longer necks. Evolution occurs through major adaptations to new environmental changes.

So, environmental changes and gene mutations are fundamental factors in evolution. We can say that evolution results from the perfect combination of these two factors, internal and external.

In our exploration of life and evolution, it becomes evident that life's complexity surpasses human capacity to manufacture a cell with DNA directly from atoms, let alone create small living organisms like ants or flies with their intricate biological processes.

Don't you think the ongoing collaboration between DNA and the changing Earth's environment that leads to the evolution of life and the emergence of humans is worth considering? The mysteries of life and evolution are boundless, suggesting that the study of life may be a pursuit without end.

Photosynthesis, Life's Only Energy Source

Photosynthesis is the sole bio-energy source across the vast biosphere, spanning a diverse array of species and fostering the evolution of life for billions of years.

One might question why there isn't a secondary form of biological energy. A key consideration lies in the necessity for organisms to absorb this energy, requiring both the energy source and the organism to share the same substance. Since photosynthesis relies on chloroplasts, which belong to carbon-based life, only such life forms can engage in this process. However, the precise origins of chloroplast formation remain a scientific mystery. Following the two principles of natural philosophy, other elements on the periodic table lack the functionality of carbon. Therefore, there is no second biological energy source.

Early photosynthesis likely occurred in seawater within an atmosphere predominantly composed of carbon dioxide, providing optimal conditions for its emergence.

This symbiotic relationship between photosynthesis and DNA sustains every living organism, forming a balanced ecosystem. Earth is the only planet capable of fostering such an ideal system, a topic we will continue to explore.

SETTING THE STAGE FOR REVERSE REASONING

While reading the preceding chapters, you likely realized the inherent intelligence in nature—an acknowledgment we must admit. This wisdom permeates every process of the Earth's development, an outlook endorsed by Einstein as "the highest wisdom" or "the Reason in nature."

This book has explained how a stochastic natural process can proceed successively to a higher material state solely under the influence of regulating factors. From this, one might infer that humanity is the most significant fruit of this highly complex process of natural development, controlled by natural Reason. In the following sections, we will employ backward reasoning to illustrate the inevitable nature of this process, starting with humans—an intentional indication of "the highest wisdom."

What is Reverse Reasoning?

Inference, or forward reasoning, belongs to formal logic and aims to get unknown knowledge from known knowledge. Conversely, reverse reasoning, also called goal-driven reasoning, is the opposite of inference. Reverse reasoning usually starts with the result and then looks at what conditions are needed to achieve the result. If those conditions are met, what subsequent conditions are required, and so on. For example, to make a ceramic vase (the result), one must first have the appropriate type of clay, plan the design, and prepare the necessary tools, supplies, and equipment. To successfully complete the project, one must be a skilled, careful, and patient artist. The project can then begin toward the completion of the envisioned piece.

The advantage of reverse reasoning is that you can understand the specific conditions required for something to happen/develop/form and their probabilities. By reasoning backward (with enough knowledge), you can go back to the origin of what happened and determine how likely it was to happen. Reverse reasoning is widely used (picture a detective in a criminal case) and has achieved compelling results. In reverse reasoning, the target of the reasoning process—the outcome—must be precise.

Before we apply reverse reasoning to the discussion of the evolution of the universe, let's first look at the human brain since the evolution of the universe would be meaningless without it. The exploration of the human brain in the following two sections is the foundation for all the arguments in this chapter. It is also the starting point of the reverse reasoning process. Take a moment to reflect on this statement.

The Human Brain—Nature's Standard of Creativity

Intelligence, as a function, is inherently tied to a specific structure: the human brain, as determined by the principles of natural philosophy. Some may perceive this assertion as overstated, pointing to the discovery of numerous distant planets in the universe, and speculate on the possibility of intelligence existing elsewhere with brains different from humans. However, dismissing the extreme complexity of the human brain based on the principles of natural philosophy is simplistic and lacks depth.

The concept of natural miracles was introduced earlier: nature always acts with the least energy, smallest space, and best effect. This "behavior" is nature's way of acting without unnecessary measures. Can you find a natural phenomenon that deviates from these characteristics? Not likely. Life is the most outstanding product of nature; you won't find a useless part in the physiology of an insect. The same is true in the inanimate world, for example, the obliquity of the ecliptic, the composition of the atmosphere, etc.

Another example of a natural miracle is precipitation, which forms fresh water on land. That is how to get the result with the least energy consumption. There is no better solution. So, if other planets could form creative brains, they must obey the laws of nature.

However, one might argue that there were giant insects in the Paleozoic era, which scientists think may have resulted from the air's high oxygen content at that time (see Chapter II.1, The Atmosphere). If the human brain evolved in air with high oxygen content, the human body and brain may also grow larger. This casual assumption is groundless because the Earth System is a perfect and precise combination of numerous factors, as you will see in the next section. It does not allow for hypotheses to be made in isolation from this intricate combination.

Building on the above, let's examine how the modern human brain evolved. Life's evolution means adding functions to meet survival needs. Evolutionary theory holds that changes in the environment are the basis of evolution. The human brain's capacity was substantially and gradually shaped by the struggle for survival (including improved diets and the use of fire) in various circumstances. Anything useless in the brain is eliminated by natural selection.

Therefore, eliminating the organs or species or physiological materials not adapted to the environment is an essential process in evolution. Natural selection makes the physical structure of life increasingly precise. Every organism is a highly sophisticated organization, and every tiny part of it is necessary. You won't find any extra components lacking function in the bodies of ants or flies. Similarly, manufactured items like clocks don't contain any useless parts.

The evolution of the human brain traces back to ancestral primates. According to archaeological evidence, the progression of cranial capacity unfolded roughly as follows: More than three million years ago, brain capacity exceeded 400 milliliters. Two million years ago, it surpassed 700 milliliters, and one million years ago, it reached approximately 900 milliliters. Around 500,000 years ago, *Homo erectus pekinensis* had a cranial capacity of about 1100 milliliters, while the Upper Cave Man in Zhoukoudian, Beijing, China, approximately 18,000 years ago, exhibited a cranial capacity ranging from 1300 to 1500 milliliters. Present-day humans typically possess a cranial capacity of around 1500 milliliters.

This data shows that evolving into a creative brain involves gradually increasing brain volume, intellectual growth, and stature development. It does not allow for redundant physiological components but is a process of constant optimization in structure and size. Atoms, molecules, cells, and brain structures, in turn, likely determine the brain's capacity. So, the brain can't grow larger or smaller unless it significantly changes its structure, which is impossible.

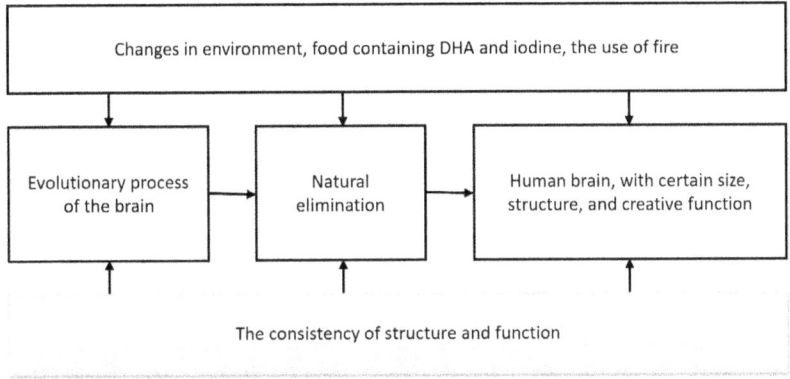

Regarding other human organs, not everyone's organs are the same size. Still, everyone's organs are made precisely in the same way, or else it is considered an abnormality. Your hands are smaller than others, but their structure and function are the same.

Likewise, the size of the human brain, with billions of cells, exhibits slight differences among individuals, which is perfectly natural. Owing to some differences in genes and environment among people, the size or the number of human brain cells may not be the same, but their brain functions are the same. Einstein's brain functioned no differently than anyone else's. However, if we were to discover a specific difference in a particular part of his brain that led to his extraordinary cognitive abilities, it would only demonstrate the complexity and intricacy of the human brain's structure.

At this point, our focus lies on the intricate structure of the brain, a significant subject of scientific research. This structure serves as the foundation for the brain's creative function. Our emotions, feelings, memories, and thoughts are inseparable from the brain's functions. According to modern science, the operation of our brain relies on complex, self-organizing, massive systems that fall within the field of chaos theory, accompanied by the occurrence of brain waves and magnetic fields.

The human brain contains over 100 billion nerve cells connected through synapses. The discovery of synapses was a major leap forward in brain research, as important as DNA in genetics. The human brain is estimated to have over a trillion synapses, an astronomical number!

Based on the argument above, the combined structure and size of the human brain is the only standard for a living body composed of cells of atomic matter and endowed with creative capabilities. In essence, any organism exhibiting creative functions necessitates an incredibly complex structure like the human brain.

The human brain is another remarkable creation of nature, alongside atoms, DNA, and photosynthesis. Researching the brain parallels the study of atoms and DNA, marked by significant milestones yet lacking definitive endpoints. Just as atoms, DNA, and photosynthesis contribute to the emergence of intelligence—the human brain—what further elements are necessary for its formation? This question guides our next inquiry, grounded in the belief that everything happens under specific conditions, to be discussed in the next section.

While we're on the subject, let's consider this question: What constitutes the value of an individual's life? One possible answer lies in the extent to which one develops their own intelligence. By honoring and effectively using one's innate wisdom, one can fulfill the potential endowed upon them by nature, thus embodying their life's value, regardless of their contributions. While everyone's brains are essentially uniform, how individuals utilize their cognitive faculties varies widely.

As emphasized throughout this book, it's worth noting that possessing a brain and maximizing its potential are distinct concepts. Nature provides us with our brains, yet our actions and choices determine how we leverage this gift.

Indeed, the distinction between animal nesting and human creation underscores the complexity of cognitive processes and the evolution of intelligence. Animal nesting primarily stems from instinct—a genetically programmed behavior to ensure survival and reproductive success. Their nests (Figure 6-2) have not changed in size or form from generation to generation, but human-made buildings have continually improved for thousands of years.

Figure 6-2: Birds and bees can build beautiful nests. They are not "designed" by their brains but credited to their genes (Source left: Tomasz Przechlewski, CC BY 2.5; right: Martin Fisch, CC BY 2.0).

Earth—A Perfect Planet for Forming the Human Brain

As asked above, what conditions are needed for forming the human brain? The most essential condition is the Earth! Earth is a perfect and mysterious planet, but we know little about its perfection. How perfect is it? Why is it so perfect? Let's explore the subtle relationship between our planet and the human brain.

According to the principles of natural philosophy, the human brain, made up of atoms, molecules, and cells, is the only physical form with a creative function. Its extreme complexity means any brain structure change will affect its creative function. Could the human brain have evolved if the Earth's gravity were different? All the arguments in this book answer this question: no, it's impossible.

The first thing to consider is that the effect of change in the Earth's gravity would be nonlinear. For example, changes in the Earth's mass would affect its gravitational balance on the solar system. These effects are too complex for modern science to describe. All we can say at this point is that the effect is certainly there, but the result would be uncertain. Below, we will explore this idea in greater detail.

Let's start with life on Earth. Aspects of life remain a mystery to modern science, and there is no doubt that life is a highly complex and precise system. Life is also a perfect combination of numerous natural laws. The birth and evolution of life on Earth, including the emergence of humans and modern civilization, is a highly complex process in which many conditions have been met in time and space. It's truly an unimaginable process that the Earth has accomplished.

One might argue that if certain conditions were not met, life could have still evolved into humans and modern civilization under different conditions, given evolution's adaptability to diverse environments. However, this perspective oversimplifies the complexities underlying the origin and evolution of life, as well as the emergence of human beings and modern civilization, and goes against the principles of natural philosophy.

If life evolved on the Moon or Jupiter, would the behaviors of molecules—including subtle liquid movements within organisms—as well as molecular mass and bonding, mirror those observed on Earth? It's important to recognize that molecular properties, including specific gravity, can be altered under varying gravitational conditions. Life requires many conditions to exist, including gravity. The intricate functions of molecules essential for life manifest fully only under optimal gravitational conditions. Consequently, alterations in gravity have the potential to change everything.

The evolution of life is complex, and we only touch on it here at its highest level. Let's take this process a step further. Now that humans and modern civilization are established, we can make the following deduction.

The whole process begins with the periodic table. According to the characteristics of the chemical elements of the periodic table, carbon-based life is the only pattern that can fulfill a large variety of highly complex physiological processes. Some have proposed the possibility of silicon-based life (see Chapter III, Understanding the element carbon), but this is mere speculation, not a practical hypothesis.

Carbon-based life can only produce the biological energy (glucose) it needs through photosynthesis. Photosynthesis brings carbon into life, forming the carbon skeleton of life. The widespread occurrence of photosynthesis requires certain astronomical conditions, such as the distance between the Sun and the Earth, the rotation period, the revolution period, and the size and mass of the Earth. The combined effect of these astronomical conditions is the proper intensity of solar radiation and the appropriate temperatures on the Earth's surface, which are required for the existence and evolution of life for more than 3 billion years (the surface temperature is affected by many factors, of which solar radiation is the primary one).

According to Milankovitch's hypothesis (see Chapter II.1, The Appropriate Distance, the Appropriate Size), even slight variations in the Earth's regular movements can trigger significant shifts in the global climate. Thus, astronomical conditions must remain consistent. Darwin's theory of evolution emphasizes that environmental changes are necessary for life to evolve. Within the existing astronomical conditions, environmental changes stem from the Earth itself, predominantly through plate tectonics. Earth's structure, based on its mass and size defined by astronomical conditions, not only enables plate movement but also influences the strength of the geomagnetic field. Consequently, life evolves through rational changes in environmental conditions over time and space, facilitated by plate tectonics under the shield of the Earth's magnetic field.

It's important to recognize that carbon-based life relies predominantly on water as well as life substances like proteins. The required conditions for life are inherently specific, such as abundant liquid water.

From this explanation, can you imagine which link is isolated or arbitrary? Which link appeared out of thin air? Not one! Beginning with the periodic table, the pattern of life is defined. Consequently, the underlying logic dictates that nature systematically generates the conditions necessary for increasingly complex natural structures with higher functions to develop, culminating in the emergence of humanity and modern civilization.

Effectively, all these conditions unfold in a chain reaction where one condition prompts the fulfillment of subsequent conditions. So, there are no arbitrary factors in this process (see Chapter I.2, The Illustration of Natural Development Chains).

As stated above, the whole process of natural development unfolded in strict accordance with the two principles of natural philosophy. There would be only one set of changes in environmental conditions for the evolution of life, leading to the eventual emergence of humans. The higher the function, the more complex the structure needed, and the more the conditions required to form the structure. Conditions matter. The two principles of natural philosophy seem to define the process of natural development.

We hold that evolving life can only adapt to natural laws and cannot change them. Evolution requires certain natural conditions, such as sufficient liquid water, but cannot change those conditions.

Let's examine the example of water to illustrate the precise relationship between life and natural conditions. Plants, which are fundamental to the survival of all life, rely on water movement driven mainly by a combination of capillarity and transpiration (similar to evaporation; refer to cohesion-tension theory). These two physical processes are closely linked, with transpiration "pulling" on capillary action.

Although modern science does not fully understand this complex and precise process, it is clear that an increase in gravity on a larger planet would weaken or even halt both capillarity and transpiration because the height of the capillary rise is inversely proportional to gravitational acceleration (g) and the density of the liquid. Conversely, if gravity decreased on a smaller planet, the process might exist, but other changes would arise, affecting the survival of life. Capillarity not only exists in plants but also widely exists in all living organisms, significantly impacting life's survival.

In addition to the unit weight of water that varies with gravity, the capacity of permeation and evaporation of water is also influenced by gravity. It can affect all kinds of physiological processes of life and cellular functions, such as body heat regulation, through the movement of water molecules. Therefore, survival requires certain natural conditions. In addition to capillaries, the change in gravity would also impact other factors at least that support survival:

- The precise dynamic balance between celestial bodies, including the Sun, the Moon, and the Earth System. Movement of the Earth System would change adversely.
- The structure and function of the Earth, including plate movement.
- The intensity of the geomagnetic field.
- The intensity of solar radiation on the Earth's surface, which would impact everything.
- The density of the atmosphere (If the Earth were smaller, it wouldn't be able to retain air molecules or even hold onto the Moon), atmospheric pressure and temperature, and air composition.
- All movement states, including molecule movement, atmospheric movement, ocean currents, surface water evaporation, global water circulation, and frictions between objects.

Again, if the Earth's gravity changed, its effects on the above and the evolution of life would be nonlinear and highly complex.

We believe that the size of an atom determines the size of molecules and life substances such as proteins. This, in turn, dictates the size of cells needed for various forms of carbon-based life, including the highly complex human brain. For the brain to function properly and for all other living organisms to survive, a planet of appropriate size and gravity is necessary. This logic is interconnected and inseparable.

The emergence of humans with their creative brains shows that the Earth, with its unique structure and gravity, meets all the relevant conditions from astronomy (such as the size of the Sun, the Moon, etc.) to geology, and perhaps many other conditions that we have yet to discover, form this perfect and precise combination of numerous natural laws conducive to life. Evolution hinges on this combination of conditions.

So, based on two natural philosophy principles and the periodic table, we may say that life on Earth is the standard pattern in the universe. No larger or smaller planet can reproduce the same states of the natural laws necessary for forming life as they are on Earth (such as the capillary action of plants described above).

Some might say all the above is a fancy of imagination. Aren't astronauts in space the same life forms as people on Earth? No! Modern science has proved that weightlessness or microgravity can bring many adverse changes to people's physiological state that become increasingly serious. However, their physiological functions do not deteriorate significantly in relatively brief forays into space.

Alternatively, if humans visited a planet with stronger gravity, such as Jupiter, their hearts wouldn't handle it, and their brains wouldn't function properly.

Some imagine that humans could grow plants on the Moon or Mars in an especially closed environment, but this is a different story from the evolution of life. Essentially, evolution is simply the process of gene expression in various environments that is very slow and complex. The more complex the structure of life, the higher the function of life, and the more conditions are required to form the structure. The human brain, as a creative living body, can be seen as the "final" fruit of the evolution of life. So, growing plants on the Moon and evolving on Earth are very different things.

The extensive impacts brought by a changed gravity are unthinkable and unacceptable. In any case, we can't underestimate the effect of gravity on life and the environment. The human brain could not have evolved under different gravity due to the extreme complexity and sensitivity of the brain's physiological tissue, as well as being subject to the principles of natural philosophy.

The creative human brain is probably the universe's most complex and sophisticated structure, composed of organic matter, and it appears to be a standard living entity within the universe. Logically, the relationship between the human brain, the Earth's size, and its gravity is not arbitrary.

THE REVERSE REASONING FOR THE EVOLUTION OF THE UNIVERSE

Building on the foundational arguments above, let's proceed with the following sequence: the human brain, the Earth System, the solar system, the Milky Way, and the Big Bang. This order will allow us to reverse-engineer the reasoning behind the universe's evolution.

Is Earth Designed for the Human Brain?

To begin, we need to clarify the word "design," which, in the context of this book, can be understood as "natural arrangement," and the "designer" means the wisdom of nature.

We've explored the human brain's complexity and subtlety and gravity's effect on life, all of which suggest that the Earth's size and mass (gravity), necessary for the evolution of life and the human brain, are not arbitrary. The two previous sections answer whether the Earth seems to be designed particularly for the human brain. Now, let's look at this question from a different viewpoint.

With sufficient knowledge, you can identify the conditions necessary for the survival of each species, and you'll likely find that humans require the most. Generally, the most demanding species tend to be the most complex in structure. The human brain, for instance, has perhaps the strictest survival requirements. Although it accounts for only about 2.1 percent of the body's weight, it consumes 25 percent of its total oxygen. It also has the shortest survival time when deprived of oxygen. Therefore, the 21 percent oxygen content in the air is far more critical for humans than other living organisms.

However, maintaining a 21 percent oxygen content in the air is highly complex. This ratio remains unchanged despite extensive human activities such as combustion, widespread deforestation (especially in tropical rainforests), and rapid population growth. What factors sustain this precise balance? Science has not yet fully understood the mechanisms behind this stability.

The emergence of the human brain marked a turning point in natural development. At this point, human will influenced natural development, altering its original trajectory and balance. We can examine this issue in three ways:

1. Human evolution, taking 3.5 to 3.8 billion years, aligns well with the lifespans of the Earth and the Sun (as will be determined in later sections).
2. This evolutionary process is intricate and subtle. Over time, the Earth's internal structure gradually reshaped its surface, creating diverse environments across geological periods and ultimately leading to the emergence of humans. The 21 percent oxygen content in the air was established during this process, despite being a complex and fluctuating journey.
3. The evolution of the human brain was the most demanding in the animal kingdom. After its initial development, further evolution required conditions such as fire, cooked food, DHA, and iodine. Nature provided and met these conditions through a long evolutionary process. Lightning helped early humans learn to use fire, which took about 500,000 years. The invention of stone tools enabled the consumption of cooked food. DHA and iodine were sourced from the Red Sea in the GRV. Lightning resulted from the planetary wind system, a long natural process. Stone tools were crafted from various natural stones, with quartzite being the best. In nature, nothing is simple!

Thus, the human brain's appearance seems to be the "final result" of life's evolution within the Earth's system. It's similar to a hen laying an egg; the hen's physiological structure is "designed" for this function. Any change in its structure would disrupt the egg-laying function, demonstrating the consistency of structure and function.

Building on the evidence, we may say that Earth was designed for humans. Over the past billions of years, the smooth functioning of the planet has been an essential condition for the birth and evolution of life. Like egg hatching, Earth requires a stable temperature and environment to hatch a chick.

The smooth functioning of the Earth depends on numerous conditions across time and space. The fact that the planet has operated seamlessly for 4.6 billion years indicates that all these conditions have been met. Without delving into specifics, it's worth highlighting the presence of the Moon. Now, let's explore this issue further.

Is the Moon Designed for Earth?

We previously determined the importance of the Moon to the Earth. The Moon is an ideal and necessary natural satellite for the planet. Its size, mass, and distance are crucial for the formation of tides and probably the stability of the obliquity of the ecliptic. But how does the Moon's motion affect the Earth? Scientists still have much to study about this. The Moon's motion is not only a simple revolution and rotation but also a constantly changing angle between the Moon's orbital plane and the Earth's equatorial plane. The Moon's axis of rotation is not perpendicular to the Moon's orbital plane either; the angle between them is making regular changes (Figure 6–3). The Moon's motion is complex.

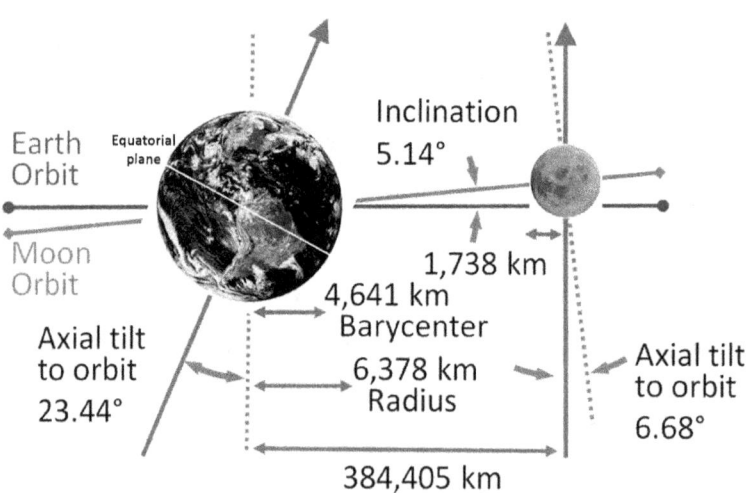

Figure 6-3: The complex motion relationship between the Moon and the Earth. The angle between the Moon's orbital plane and the planet's equatorial plane is not constant but does periodic reciprocating motion. The Moon's axis of rotation is not perpendicular to the Moon's orbital plane; the angle between them also constantly changes. All these changes may impact the Earth's environment (Source: NASA, public domain).

Since the Moon and Earth form a system of motion, and neither is a standard sphere nor homogeneous, the Moon's motion may have subtle effects on the Earth's environment. We can naturally imagine that the gravity field will change with the state of motion between the two bodies (gravity varies with distance). Some impacts on the Earth's environment may be inevitable. For example, how do these gravitational changes affect the seawater surface and atmospheric movement? We know little about these changes, but they must be cyclical and conducive to humans and other life; otherwise, this periodic influence, over time, would be adverse to life on Earth. If so, could life have evolved to this day?

Looking back at the Moon's history, it becomes even more impressive. Modern science believes that the Earth's initial rotation rate was much higher than today, which probably helped to create the geomagnetic field. Still, the Earth's rotation rate gradually slowed down due to the Moon's presence. Today, the Earth's rotation rate and the obliquity of the ecliptic are perfect for humans and all life. However, the Moon's orbit has been moving away from the Earth so slowly that the Earth-Moon system will no longer exist in billions of years. This process naturally makes people feel as if the Moon came to the Earth with a particular natural mission necessary to establish the Earth's environment, the evolution of life, the survival of humans, and the development of civilization. After completing these natural missions, the Moon will move away, forcing the Earth's motion into a new state and triggering a major change in the Earth's environment—deterioration. This is probably the stage humans must go through, but that's a long, long way off.

From the Moon's formation to its significant role on the Earth and its gradual moving away, the whole process sounds fascinating and seems precisely calculated for its natural mission. Isn't that the case?

Further, if indeed a collision caused the Moon's formation, the resulting Earth and Moon system is so precise in mass, volume, and motion that it's tough to imagine. Overall, the Moon benefits the Earth and is critical for the planet's environment and human existence. We can look forward to more discoveries in this field.

Given the Moon's crucial role for the Earth, humans, and life, it's not far-fetched that the Moon seems designed for the Earth. The formation of the Earth-Moon system is a continuous and highly complex process requiring precise timing that couldn't have occurred solely through random chance within a specific timeframe.

Is the Solar System Designed for Earth?

What kind of celestial system allows the Earth to function stably? Our solar system.

Undoubtedly, Earth could not exist in isolation in the universe and must rely on a celestial system to keep it moving steadily. In fact, for the past five billion years, our solar system has not only kept the Earth moving steadily in dynamics, but it has also met the Earth's particular requirements as follows:

- The solar system is in a suitable location within the Milky Way galaxy, protecting life on Earth from the dangers emanating from the galactic center.
- Solar radiation is the essential energy source for Earth. Its strength fluctuates over long periods, significantly impacting life. Currently, Earth is in an optimal period of solar radiation, providing the ideal intensity required for life. The size and age of the Sun, along with Earth's position and size, are perfectly aligned to support this balance.
- The presence of Jupiter and Saturn may protect the Earth from possible celestial collisions, keeping life safe. In fact, during the past 3.5 billion years, just one or two massive celestial collisions would be enough to wipe out all life on Earth, but this has not happened.
- As stated above, the Moon's presence has an indispensable, irreplaceable value.
- Modern science believes that Earth also needed comet impacts to deliver the large amounts of water and organic molecules essential for life.

All these conditions seem to be for one bright spot: life on Earth. The solar system has been running steadily for five billion years while meeting the above conditions. Given that these conditions are satisfied, the rest of the celestial bodies of the solar system, such as Mars and Neptune, both in size and in quantity, etc., may be the inevitable products of the solar system's formation and may also be necessary for the mechanical equilibrium or the stability of the vast system. Modern science shows that in addition to the eight planets, there are the distant Kuiper Belt and even more distant Oort Cloud. Although their origins and significance are still being explored, it is safe to say that they are also inevitable products of the birth of the solar system, not redundant parts.

But the complexity and stability of the solar system don't seem to matter much to other planets where there's no life, no evolution, and no humans. Only Earth needs a high level of stability. Our planet exists under the thoughtful protection of the solar system.

Could the Solar System Remain Stable for Another Five Billion Years?

Earth's stability depends on the solar system, so will the solar system remain stable for another five billion years? It depends first on the stability of the Sun.

The Sun is a single star (there are also binary star systems in the Milky Way). The Sun accounts for 99.86 percent of the solar system's total mass, and any change in its condition would have a massive impact on the entire solar system. Our Sun has performed an ideal role in the solar system, with its perfect mass, size, rotation speed, rate of nuclear fusion, and combustion stability. All these determine the possibility of stabilizing the solar system for another few billion years.

The orbit of our single-star system appears simple, with the eight planets moving in roughly the same plane, facilitating the stable operation of the solar system. However, it's crucial to acknowledge that our solar system is complex and nonlinear. As we understand, nonlinear systems lack stable solutions along the time axis and can generate chaos under specific conditions. Chaos has always been present within our solar system. For instance, individual asteroids in the asteroid belt "suddenly" veered out of orbit.

In general, the conditions outlined above are unlikely to change easily unless a powerful external factor impacts the system, though the likelihood of this occurring in a stable galaxy is minimal. Over the past five billion years, our solar system has remained largely stable, which can be considered a miraculous feat in theory.

Given this stability, we can be optimistic about the solar system's continued stability in the next few billion years. It is a perfect and precise star system, a rarity not found elsewhere in the Milky Way. Like Earth, the formation of our solar system has been a continuous and highly complex process requiring specific timing. Such intricate processes couldn't have occurred within a particular timeframe solely through random chance.

Is the Milky Way Designed for the Solar System?

The next step in our reverse reasoning is considering our Milky Way galaxy. What type of galaxy does the solar system require to maintain its stability and support the evolution of life? The five-billion-year history of the solar system indicates that the Milky Way is precisely the type of galaxy capable of meeting the requirements for the solar system's motion.

The Milky Way is an ordinary galaxy in the vast universe, with more than an estimated 200 billion stars. Scientists conjecture that the center of the Milky Way (the galactic center) has extreme radiation. Our solar system has been sitting steadily in the "suburb" of the Milky Way, far from the galactic center.

The Milky Way is a highly complex structure filled with various forms of matter and energy. Though poorly understood, we can offer a brief examination of the relationship between the evolution of the Earth and the Milky Way (Figure 6–4).

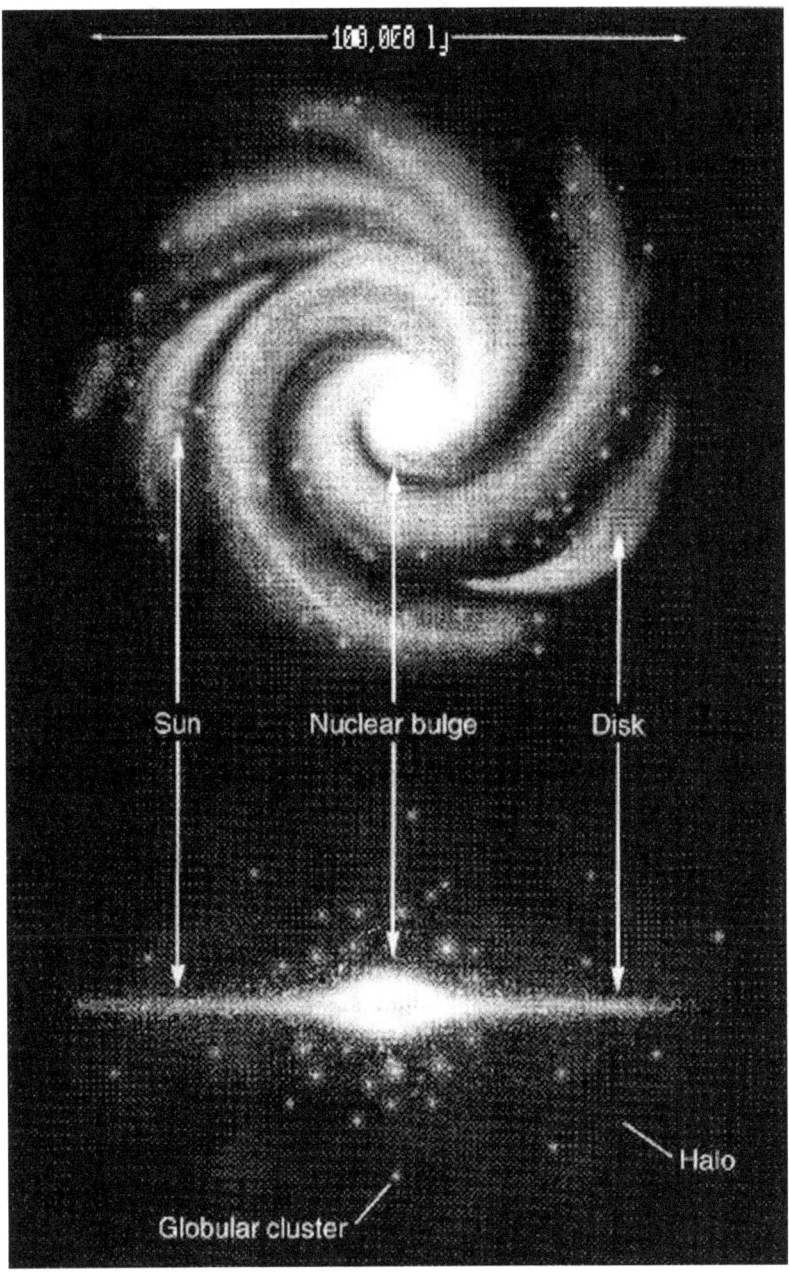

Figure 6-4: The Milky Way galaxy. There is no way to truly map the Milky Way (Source: Smith, *The Milky Way Galaxy*).

In the 160 years since C. R. Darwin published his theory of evolution in 1859, our understanding of evolution has been deepening with fresh discoveries. Modern science has come to understand that evolution is a highly complex process during which the change in environmental conditions matters.

Two contrasting theories emerge in studying life's evolution: progressive evolutionary processes (phyletic gradualism) and abrupt evolutionary processes (catastrophism or convulsionism). There's also the notion that evolutionary processes take less time as time progresses (evolution accelerating over time). However, we consider both theories for reference in this chapter.

There are five recognized mass extinctions in evolutionary history in which nature "rapidly" eliminated the species that did not adapt to the changed environment while new species emerged that successfully adapted. For example, after the completion of the "natural mission" of dinosaurs, the changed environment drove them to extinction and allowed the increased development of mammals. Without these catastrophes, gradual evolution would not be accomplished within the Earth's lifespan. Besides the mass extinctions, there was the Cambrian explosion, which laid out the "blueprint" for animal evolution.

But what caused the great change in environmental conditions which impacted the evolution of life? Some scientists are looking at the Milky Way for answers because the conditions around our solar system have not been unchanged. During the revolution of the solar system around the Milky Way, the different interstellar regions it passes through may significantly influence the Earth's environment (see below).

Scientists now believe the Milky Way to be about 98,000 light-years in diameter and to have a rod-like center from which four spiral arms extend outward. Our solar system lies within the Orion cantilever, about 26,000 light-years from the galactic center (Figure 6-4). The motion of the solar system around the Milky Way has the following features (Figure 6-5):

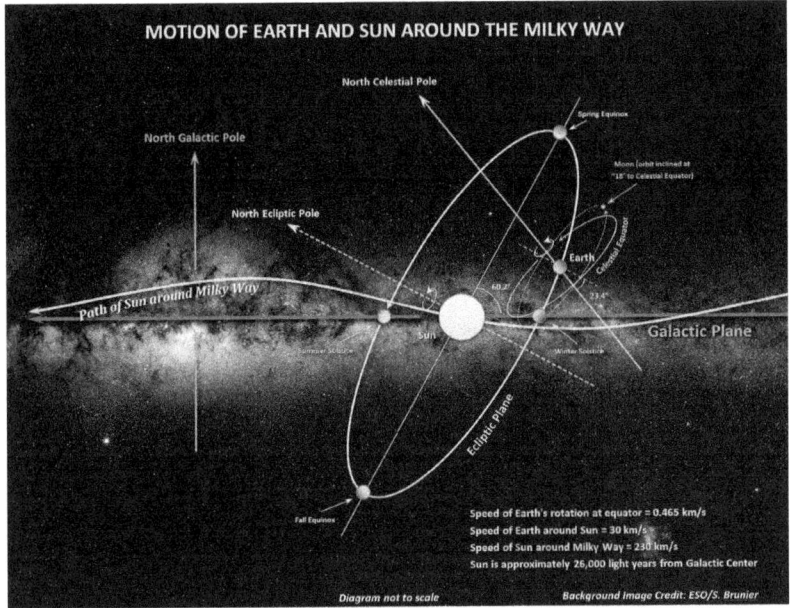

Figure 6-5: How the solar system moves around the Milky Way. The celestial system's complex, precise, and regular motion probably has no similarities outside our solar system (Source: Jim slater307, CC BY-SA 4.0).

The orbit of the Milky Way takes about 220 to 250 million years, or one galactic year (revolution). Our solar system has been around for at least 20 galactic years. Instead of following a simple plane motion, it moves up and down through the galactic plane in a wavy movement (Figure 6-5), a bizarre behavior! Current research also suggests that the Galactic Disc might not be a flat plane but rather a wavy one along the galaxy's radial direction and that the density of stars may vary significantly throughout the galaxy.

In addition, the orbit passes through the four spiral arms. All of these orbital features inevitably increase the possibility of periodic changes in the surrounding conditions of our solar system.

Note that the Milky Way's intense radiation field originates from numerous stars, each undergoing its unique developmental stages and emitting varying intensities of radiation over time. Consequently, the interior of the Milky Way is far from tranquil; it could be characterized by disorder or give rise to potent combinations, including supernova explosions. Such phenomena not only directly affect life at a genetic level but also have the potential to influence the Earth's environment, as radiation is a form of energy.

On the other hand, Earth experiences periodic natural events such as the Great Ice Age, mass extinctions, large-scale transgression and regression of the sea, geomagnetic reversals, and so on. Is there a close relationship between these events and the periodicity of the Milky Way? This question remains unresolved. Modern science has a limited understanding of the highly complex structure of the Milky Way and its impact on the solar system during its revolution. We can only assert broadly that the Milky Way's influence on the Earth exists, but the precise relationship remains to be studied.

Let's expand on this point. On the one hand, the solar system must maintain stability to ensure the Earth's motion remains stable; on the other hand, dramatic changes in the planet's environment are necessary for life to evolve. This presents a paradox. The only apparent solution is to delegate the latter "task" to the Milky Way galaxy, suggesting a "division of labor" between our solar system and the Milky Way, wherein the stability of the Earth's motion is ensured within the solar system, while significant environmental changes on the planet allow life to continue evolving.

We emphasize the change in the radiation intensity of the galaxy rather than specific celestial objects. The impact of radiation intensity on the Earth may be invisible, yet it can influence genes and the environment.

We may not know why the solar system's orbit is so strange or why spiral arms exist in the Milky Way. Still, we can say for sure that if the motion of our solar system were confined entirely to the galactic plane, with no undulating motion, and if the Milky Way were not cantilevered, then the interstellar conditions through which the solar system passes in a galactic year would be roughly the same. What would happen to the Earth's environment in this case? The story of the Earth's evolutionary process would undoubtedly be different.

At any rate, the Milky Way gave birth to the solar system, the Earth, and human beings. Given the structure and function of the Milky Way galaxy, this highly complex formation could be neither random nor arbitrary because everything happens conditionally (see Chapter II.1, The Mysterious Solar System and the Galaxy).

So, it is not fantasy to say that the Milky Way is designed for the solar system. Of course, this is only a hypothesis in need of scientific proof.

Continuing our thought process, how did the universe construct the Milky Way? The answer is of intense speculation in modern science. The Milky Way is typically estimated to be at least 12.2 billion years old, emerging shortly after the Big Bang.

The Big Bang

Our final reverse reasoning aims to uncover how the universe constructed our galaxy. As previously mentioned, nature initially formed atoms as the fundamental building blocks of matter during the Big Bang. Atoms possess immense versatility, capable of manifesting as the Milky Way, the solar system, the Earth, human beings, and more. Atoms provide the matter for everything! Without atoms, there would be no universe. However, atoms are not exclusive to the Milky Way; they are a universal tool, constituting countless other galaxies in the vast universe.

We might ask about the significance of the countless galaxies in the universe for the emergence of human beings. Are they also essential for the universe? This profound question remains unanswered by modern science. However, logically, the following points are noteworthy:

- The universe could only be an interconnected, interdependent whole from which no individual part could be isolated.
- All material forms and properties are derived from the atomic structure. In other words, the substances of which atoms are made and the substances of which the universe is made are interchangeable.
- The universe has continually been expanding, and only an expanding universe could support life (see the next section).
- The Milky Way results from the universe's expansion at a specific stage. Its notable feature is its role in the creation of our solar system and human beings, along with the solar system's complex and peculiar movement patterns. As previously mentioned, all of this contributes to the galaxy's mystery. Undoubtedly, its formation isn't in isolation; it could only form within an expanding universe filled with numerous galaxies. Also, it must be intricately linked to its surrounding conditions in time and space, unveiling the mysterious relationship between the universe and the Milky Way.

- Based on the above, the universe as an interconnected whole "must" be big enough, and there "must" be enough galaxies to give birth to our Milky Way with our solar system.
- As for theories about dark matter and dark energy that have yet to be confirmed, if true, their existence must be tied to visible matter. Without atomic matter, these invisible things would lose their meaning. In theory, dark matter and dark energy are essential components of the universe, making modern theories about the universe more plausible.

Summary

Our universe ultimately gave birth to humans and modern civilization. We can divide our reasoning above into five stages:

1. The human brain (the starting point) ↔
2. The Earth System stage ↔
3. The solar system stage ↔
4. The galaxy stage ↔
5. The Big Bang stage ↔

Based on the two natural philosophy principles and the periodic table, it is easy to understand that only a hen can lay a chicken egg because it possesses all the physiological structures and functions for producing a chicken egg. Similarly, only the Milky Way can make humans and their civilizations because it has all the structures and functions for the emergence of human beings in time and space.

We do not know how the Milky Way produced the solar system nor how the solar system nurtured the Earth's formation. Still, we might say that without the precise structure and corresponding function, the universe could not create a civilized planet. However, the conditions required to form this series of structures and functions in time and space might be highly complex and fixed.

It's worth noting that modern science primarily focuses on studying natural structures themselves, such as life, while the conditions leading to the formation of these structures have received less attention. Exploring these conditions tends to be abstract and lacks standard answers, posing numerous challenges to research. Yet, these conditions remain the most mysterious aspect of nature.

So, logically, from establishing conditions to structure to function, it must be an incredibly intricate process, with no room for alternatives because the conditions are unchangeable.

The human brain and modern science seem to be the "final result" of 13.7 billion years of natural development, which shows that since the Big Bang, the universe appears to have focused on the formation of the human brain and the emergence of modern civilization.

The following diagram illustrates the logic of backward reasoning.

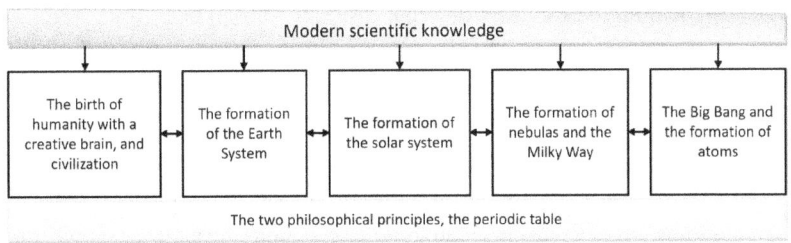

↔ Requires/results in

At each stage, the farther one moves from the Earth (left to right), the less scientific knowledge is available. This is understandable.

The time elapsed at each stage in the chart is as follows:

- From the Big Bang (13.7 billion years ago, marking the formation of atoms)
- to the formation of nebulas and the Milky Way (12.2 billion years ago)
- to the formation of the solar system (5 billion years ago)
- to the formation of the Earth (4.6 billion years ago)
- to the emergence of life through evolution (3.5 billion years ago)
- to the appearance of hominids (3 million years ago)
- and finally to modern civilization (300 years ago).

This series of figures shows that the evolution of the universe has the following characteristics:

- The initial stage is lengthy and shortens over time, following the general law of developmental progression. Therefore, it represents an organized process of evolution rather than an arbitrary process.
- The evolution of the universe requires strict timing. All these processes must occur during the universe's expansion, and life must occur during the optimal period of solar radiation.

Regarding the scale of each of these stages, logically, we can arrive at the following understanding:

The size of hydrogen atoms sets the scale for all other elements in the periodic table, influencing the sizes of various molecules, cells across species, the human brain (as argued above), the human body, and even the Earth (as mentioned regarding its gravity). Subsequently, this influences the size of the Sun and the entire solar system (as argued previously in this book), potentially extending to the size of the Milky Way (as argued above). This progression represents a harmonious combination devoid of arbitrariness; instead, it unfolds as a precise, subtle, and logical process.

Which stage or sub-stage do you believe could be altered? Could the wavy movement of the solar system be modified? Could the size of the Sun be changed? Could the distance between the Sun and the Earth be altered? Could the Moon be removed? Could the distribution of land and sea on Earth change? Could the structure of the human brain be modified?

There is only one way for the universe to evolve, from the appearance of atoms to the formation of chemical elements to the emergence of a planet of civilization. This process is a logical, rational evolutionary system that does not allow for change (please review this chapter).

BACK TO EINSTEIN AND HAWKING

Einstein's Natural Philosophy

Einstein always believed that nature has "the Reason" and has a profound wisdom that humans can only glimpse occasionally.

> Quantum mechanics is certainly imposing. But an inner voice tells me that it is not yet the real thing. The theory says a lot, but does not really bring us any closer to the secret of the 'Old One.' I, at any rate, am convinced that He is not playing at dice.[1]

---

1. Einstein, *Letter to Max Born*

> I am satisfied with the mystery of the eternity of life and with the awareness and a glimpse of the marvelous structure of the existing world, together with the devoted striving to comprehend a portion, be it ever so tiny, of the Reason that manifests itself in nature.[2]
>
> To know that what is impenetrable for us really exists and manifests itself as the highest wisdom and the most radiant beauty, which our dull faculties can comprehend only in their most primitive forms—this knowledge, this feeling, is at the center of true religiousness. In this sense, and in this sense only, I belong in the ranks of devoutly religious men.[3]

His words convey this fundamental idea. He expressed similar thoughts numerous times, which we will not quote here.

Even ordinary people, if they observe and reflect carefully, will find nature's wisdom everywhere. Isn't the wonderful world, including the mysterious atom, created by nature? Therefore, it is evident that nature itself possesses profound wisdom.

Hawking's Ideas and R. Penrose's Theory

Stephen Hawking described the universe (space-time) as follows:

> If the classical theory of general relativity was correct, the singularity theorems that Roger Penrose and I proved show that the beginning of time would have been a point of infinite density and infinite curvature of space-time. All the known laws of science would break down at such a point.[4]
>
> On the other hand, the quantum theory of gravity has opened up a new possibility, in which there would be no boundary to space-time and so there would be no need to specify the behavior at the boundary. There would be no singularities at which the laws of science broke down, and no edge of space-time at which one would have to appeal to God or some new law to set the boundary conditions for space-time.[5]

---

2. Einstein, *Letters to Solovine*, 102
3. Einstein, *The World*
4. Hawking, *A Brief History*
5. Hawking, *A Brief History*

THE RATIONAL UNIVERSE EVOLVING FOR HUMANS

Space-time would be like the surface of the earth, only with two more dimensions. The surface of the earth is finite in extent but it doesn't have a boundary or edge; if you sail off into the sunset, you don't fall off the edge or run into a singularity.[6]

Note: The 2020 Nobel Prize in Physics was awarded to Roger Penrose for his work on the Singularity Theorem.

So, according to his quantum theory of gravity and imaginary time, Hawking likened space-time to the Earth's surface (Figure 6–6, left), correspondingly from the Big Bang to the Big Crunch (Figure 6–6, right).

The universe starts at the North Pole as a single point. As one moves south, the circles of latitude at constant distance from the North Pole get bigger, corresponding to the universe expanding with imaginary time. The universe would reach a maximum size at the equator and would contract with increasing imaginary time to a single point at the South Pole. Even though the universe would have zero size at the North and South Poles, these points would not be singularities, any more than the North and South Poles on the Earth are singular. The law of science will hold at them, just as they do at the North and South Poles on the Earth.[7]

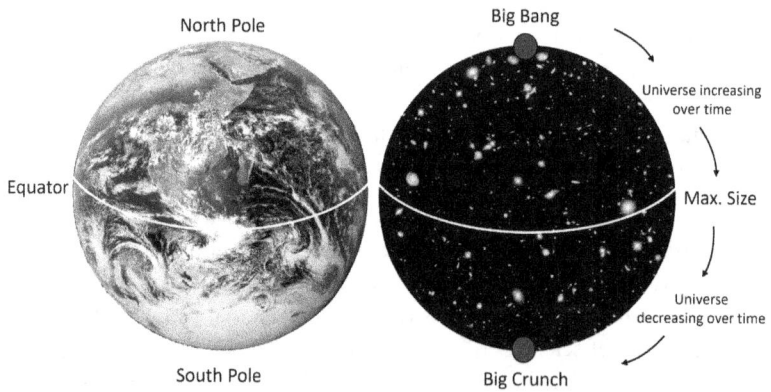

Figure 6–6: The Big Bang to the Big Crunch (Source: Hawking, Stephen. *A Brief History of Time*. New York: Bantam, 1988).

6. Hawking, *A Brief History*
7. Hawking, *A Brief History*

Hawking goes on to say, "I'd like to emphasize that this idea that time and space should be finite 'without boundary' is just a *proposal*: it cannot be deduced from some other principle."[8]

However, logically, if space-time were infinite, there would be no time requirement for civilizations to emerge in the universe. This implies that life could only appear at a point infinitely distant on the timeline, which contradicts the reality of human existence. Thus, the concept of infinite space-time is illogical: ". . . intelligent life could not exist in the contracting phase of the universe . . ." and ". . . the conditions to be suitable for intelligent life only in the expanding phase . . ."[9]

Thus, life could only arise and evolve during the expansion of the universe—from the North Pole to the Equator.

Having no boundary or edge and neither beginning nor end, the evolution of the universe from the Big Bang to the Big Crunch could be called a cosmic cycle.

To learn more about the cosmic cycle, refer to the works of the scientist Sir Roger Penrose[10] and his conformal cyclic cosmology (CCC) theory. This book is not positioned to explore this profound topic further.

The Characteristics of Cosmic Evolution

There are two fundamental characteristics of the natural process, from the Big Bang to the appearance of modern science on Earth, which are consistent with the thoughts of Einstein and Hawking.

The first is the requirement for timing.

In the preceding chapters, we have demonstrated the time effect in the Earth's development. The requirement for timing is widespread in the natural development from the Big Bang to the emergence of modern civilization. The evolution of life on Earth must occur during the optimal period of solar radiation, and the solar system and the Milky Way galaxy must exist during the expanding phase of the universe. Timing is an essential element in natural development. Without the requirement of timing, natural development would be disordered, with no meaningful results. The point here is that natural development has a finite length of

---

8. Hawking, *A Brief History*
9. Hawking, *A Brief History*
10. Penrose, *Cycles of Time*

time. Each stage of natural development has a time requirement, so the timing of each stage is critical.

The universe has completed the evolution from the Big Bang to the appearance of humans and civilization within the expansion phase. We may call it "cosmic evolution" (natural development).

The second fundamental characteristic of the natural process is the extreme complexity of cosmic evolution.

According to the two principles of natural philosophy, the universe's evolution must be of extreme complexity. If you have attentively read all the above chapters, you may realize the complexity of cosmic evolution. The only way to accomplish this highly complex cosmic evolution in a limited time is to make intentional rather than arbitrary selections.

Let's describe this issue further with the following two points:

1. In the process of natural development described in this book, natural selection at each stage leads in one direction: the emergence of humans and modern civilization.

- Planetesimal stage: Earth's appropriate position, size, and mass were established, meanwhile absorbing a large amount of silicon, aluminum, and iron according to its own needs, forming different element abundance than the nebula.

- Formation stage: Earth formed a layered structure, which led to plate movement and a strong geomagnetic field.

- Development stage: Over 4.6 billion years, the Earth formed the layered atmosphere required for the survival of life and the development of civilization, and through the plate movement, formed the global land/sea distribution with various geographical conditions and a large number of natural resources (freshwater, various minerals) required for the birth and development of humanity.

The above processes are undoubtedly stochastic, but nature's selection in those processes is always in one direction: forward to the birth and development of humans and their civilizations. We may call it the unidirectionality of cosmic evolution (see Chapter IV, What Exactly is the Earth? (5)).

The detailed process of the above stages has been explored throughout the chapters of this book.

2. The progression from the Big Bang to the emergence of humans and their civilization hinges on numerous precise conditions, though

THE UNIVERSE EVOLVED FOR HUMANS

their precision is relative and can vary within a narrow range. For instance, astronomical values, such as the distance between the Sun and the Earth, can only change within a minimal allowable range. This is understandable. Similarly, there must be a small permissible range of variation in the composition of the air, particularly concerning oxygen and carbon dioxide levels. While the allowable range of change in the distribution of land and sea might be slightly more extensive, no changes can occur that would disrupt the global climatic patterns, including ocean currents, that humanity has relied on for thousands of years.

According to Hawking:

> The remarkable fact is that the values of these numbers seem to have been very finely adjusted to make possible the development of life. For example, if the electric charge of the electron had been only slightly different, stars either would have been unable to burn hydrogen and helium, or else they would not have exploded.[11]

Of course, if you're a theoretical physicist, you can provide numerous examples demonstrating that nature's composition (structure) is precise and rigid. Since the Quaternary Period, various values that impact the global environment and climate appear to fall within an allowable range (see Chapter I.1, The Timely Appearance of the Great Ice Age during the Quaternary Period).

The two points above further illustrate the intent of cosmic evolution. Let's consider the relationship between cosmic evolution's complexity, intentionality, and unidirectionality. In this book, we have often illustrated the extreme complexity of cosmic evolution. In essence, this complexity is determined by two principles of natural philosophy. Given the requirement for timing, extreme complexity must result in intentionality in cosmic evolution, and intentionality must lead to unidirectionality in that process. Hence, complexity → intentionality → unidirectionality are indivisible, based on the extreme complexity and the requirement for timing.

Given this conclusion, complete randomness couldn't result in such extreme complexity. We are not denying the existence of randomness, which is inevitable in cosmic evolution, but if cosmic evolution occurred purely by random selection, it would have to go on indefinitely with limited results. We emphasize that "the highest wisdom" or "the Reason" manifests itself in numerous stochastic natural processes.

11. Hawking, *A Brief History*

Broadly speaking, there are only two kinds of natural selection in natural development: random and intentional. So far, which one could support cosmic evolution?

Again, considering the time constraints of cosmic evolution and the extreme complexity of the universe, it could only be the result of intentional selection among numerous potential random choices or a selection process oriented in a specific direction.

The universe is neither magical nor blind; on the contrary, it is rational. Its unfathomable wisdom has manifested itself in the countless random choices of cosmic evolution, leading to humans' emergence and civilization's continual development. Though a logical inference from popular science, we don't really know what Nature is. Therefore, we cannot underestimate "the Reason in nature" proposed by Einstein, a profound idea reflecting his view of nature.

One might wonder if the universe, evolving for humans, could have also created another world for humans, which doesn't necessarily contradict the viewpoint of this book. To address this question, one may refer to the backward inference section in this chapter. The process is so intricate that the universe appears to have had to focus on it from the outset or exhibit a unidirectional nature. It's akin to a magnifying glass with only one focal point. The universe couldn't concentrate on two identical processes simultaneously, nor does it require two groups of humans on different planets to fulfill the same natural mission (see the next section).

Here, we must also note that the universe is a nonlinear system, which cannot produce two natural processes with precisely the same detail, such as the astonishing Moon and its complex movement. This is a natural law, created by nature, and cannot be violated by her. It is also impossible for the universe to use the same periodic table to create two life forms from different chemical elements.

We could assert that the universe has evolved over a long period to shape the Earth System we cherish, which is full of "natural miracles" guiding humanity from ancient times to the present and beyond into the distant future. The notion of "natural miracles" proposed in this book implies that this is the inherent manner in which nature operates. Cosmic evolution appears to represent an overarching super-design that, once initiated, progresses toward its ultimate culmination.

A SIMPLE SUMMARY OF COSMIC EVOLUTION

The Chart of Cosmic Evolution

The chart below simplifies the process of cosmic evolution. From this chart, you can understand that for the human brain to emerge in the universe, which is essential for civilization, this is the only path determined by the principles of natural philosophy. Given the complexity of the human brain, its formation requires countless natural conditions in time and space. Nature created these conditions to gradually form more complex structures with higher functions, allowing the human brain to evolve step by step.

The universe began in its simplest state, all energy. Next, conditions arose to form a natural structure with specific functions (atoms). Then, more conditions emerged to create increasingly complex structures with higher functions (elements, molecules). This process continued until the formation of the brain and the advent of modern civilization. It is a highly complex, gradual, and logical process that requires sufficient time and intelligence, or "the highest wisdom" to complete. There is no other alternative. Einstein's concept of "the Reason" is the only guidance capable of overcoming randomness.

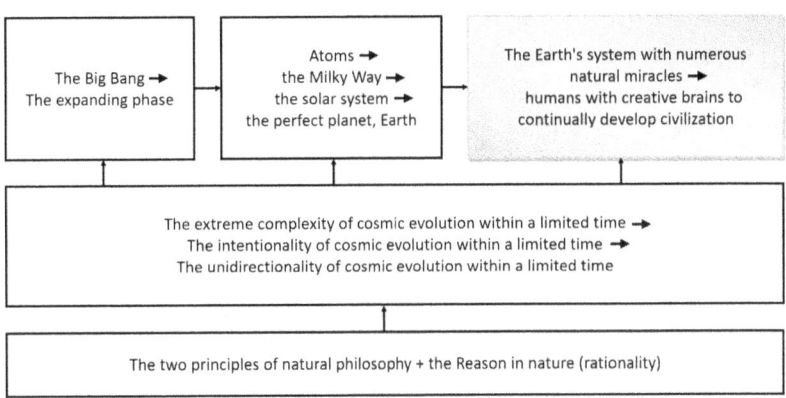

In this chart, we would like to emphasize again the "the highest wisdom" and "the Reason in nature." Cosmic evolution is a stochastic process filled with random selections, but how did nature know its selection every time (such as the Earth's conditions of astronomy) would be the best one for the eventual emergence of human beings in the universe? The two principles of natural philosophy, like two rails, form the track of the process of natural development. The guidance of this track is natural rationality, or "the Reason," which has always guided natural development towards the eventual emergence of humans and modern civilization within a limited time (see Chapter IV, What exactly is the Earth? (4)).

Random selection cannot produce meaningful results within a limited time, or it would require an infinite amount of time.

Further Understanding the Element of Carbon

Carbon, the Initiator

We've argued that carbon is the only element in the periodic table capable of fulfilling two vital missions: supporting life and enabling civilization. This is due to the unique properties nature has bestowed upon carbon. Its oxide, carbon dioxide, is a gas, a crucial and exceptional property that allows it to be fixed into organic molecules through photosynthesis to form organic matter. Forest fires caused by lightning strikes (the combustion of carbon) introduced early humans to the value of fire, setting the stage for the emergence of civilization. Humans used plants rich in carbon as fuel to make fire, which is considered the father of civilization (see Chapter II.1, The Mystery of Fire).

As humans increasingly used combustion, more carbon stored in ancient life forms was released, resulting in higher atmospheric carbon dioxide levels. The continuous accumulation of carbon dioxide and other extensive human activities have led to environmental deterioration. This process is inevitable. The melting ice sheet at the poles will not recover, but the rate of melting may slow down in the future if humans can reduce carbon dioxide emissions. The worsening environment will significantly impact the survival of humans and all other life, accelerating the development of science and technology in response. All chemical elements will be utilized to their fullest potential.

We can logically imagine that highly intelligent robots will eventually prevail. With their help, cellular humans may create intelligent energy beings (see Chapter V, The Remote Future of Humanity (4)). However, cellular humans will eventually need to leave the natural world due to the severe deterioration of the Earth's environment—exacerbated by long-term astronomical changes and continental drift. As carbon-based life forms, cellular humans are merely a transitional phase in the long cosmic evolution. Ultimately, carbon will complete its natural mission. This is, of course, an extremely long process, marking the dual mission of carbon in this grand evolutionary narrative.

The Entire Process is Extremely Slow

Firstly, macromolecules formed from carbon skeletons evolve so slowly, on a time scale of one hundred million years, they lived in the sea for the first three billion years of life on Earth. They evolved extremely slowly because the structure of life was very simple, with few functions, and the marine environment also changed very slowly.

As life extended into terrestrial environments, evolution relatively accelerated. Here, the rate of molecular evolution, determined by the accumulation process of neutral mutations (neutral gene mutations), is very slow.[12] Note that the main elements that make up the DNA molecule are still carbon, hydrogen, oxygen, nitrogen, and phosphorus, and its main role is carbon-based because only carbon can form macromolecules. Gene mutation is an inevitable property of DNA molecules and the only way carbon-based life can continue from generation to generation. Therefore, the evolution of carbon-based life is necessarily extremely slow.

Secondly, global environmental changes (such as changing global atmospheric temperatures, continental drift, the transformation of sea into land, orogeny, etc.) are also relatively prolonged processes.

Molecular evolution can be regarded as the internal factor of life's evolution (internal material basis), while global environmental change serves as the external condition for life's evolution. The global environment must not change too quickly; its rate of change must roughly match the rate of molecular evolution. This alignment ensures that the evolution of life can proceed seamlessly.

What amazes us is that these two processes were well-matched, successfully realizing the long evolutionary process of life, ultimately leading to the appearance of humans and their civilization. This process required a planet of specific size and mass (see Chapter VI, Earth—A Perfect Planet for Forming the Human Brain), and the optimal period of solar radiation had to be sufficiently long to allow for the time needed for life to evolve, which in turn determined the necessary size and mass of the Sun.

12. Kimura, *The Neutral Theory*

So, the element of carbon determines not only the material process from life to civilization but also the time required for natural development and the conditions needed, which includes the size and the mass of the Earth, the Sun, and even the Milky Way. In this sense, our universe might even be called a carbon universe.

That cosmic evolution seems well defined (arranged) by the highly mysterious periodic table is not nonsense but a scientific fact. Don't you think so?

But why does carbon have such magical properties? Note that carbon is only the sixth element in the periodic table in terms of its electron arrangement; however, its enormous potential is incredible. So, this question is beyond the scope of modern science and can only be explained by natural philosophy (including the natural philosophical principles). But this explanation is no more than another way of showing nature's great mystery, just as Einstein said, "The most beautiful experience we can have is the mysterious."[13]

Is it "the most profound reason" and "the highest wisdom" that manifests itself in nature? Philosophically, it is!

The Earth, Only a Tool

Given the extreme complexity of the human body, "the processing" of chemical elements into humans must be an extremely complex evolutionary process, which requires a special planet in time and space to complete. The universe did its best to build this perfect planet, Earth. But when the Earth has completed its natural mission, it and the solar system will wither away. Therefore, Earth is only a tool of the universe, and its existence is relatively short-lived. This is an inevitable process.

Humans are the value of nature in the universe, and we will coexist with the universe through evolution to fulfill our natural mission given to us by the universe (see Chapter V, The Remote Future of Humanity (4), and the last section).

Of course, you can also imagine that humans will disappear from nature with the destruction of Earth. In other words, humans are only a temporary "tragedy" in the universe. So, what is the significance of this tragedy? It does not make sense that Nature took 13.7 billion years to

---

13. Einstein, *The World*

create humans for no purpose, nor is it consistent with the logic of Nature, which has "the highest wisdom" and "the Reason."

A Focusing Process

Now, let's look at the evolution of the universe from another perspective.

In essence, the universe's evolution is a process of increasing the function of matter and energy (ME), which requires increasing the density of ME. Otherwise, is there a need for the existence of the universe?

So, continually increasing ME density in a region is a primary condition for the universe to produce anything meaningful. This is mentioned by Hawking:

> In an expanding universe in which the density of matter varied slightly from place to place, gravity would have caused the denser regions to slow down their expansion and start contracting. This would lead to the formation of galaxies, stars, and eventually even insignificant creatures like ourselves.[14]

Let's consider the density of matter as a representation of the density of ME. We conjecture that the density of matter in the solar system is greater than that of the Milky Way (the closest star to our solar system is Proxima Centauri, which is 4.22 light-years away) and that the Earth's density surpasses that of the Sun, making Earth the densest planet in the solar system. Although the human brain is less dense overall, this does not contradict the trend of increasing matter density from the Milky Way to the Earth.

The formation of intelligence depends not on matter density but on the density of material organization (material structure). The higher the function, the more complex the structure required. As previously mentioned, the human brain evolved under Earth's gravity and may be the most densely organized structure in the universe. It contains billions of neurons (nerve cells) interconnected by synapses, with unimaginably vast numbers of synapses. This suggests an extraordinary density and sophistication of the brain's organization. The discovery of the function and structure of synapses marks a significant breakthrough in brain research. Therefore, it can be said that from the Milky Way (12.2 billion years ago) to the solar system (5 billion years ago) to the Earth (4.6 billion years

14. Hawking, *A Brief History*

ago) to the human brain (about 30,000 to 10,000 years ago), there has been a progressive focusing process of increasing ME density.

This process (from the Big Bang to the emergence of the human brain) should conform to the theory of dissipative structure of nonlinear systems (see Chapter IV, No Other Identical Natural Process in the Universe). We may well logically imagine that soon after the Big Bang, energy gradually formed an orderly system of ME from the initial chaotic state. The first ordered ME system was galaxies, including the Milky Way. Then, the Milky Way gradually generated the ordered solar system from its chaotic state. In the same process, the solar system gradually generated the solid Earth from its chaotic state. With its unique functions, Earth gradually processed the chemical elements into the human brain, leading to modern science.

The mystery lies in the fact that at each stage, where chaos transitions into order, events occur at positions crucial for the eventual emergence of human beings. For instance, our solar system is positioned far from the center of the Milky Way, our planet occupies the right spot within the solar system, and the GRV is situated precisely on the equator of the African Continent. These occurrences may illustrate "the Reason that manifests itself in nature." Every stage of the process is necessary, and you couldn't find a better way to achieve it.

Here, we must note that the formation of the Milky Way must have required many conditions, and creating these conditions must have been an extremely complex process starting from the Big Bang. So, given the extreme complexity and precision in time and space of the process from the Big Bang to modern civilization's appearance, the principles of natural philosophy, and nature's nonlinearity, the universe could not focus on two identical Earth-brewing processes from the start. As mentioned above, it is like a magnifying glass that cannot produce two focal points. It's a bit like a chicken not having two biological systems for laying eggs. It is also easy to understand that according to the two principles of natural philosophy, the same periodic table could not produce two different types of life made of various elements using different properties.

According to all the arguments above, we might say that the universe without intelligence did not necessarily come into being or needed to exist. It is groundless to imagine that there is another human world in the universe with a higher civilization than Earth. It is also unfounded to imagine another civilization universe without atoms.

Maintaining Intelligence Through Cosmic Cycles

Let us first review the logic of natural development (cosmic evolution). From all the arguments in this book, we can see the logic of natural development to create human intelligence and bring it into full play. From the Big Bang of the universe to the birth of Earth, to the birth of humanity in the GRV, to the Mediterranean Sea, to the Eurasian continent, to the global distribution of land and sea, to the distribution of various natural environments and natural resources, through this series of natural development, on the one hand, nature created climate conditions suitable for the development of civilization. On the other hand, nature created environmental conditions that constantly stimulated human intelligence. Then, based on the characteristics of human nature with human activity on a large scale and the periodic changes in geology or astronomy, such as the ice age and global high temperatures, the global environment inevitably goes into deterioration, forcing humanity towards extinction while entering into intelligence in the form of energy (see Chapter V, The Remote Future of Humanity (3) and (4)).

So, what is the significance of intelligence in the form of energy? From the perspective of the universe, we can further understand the problem through Hawking's and Penrose's theories.

According to Hawking's theory, the universe undergoes a cosmic cycle from the Big Bang to the Big Crunch. He argued that the universe is self-contained, with no beginning or end. The Big Crunch would lead to the next Big Bang, creating a continuous cycle (see Roger Penrose's theory mentioned earlier). In this process, the emergence of human beings reflects the wisdom of the universe, seemingly demonstrating the significance of the universe's existence. We might wonder: if no humans ever existed in a universe, would that universe have any meaning? Accordingly, humans are likely to appear in the next cosmic cycle. If you ask, what's the purpose of human existence? Humanity cannot yet answer this question definitively; we can only say that humans are essential for the universe's evolution.

According to general philosophy, everything in the world, except the subatomic world, develops continuously, so the universe's evolution would also be continuous. The next universe would still need humans. In other words, an intelligent universe must ensure the presence of humans in every cycle. Achieving this continuity requires the existence of intelligent energy beings, which return to the universe's original energy state.

THE RATIONAL UNIVERSE EVOLVING FOR HUMANS

Cellular humans, having evolved from atoms, face demanding conditions for their existence and cannot last forever. As the global environment deteriorates, cellular humans will inevitably decline and eventually disappear from nature (see Chapter V, The Remote Future of Humanity (4)).

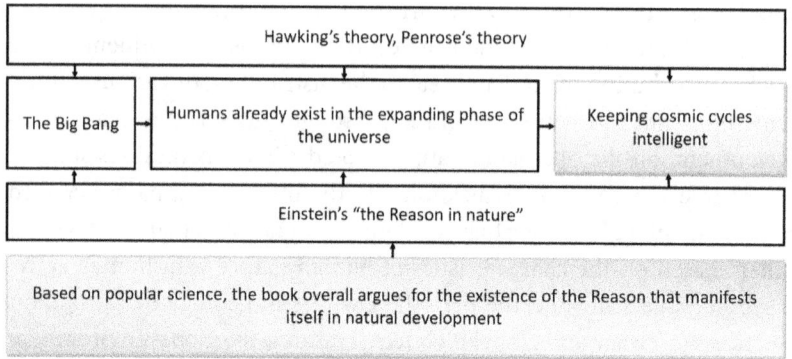

As the chart above shows, the ideas of Einstein, Hawking, and Penrose are connected in the sense of natural philosophy. Einstein's "the Reason in nature" and "the highest wisdom" are also on full display in Hawking's and Penrose's theories. Without "the Reason in nature," the three theories could not be philosophically integrated. Although this chart is only inferential and can never be verified, it makes sense scientifically and logically.

Since the logic of natural development is that the universe evolved to create intelligence, the human world should also follow this logic, fulfilling the mission of human intelligence in the evolution of the universe. Fortunately, since the Renaissance, humanity has begun to respect human rights and constantly liberate human wisdom.

However, one may ask a deeper question: although human intelligence is invisible, it can always manifest through human actions, such as innovation. The birthplace of human intelligence is the brain. So, since "the highest wisdom" is embodied in natural development, as described in this book, where does it come from? It's a question that can never be answered scientifically, and it's also a question without any significance. Einstein's natural reason is a property of nature (the universe), according to which it is sufficient for people to recognize their natural mission. Do you have to find an "entity" that can command everything? Einstein always disagreed with this idea. Einstein's views on religion are best expressed in his writings and letters, particularly in his letter to philosopher

Eric Gutkind in 1954 (known as the "God Letter"). In it, he makes clear that he did not believe in a personal God. He described the idea of a personal God as a "product of human weakness."[15]

So, "the Reason in nature," as a property of the universe, is more reasonable, logical, and meaningful than an entity. Now, let us end the book with Professor Roger Penrose's religious views:

"I'm not a believer myself. I don't believe in established religions of any kind. I would say I'm an atheist."[16]

"I think I would say that the universe has a purpose, it's not somehow just there by chance. . . some people, I think, take the view that the universe is just there and it runs along—it's a bit like it just sort of computes, and we happen somehow by accident to find ourselves in this thing. But I don't think that's a very fruitful or helpful way of looking at the universe, I think that there is something much deeper about it."[17]

However, Penrose did not specify what purpose the universe has. This book seems to answer that big question, which is based on the philosophical ideas of Einstein, Hawking, and Penrose.

15. Einstein, *Der Einstein-Gutkind Brief*
16. BBC News, *Big Bang Follows*
17. A Brief History of Time, Morris, Dir.

Epilogue
(On the human brain)

It sounds absurd that the universe evolved for humanity or our brains. But if you follow all the arguments in this book, perhaps you'll find it is a grand project and makes sense.

Don't you find your brain amazing? It can not only remember but also bring forgotten scenes back to life in your mind. It can produce a variety of complex feelings and human nature. It can analyze, understand, and solve a problem, which is the creative function. So, what is this physiological structure? What is this biochemistry that gives it these fantastic functions? The formation of the human brain in the universe is far from simple!

The human brain is the product of a long evolutionary process of constantly combining the physical characteristics of the Earth with the physiological characteristics of life. Gravity and DNA molecules are two primary factors in this combination. According to the two principles of natural philosophy, the process is so complicated and subtle that the human brain couldn't have been created on any other planet. Though humans differ in height, their bodies do the same job of effectively supporting their brains. The value of the human brain lies in its creativity, which enables the endless development of civilization, which might eventually connect with the cosmic cycle. So, the creativity of the human brain consists of the logic of the universe's evolution; therefore, disrespecting the human brain and thoughts, no matter what the excuse may be, is against the logic of the development of nature.

However, to create the brain, the universe had to start with the formation of atoms, then the Milky Way, the solar system, the Earth, and finally life, each stage nurturing the next. In this cosmic evolution process, the principles of natural philosophy define that it had to be time-consuming and complicated. The extreme complexity and precise timing requirements suggest that the evolution could only be intentional and unidirectional. This approach ensures the process can be completed within the universe's expansion phase. If each step were determined entirely by random selection, the process would stretch indefinitely without yielding meaningful results.

All this indicates that since the Big Bang, the universe's unidirectional evolution has inevitably led to the birth of humans, allowing no room for deviation. It's like the development of a fetus from a fertilized cell, where every part of the process is necessary and purposeful.

It would be illogical and impossible for a universe governed by the highest wisdom to evolve without a purpose. As for the countless other galaxies, they may serve as components of the entire universe in supporting our unidirectional evolution.

Einstein proposed the profound idea of studying "God's thoughts," which we can interpret as follows:

"God's thoughts" may be understood as God's wisdom and logic. Throughout this book, we've explored the wisdom of God, exemplified by various natural miracles. Nature comprises countless such miracles, and we've explored the logical relationships between different natural factors, often represented with block diagrams. There may be no better way to convey this wisdom.

God's logic stems from God's wisdom, an inherent attribute rather than a supernatural power beyond time and space. God is Mother Nature, and understanding nature involves not only addressing the physical question of what but also the philosophical question of why nature does what it does. This dual inquiry reflects the division of labor between natural science and natural philosophy. Given that the universe possesses "the highest wisdom" and has evolved in an extraordinarily complex manner over billions of years, we must ask: what is its purpose or intention? This question is inevitable since nature has created everything, including humans.

The author believes that "the religion of the future will be a cosmic religion," representing the pinnacle of humanity's development. Through this religion, humanity will advance to a distant and glorious future. While we may seem insignificant relative to the universe, our intelligence is profound enough to impact its future.

EPILOGUE

Human society must continually progress to unlock and improve the level of human wisdom. This can only be achieved through a nonlinear system of society, as opposed to a linear one. Therefore, the advancement of our social system is crucial.

You are fortunate to be a member of the human race, and the universe has great hopes for humanity. Treasure your own wisdom, for humanity is not merely a fleeting occurrence in the universe but a significant part of its grand design.

Studying the ultimate natural mission of humanity is a profound and exciting subject in natural philosophy. As modern science advances, we can philosophically comprehend our planet and the universe, leading to a deeper understanding of human history and the value of our existence within the vast, mysterious cosmos. This understanding also allows us to predict future trends.

Comprehension Exercises

This book is highly informative and introduces many new concepts and ideas presented by the author. Below are just a few of the most fundamental ones. If you have different perspectives, please feel free to share them.

1. *Two principles of natural philosophy governing the development of nature: "The consistency of structure and function" and "Everything happens conditionally".* Their relationship is such that a specific function requires a particular structure that, in turn, requires certain conditions for its formation. Essentially, the development of nature involves continually generating new conditions to form more complex structures with higher functions. These principles define the means of natural development, which must be an extremely complex process that unfolds over a long period. Earth has successively created new geographical conditions that allowed life to evolve into more complex structures with higher functions, ultimately forming a civilization. This process makes Earth so thought-provoking. The reason nature continuously creates new conditions to form more complex structures with higher functions seems to be explained only by Einstein's concept of "the Reason in nature."

2. *Limitation of time and time effect.* The evolution of carbon-based life is an extremely time-consuming process that could only occur within the optimal period of solar radiation. The lifespans of the Sun and the Earth are limited to the expanding phase of the universe, an essential precondition for natural development. The effective lifespans of the Sun and the Earth must align with the time

COMPREHENSION EXERCISES

requirements for the evolution of life. While natural processes are inherently random, modern science tells us that timeliness is also a factor in the development of the Earth System. The progression from the birth of the Earth to the emergence of human beings and modern civilization is governed by time effect (timeliness), making it a logical process. These two temporal characteristics—limitation of time and time effect—are inextricably linked in natural development. The limitation of time necessitates that natural development must incorporate time effect, ensuring that evolution is completed within the available timeframe.

3. *The Reason in nature.* If natural development did not require time, it would be illogical, irrational, and inconceivable, which is easily understood. So, the two time characteristics of natural development, as stated above, are the essential and inevitable conditions. "The Reason in nature" made every natural selection necessary in time and space for the eventual emergence of human beings and modern civilization on Earth, meeting the time requirements of natural development. The value of this book lies in the detailed demonstration of the existence of the Reason (rationality) in the process of natural development since the Big Bang. Without the control of the Reason, a completely random process could not produce any meaningful results.

4. *Natural mission.* In the immense network of natural development, each natural factor (node, link) has its own natural mission, and there is no isolated natural factor; that is, there is no natural factor that has no natural mission, but humans have not yet discovered it. Note that natural factors refer to a concept, such as a natural phenomenon, rather than an individual (an ant, a wind, etc.).

5. *Nonlinear properties of nature.* Modern science believes that nature is essentially nonlinear, composed of numerous nonlinear systems. Only nonlinear systems can evolve, and human society is no exception. However, early human society was a simple linear system controlled by absolute power. The brain of the whole people had to follow the brain of the ruler, and it was inevitable for this absurd system to be eliminated by history.

6. *The logic of natural development.* Briefly, the logic is that the universe evolved for humans. It can also be called the logic of the evolution of the universe. From all the arguments in this book, we can see that

the logic of natural development is to create human intelligence and bring it into full play. Before humanity finally completes its natural mission, all institutions and cultures that go against the logic of natural development will be eliminated.

7. *The planetary wind system.* The planetary wind system did not form simultaneously with the Earth. Instead, the conditions for its development gradually emerged in the late Cenozoic era, illustrating the principles of natural philosophy. The planetary wind system had a profound impact on the global environment. One of the most significant impacts is in the seasonal precipitation patterns, which played a crucial role in the formation of rivers, soil, alluvial plains, and the salinization of seawater, all of which created the necessary conditions for the emergence of early civilizations. By highlighting the formation of planetary wind systems, the author outlines natural development chains. Had the planetary wind system appeared during the Paleozoic or Mesozoic eras, it would have posed significant challenges to the evolution of life.

8. *The global land-sea distribution.* The current distribution is remarkably logical and conducive to human development, with the Mediterranean Sea playing a particularly crucial role in the emergence of modern science. From the GRV and the Mediterranean to the Eurasian and American continents, this distribution appears purposeful rather than arbitrary. It resembles a roadmap with profound meaning tailored to human nature and the advancement of human intelligence. Without any stage in between, human history would have changed so much that it would not be where it is today (such as without the North American continent). This is thought-provoking.

9. *The Two Primary Substance Chains.* The two primary substance chains are natural metal/energy chains serving civilization. The series of links in the two chains in the natural world seem to be entirely unrelated and randomly distributed. Yet all the links in either chain seem to be naturally arranged in such a way from being easily obtained but of fewer uses (say, dry leaves on the ground) to being difficult to obtain but of more uses (say, oil deep below the ground). Thus, early humans could discover and use them step by step, from simple to complex, over the last few thousand years to the present.

COMPREHENSION EXERCISES

10. *Human nature.* A fundamental characteristic bestowed by nature to humanity, human nature is pivotal in understanding human development. The author has frequently emphasized the distinction between possessing intelligence and fully utilizing it. Nature has intricately shaped various geographical conditions, facilitating humanity's journey toward modern scientific advancements by enabling the full expression of human intellect. Therefore, nature has created human nature and given full play to human intelligence. The strategic global distribution of land and sea, aligned with the requirements of human nature, actively fosters the utilization of human intelligence, underscoring the impossibility of humanity's progression from primitive existence to modern civilization solely on an isolated and unique continent over thousands of years.

11. *The human brain.* The human brain is the only way the universe created intelligence. It is also an indispensable part of the evolution of the universe and will impact its fate. Remarkably, Earth is the only planet where life has evolved to produce humans with such advanced cognitive capabilities. This distinction underscores Earth's and humanity's unparalleled distinctiveness in the cosmic expanse.

12. *Abundance of chemical elements.* The solar system began as a nebula, but the chemical composition of the Sun is markedly different from that of the Earth. This necessary difference, along with Earth's precise astronomical conditions, laid the foundation for the emergence of life and human intelligence. Such conditions could not have arisen through purely random processes. Among Earth's chemical elements, carbon plays a unique role—not only activating all other elements in the periodic table to make life and civilization possible but also guiding nature's developmental process and its duration. Carbon is, therefore, the cornerstone of the periodic table.

13. *The extreme complexity of natural development.* This is an exciting story of natural development from the Big Bang to the emergence of humanity and modern civilization. Based on time limitations and principles of natural philosophy, natural development is bound to be an extremely complex nonlinear system in time and space. It can only be regulated by "the Reason in nature" that manifests itself as "the highest wisdom" in the cosmic cycles.

COMPREHENSION EXERCISES

14. *The focusing process of matter and energy.* Continuously increasing the density of matter and energy in a region is a primary condition for the universe to produce anything meaningful. This "focusing process" occurred soon after the Big Bang and gave rise to the Milky Way, the solar system, the Earth, and humans. All factors involved in the process, from astronomy to geography, seem to have been precisely adjusted in time and space to make the evolution of life and the emergence of humans and civilization possible. This focusing process is unique in the universe, indicating that the universe's evolution is for human beings. Earth, however, is nothing but a tool of the universe, which can process all the chemical elements into humans and their civilization.

The concepts above convey the idea that the universe operates with intelligence and rationality. This perspective is grounded in human understanding; if nature exhibits rationality for humans, then the universe's evolution towards humanity is logical and inevitable.

Contemplating humanity's distant future is rooted in the logic observed throughout the universe's evolution thus far. Humanity cannot be a senseless, self-destructive momentary tragedy in the universe's evolution because such a fate doesn't logically fit within the scientifically understood universe.

While this viewpoint may seem absurd at first glance, its validity is not based on its initial impression but on its foundation. It stems from profound and extensive research into nature's current and past states, unveiling the intricate complexity and logic behind natural development. If this perspective appears somewhat akin to religious ideology, it's only Einstein who can define such a belief.

This book primarily delves into "the Reason" of nature, revealing the rationality inherent in the natural world and its formation processes. Without this guiding principle, the intricate natural world humanity inhabits could not have come into existence. Understanding the profound complexity of nature and its formation is vital to grasping its inherent rationality. The author contends that "the Reason" is an intrinsic attribute of nature itself, not something transcendent beyond time and space. It embodies the universe's nature, inspiring awe. As children of the universe, only through a deep understanding of this remarkable attribute can humans navigate the universe's logic and embark on a promising future, fulfilling our sacred natural mission!

COMPREHENSION EXERCISES

Do you believe that random processes alone could have led to the emergence of humans and their civilization in the universe? Please convey your reasoning based on the principles of natural philosophy. Remember the crucial question of whether the necessary natural conditions for universal evolution could have arisen.

After reading this book, you may understand why the logic of natural development is always consistent. From the atom to the human brain to intelligence in the form of energy, it's all about keeping intelligence in the cosmic cycles. We reiterate that a universe without humans would be meaningless. Humanity, you must respect your sacred natural mission!

Human awakening and progress will last forever!

Reflection Questions

Having completed this book, please take a moment to reflect on and answer the following questions:

- Why does the author argue that, once all chemical elements were present in cosmic evolution, the process of the birth and development of civilization was logically defined? How do natural philosophical principles contribute to understanding this process? Considering the complexity and subtlety involved, do you agree that Einstein's natural philosophy provides the best framework for understanding this process?

- Why does the author dedicate significant attention to demonstrating the rationality of natural development in both time and space? What implications does this rationality have for humanity? How does the author demonstrate the strong logic of cosmic evolution, which governs not only the birth and distribution of chemical elements but also the evolution of matter and energy?

- The universe's conception of civilization through the Earth and the development of fertilized eggs share the same principles and logic. Both are extremely complex nonlinear systems requiring a "womb and placenta," regulatory factors, and the influence of time. With this analogy in mind, how do you understand the nature of the Earth and nature (the universe) in this process?

- Is the remote future of humanity, as described in the book, a mere fantasy or an inevitable outcome of the logic of cosmic evolution? In this context, how do you interpret the significance of human existence in the universe and the direction of humanity's historical

REFLECTION QUESTIONS

development? To fully grasp the value of humanity and the meaning of social progress, we must consider them from the perspective of cosmic evolution. As humans are part of nature, is there any criterion beyond nature by which we can measure them?

- Can you grasp the profound relationship between human nature and the natural world? The progress of civilization cannot be driven by an intelligent being lacking human nature. Does this suggest that human nature is a purposeful design of nature, corresponding to the natural conditions of the Earth? How do these two aspects align with each other?

The questions above can only be thoughtfully addressed after a thorough reading and deep reflection on the book. They explore the essence of nature (the universe) and represent fundamental inquiries that are unavoidable in the pursuit of understanding nature.

At any rate, the relationship between the universe and human beings is a very exciting topic!

Bibliography

A Brief History of Time. Directed by Errol Morris. New York: Maysles Films and Warner Bros., 1991.
Benpei Liu, et al. *Geohistory*. 3rd ed. Beijing: Geology Press, 1996.
BBC News. "Big Bang Follows Big Bang Follows Big Bang." *Today,* September 25, 2010.
Brownlee, Don. "Stardust: A Mission With Many Scientific Surprises." *NASA Solar System Exploration*, October 29, 2009. https://solarsystem.nasa.gov/stardust/news/news116.html.
Callahan, M. P., et al. "Carbonaceous Meteorites Contain a Wide Range of Extraterrestrial Nucleobases." *Proceedings of the National Academy of Sciences of the United States of America* 108, no. 34 (2011) 13995–98. https://doi.org/10.1073/pnas.1106493108.
Einstein, Albert. *Albert Einstein: The Human Side*. Edited by Helen Dukas and Banesh Hoffmann. Princeton, NJ: Princeton University Press, 1979.
———. "Der Einstein-Gutkind Brief." (1954) Richard Dawkins Foundation for Reason and Science, trans. https://de.richarddawkins.net/articles/der-einstein-gutkind-brief-mit-transkript-und-englischer-ubersetzung.
———. "Einstein Believes in 'Spinoza's God.'" *The New York Times*, April 25, 1929.
———. *Ideas and Opinions*. Crown Publishers, 1954.
———. "Letter to Max Born, December 4, 1926." *Physics Today* 58, no. 5 (2005): 16. https://doi.org/10.1063/1.1995729.
———. *Letters to Solovine, 1906–1955*. Translated by Wade Baskin. New York: Philosophical Library, 1987. https://archive.org/details/letterstosolovinooooeins_r6g8.
———. *The World as I See It*. Translated by A. Harris. London: John Lane The Bodley Head, 1935.
Hanslmeier, Arnold. *Habitability and Cosmic Catastrophes*. Berlin: Springer, 2008.
Hawking, Stephen. *A Brief History of Time*. New York: Bantam, 1988.
Hermanns, William (1983). *Einstein and the poet: in search of the cosmic man*. Brookline Village: Branden.
Kimura, Motoo. *The Neutral Theory of Molecular Evolution*. Cambridge: Cambridge University Press, 1983.
Lenton, Timothy M., et al. "Earliest Land Plants Created Modern Levels of Atmospheric Oxygen." *Proceedings of the National Academy of Sciences* 113, no. 35 (August 30, 2016) 9704–9.

BIBLIOGRAPHY

Meech, Karen. "1997 Apparition of Comet Hale-Bopp: What We Can Learn from Bright Comets". *Planetary Science Research Discoveries*, February 14, 1997 (updated March 24, 1997). http://www.psrd.hawaii.edu/Feb97/Bright.html.

Mei, Yuan, et al. "Study on Comprehensive Technology of Preventing Mud Cake of Large Diameter Slurry Shield in Composite Stratum." *Buildings* 12, no. 10 (2022) 1555. https://doi.org/10.3390/buildings12101555.

NASA. "Comet Stardust Findings Suggest Comets More Complex Than Thought." *NASA Solar System Exploration*, December 14, 2006. https://solarsystem.nasa.gov/stardust/news/news110.html.

Newton, Isaac. *The Mathematical Principles of Natural Philosophy*. Translated by Andrew Motte. New York: Published by Daniel Adee, c1846. https://archive.org/details/newtonspmathema00newtrich/page/n7/mode/2up.

Penrose, Roger. *Cycles of Time: An Extraordinary New View of the Universe*. New York: Alfred A. Knopf, 2011.

Prigogine, Ilya, and Isabelle Stengers. *Order Out of Chaos: Man's New Dialogue with Nature*. New York: Bantam, 1984.

ScienceDaily. "How Openings in Antarctic Sea Ice Affect Worldwide Climate." *Science News*, September 11, 2017. https://www.sciencedaily.com/releases/2017/09/170911092039.htm. Credit: University of Pennsylvania.

Smith, H. E. *The Milky Way Galaxy*. University of California, San Diego, April 28, 1999. https://casswww.ucsd.edu/archive/public/tutorial/MW.html.

Salisbury, Steven W., Anthony Romilio, Matthew C. Herne, Ryan T. Tucker, and Jay P. Nair. "The Dinosaurian Ichnofauna of the Lower Cretaceous (Valanginian–Barremian) Broome Sandstone of the Walmadany Area (James Price Point), Dampier Peninsula, Western Australia." *Journal of Vertebrate Paleontology* 36, no. S1 (2016): 1–152. https://doi.org/10.1080/02724634.2016.1269539.).

Stavrianos, L. S. *The World to 1500: A Global History*. Englewood Cliffs, NJ: Prentice Hall, 1982. https://archive.org/details/worldto1500globa00stav.

United Nations Intergovernmental Panel on Climate Change. *Global Warming of 1.5°C: An IPCC Special Report*. Geneva: IPCC, 2018. https://www.ipcc.ch/sr15/.

Waugh, David. *Geography: An Integrated Approach*. Cheltenham, UK: Nelson Thornes, 1995.

Zhang, Guowen. "Can Solar Neutrinos Heat the Earth?" *HANS Preprints*, May 15, 2020. https://pdf.hanspub.org/hanspreprints20200100000_41404959.pdf.